HIGHER-ORDER GROWTH CURVES AND MIXTURE MODELING WITH M*PLUS*

This practical introduction to second-order and growth mixture models using M*plus* introduces simple and complex techniques through incremental steps. The authors extend latent growth curves to second-order growth curve and mixture models and then combine the two. To maximize understanding, each model is presented with basic structural equations, figures with associated syntax that highlight what the statistics mean, M*plus* applications, and an interpretation of results. Examples from a variety of disciplines demonstrate the use of the models, and exercises allow readers to test their understanding of the techniques. A comprehensive introduction to confirmatory factor analysis, latent growth curve modeling, and growth mixture modeling is provided so the book can be used by readers of various skill levels. The book's datasets are available on the web.

Highlights include:

- Illustrative examples using M*plus* 7.4 include conceptual figures, M*plus* program syntax, and an interpretation of results to show readers how to carry out the analyses with actual data.
- Exercises with an answer key allow readers to practice the skills they learn.
- Applications to a variety of disciplines appeal to those in the behavioral, social, political, educational, occupational, business, and health sciences.
- Data files for all the illustrative examples and exercises at www.routledge.com/9781138925151 allow readers to test their understanding of the concepts.
- *Point to Remember* boxes aid in reader comprehension or provide in-depth discussions of key statistical or theoretical concepts.

Ideal as a supplement for use in graduate courses on (advanced) structural equation, multilevel, longitudinal, or latent variable modeling, latent growth curve and mixture modeling, factor analysis, multivariate statistics, or advanced quantitative techniques (methods) taught in psychology, human development and family studies, business, education, health, and social sciences, this book's practical approach also appeals to researchers. Prerequisites include a basic knowledge of intermediate statistics and structural equation modeling.

Kandauda A. S. Wickrama is a Georgia Athletic Association Endowed Professor in the Department of Human Development and Family Science at the University of Georgia.

Tae Kyoung Lee is a Senior Research Associate at the University of Miami in the Department of Public Health Sciences.

Catherine Walker O'Neal is an Assistant Research Scientist at the University of Georgia in the Department of Human Development and Family Science.

Frederick O. Lorenz is University Professor of Statistics and Psychology at Iowa State University.

"This book goes way beyond the basics of growth curve modeling. The authors manage to explain complicated and potentially confusing models like factor-of-curves and curve-of-factors very well. I like the way they explain how to interpret the models, including substantive interpretations of real world examples."

—Joop Hox, Utrecht University, The Netherlands

"This timely work gives clear statistical advice and offers step-by-step coverage for M*plus* users in the analysis of many kinds of growth curve models and also mixture models. A wealth of syntax examples and available data sets give additional opportunities for practice."

—Rex Kline, Concordia University, Canada

"This book would be an excellent addition to graduate courses on the analysis of longitudinal data, advanced courses on structural equation modeling and multi-level regression, or for a workshop on how to conduct growth curve modeling analysis. This would also be an excellent resource for researchers conducting analyses of longitudinal data."

—Daniel W. Russell, Iowa State University, USA

"The authors' approach to explaining statistical analysis is one of the best I've seen, and each chapter is clear and easy to read. ... I would recommend it to people attending courses I run in the Quantitative Research Methods Training Unit. ... This project fills a significant gap ... by providing step by step procedures in the application of latent growth modelling techniques and translating complicated statistical language into simple English."

—D. Daniel Boduszek, University of Huddersfield, UK

"This book has the potential to contribute greatly to the field. ... A resource that integrates ... M*plus* with the analysis of different kinds of growth models will be widely used. ... The style is straightforward and easy to follow. ... I would consider adopting it for my graduate course ... Longitudinal Research Methods & Analysis.... I would seriously consider ... using it for both research and teaching needs."

—Joel Hektner, North Dakota State University, USA

"Examples are provided to help researchers see how to apply the methodology with actual data, and interpret the results. ... The book will be of interest to students, faculty, and researchers working with longitudinal data in such areas as behavioral science, business, social sciences, and human development and family studies. ... The book can be used ... in advanced undergraduate or graduate level statistics courses, or a welcome addition to the methodological reference resources for researchers."

—Lisa L. Harlow, University of Rhode Island, USA

"I find the writing style simple and straightforward. ... I would recommend this book to my colleagues. ... It is appropriate ... for an advanced SEM course. ... The Mplus codes that accompany the models ... would be very helpful to researchers."

—Wen Luo, Texas A & M University, USA

Multivariate Applications Series

Sponsored by the Society of Multivariate Experimental Psychology, the goal of this series is to apply statistical methods to significant social or behavioral issues, in such a way so as to be accessible to a nontechnical-oriented readership (e.g., non-methodological researchers, teachers, students, government personnel, practitioners, and other professionals). Applications from a variety of disciplines such as psychology, public health, sociology, education, and business are welcome. Books can be single- or multiple-authored or edited volumes that (1) demonstrate the application of a variety of multivariate methods to a single, major area of research; (2) describe a multivariate procedure or framework that could be applied to a number of research areas; or (3) present a variety of perspectives on a topic of interest to applied multivariate researchers.

Anyone wishing to submit a book proposal should send the following: (1) author/title; (2) timeline including completion date; (3) brief overview of the book's focus, including table of contents and, ideally, a sample chapter (or chapters); (4) a brief description of competing publications; and (5) targeted audiences.

For more information, please contact the series editor, Lisa Harlow, at Department of Psychology, University of Rhode Island, 10 Chafee Road, Suite 8, Kingston, RI, USA. 02881-0808; phone (401) 874–4242; fax (401) 874–5562; or e-mail LHarlow@uri.edu.

- *What if there were no significance tests?* co-edited by Lisa L. Harlow, Stanley A. Mulaik, and James H. Steiger (1997)

- *Structural Equation Modeling with LISREL, PRELIS, and SIMPLIS: Basic Concepts, Applications, and Programming,* written by Barbara M. Byrne (1998)

- *Multivariate Applications in Substance Use Research: New Methods for New Questions,* co-edited by Jennifer S. Rose, Laurie Chassin, Clark C. Presson, and Steven J. Sherman (2000)

- *Item Response Theory for Psychologists,* co-authored by Susan E. Embretson and Steven P. Reise (2000)

- *Structural Equation Modeling with AMOS: Basic Concepts, Applications, and Programming,* written by Barbara M. Byrne (2001)

- *Conducting Meta-Analysis Using SAS,* written by Winfred Arthur, Jr., Winston Bennett, Jr., and Allen I. Huffcutt (2001)

- *Modeling Intraindividual Variability with Repeated Measures Data: Methods and Applications,* co-edited by D. S. Moskowitz and Scott L. Hershberger (2002)

- *Multilevel Modeling: Methodological Advances, Issues, and Applications,* co-edited by Steven P. Reise and Naihua Duan (2003)

- *The Essence of Multivariate Thinking: Basic Themes and Methods,* written by Lisa Harlow (2005)

- *Contemporary Psychometrics: A Festschrift for Roderick P. McDonald,* co-edited by Albert Maydeu-Olivares and John J. McArdle (2005)

- *Structural Equation Modeling with EQS: Basic Concepts, Applications, and Programming, Second Edition,* written by Barbara M. Byrne (2006)

- *A Paul Meehl Reader: Essays on the Practice of Scientific Psychology,* co-edited by Niels G. Waller, Leslie J. Yonce, William M. Grove, David Faust, and Mark F. Lenzenweger (2006)

- *Introduction to Statistical Mediation Analysis,* written by David P. MacKinnon (2008)

- *Applied Data Analytic Techniques for Turning Points Research,* edited by Patricia Cohen (2008)

- *Cognitive Assessment: An Introduction to the Rule Space Method,* written by Kikumi K. Tatsuoka (2009)

- *Structural Equation Modeling with AMOS: Basic Concepts, Applications, and Programming, Second Edition,* written by Barbara M. Byrne (2010)

- *Handbook of Ethics in Quantitative Methodology,* co-edited by Abigail T. Panter & Sonya K. Sterba (2011)

- *Longitudinal Data Analysis: A Practical Guide for Researchers in Aging, Health, and Social Sciences,* co-edited by Jason T. Newsom, Richard N. Jones, and Scott M. Hofer (2011)

- *Structural Equation Modeling with MPlus: Basic Concepts, Applications, and Programming,* written by Barbara M. Byrne (2012)

- *Understanding the New Statistics: Effect Sizes, Confidence Intervals, and Meta-Analysis,* written by Geoff Cumming (2012)

- *Frontiers of Test Validity Theory: Measurement, Causation and Meaning,* written by Keith A. Markus and Denny Borsboom (2013)

- *The Essence of Multivariate Thinking: Basic Themes and Methods, Second Edition,* written by Lisa L. Harlow (2014)

- *Longitudinal Analysis: Modeling Within-Person Fluctuation and Change,* written by Lesa Hoffman (2015)

- *Handbook of Item Response Theory Modeling: Applications to Typical Performance Assessment,* co-edited by Steven P. Reise & Dennis Revicki (2015)

- *Longitudinal Structural Equation Modeling: A Comprehensive Introduction,* written by Jason T. Newsom (2015)

- *Higher-Order Growth Curves and Mixture Modeling with Mplus: A Practical Guide,* written by Kandauda A. S. Wickrama, Tae Kyoung Lee, Catherine Walker O'Neal & Frederick O. Lorenz (2016)

- *Structural Equation Modeling with AMOS: Basic Concepts, Applications, and Programming, Third Edition,* written by Barbara M. Byrne (2016)

HIGHER-ORDER GROWTH CURVES AND MIXTURE MODELING WITH M*PLUS*

A Practical Guide

Kandauda A. S. Wickrama,
Tae Kyoung Lee, Catherine Walker O'Neal, &
Frederick O. Lorenz

Routledge
Taylor & Francis Group

NEW YORK AND LONDON

First published 2016
by Routledge
711 Third Avenue, New York, NY 10017

and by Routledge
2 Park Square, Milton Park, Abingdon, Oxon, OX14 4RN

Routledge is an imprint of the Taylor & Francis Group, an informa business

Library of Congress Cataloging in Publication Data
A catalog record for this book has been requested.

ISBN: 978-1-138-92514-4 (hbk)
ISBN: 978-1-138-92515-1 (pbk)
ISBN: 978-1-315-64274-1 (ebk)

Typeset in Bembo
by Out of House Publishing

CONTENTS

PART 2
Growth Mixture Modeling 189

PREFACE

Researchers in a variety of disciplines have used latent growth curve modeling (LGCM) in a structural equation modeling (SEM) framework to describe and analyze change in individual attributes, including personal characteristics, behaviors, aspects of intimate relationships, and health outcomes over time. However, as individuals' developmental patterns become more complex various extensions of conventional LGCMs, such as second-order growth curve models with higher-order factor structures and growth mixture models capturing unobserved heterogeneity, are necessary in order to properly investigate substantively important, yet complex, research questions.

Many of these investigations are likely to involve multidimensional data structures with higher-order factor structures, and a conventional LGCM does not allow for an assessment of higher-order factor structures. To address this deficiency, a conventional LGCM can be extended to a second-order latent growth model to capture these potential higher-order factor structures in a longitudinal context. Additionally, a conventional LGCM does not take into account the potential heterogeneity, or clustering, that may exist among individual trajectories because a conventional LGCM assumes that all individuals come from a single, homogenous population and the same pattern of growth can approximate the entire population. To address this limitation, a growth mixture model (GMM), an extension of a LGCM, can be employed to identify potential unobserved heterogeneity in individual trajectories. GMM is an emerging statistical approach that models such heterogeneity by classifying individuals into "groups" or "classes" with similar patterns of trajectories. More importantly, these two LGCM extensions (a second-order latent growth model and a growth mixture model) can be combined to form various types of higher-order growth mixture models assessing *both* higher-order factor-structures and heterogeneous trajectories simultaneously.

The M*plus* statistical program is gaining popularity for these advanced statistical analyses, and although these LGCM extensions are increasingly being used in various fields of research, particularly psychosocial and development areas of research, a practical guide for these advanced applications is not readily available. Consequently, the present book intends to fill this gap by providing:

- a brief conceptualization of these models with their equations,
- illustrative examples using M*plus* Version 7.4 software for both second-order growth curves and growth mixture models as well as combination models incorporating both of these cutting-edge methods,
- exercises (and their answers) to give readers the opportunity to actively practice analyzing these models.

For each illustrative example model, we include:

- conceptual figures,
- M*plus* program syntax,
- an interpretation of results.

The data files for all the illustrative examples and exercises are available at www.routledge.com/9781138925151.

For each model, the measurement and structural components of the model are clearly shown in the figures and also presented as regression equations. M*plus* syntax is provided with accompanying labels identifying key components of the model. Throughout the chapters, *Point to Remember* boxes provide key points to aid in reader comprehension or more in-depth discussions of statistical or theoretical concepts that are unique to these types of models.

Intended Audience

It was our intention to create a practical guide that could be used by researchers of various skill levels. Consequently, advanced knowledge of conventional approaches for analyzing longitudinal data is not necessary for comprehending the information presented in this book because we provide a comprehensive introduction to confirmatory factor analysis, latent growth curve modeling, and growth mixture modeling before discussing the second-order alliances of these first-order models. In this book, we have attempted to take readers from simple models to complex models in incremental modeling steps. To avoid unnecessary complexity, all of the example models use the same datasets and are incrementally related.

This book is designed to be used as a supplementary practical guide along with other available books on advanced longitudinal analysis for graduate level courses or self-study. Students who have completed intermediate statistics and

SEM courses have the basic knowledge and skills necessary to use this book. Although the illustrative examples are from the authors' particular field of study (namely, family and human development) the applications presented in this book are applicable to a wide variety of disciplines, including behavioral, social, political, educational, occupational, business, and health disciplines.

Content

The chapters of this book can be divided into two sections. The first part contains Chapters 1 through 6 and introduces basic structural equation modeling as well as first- and second-order growth curve modeling. The second part of the book contains Chapters 7 through 9. This second part provides a detailed guide for growth mixture modeling and second-order growth mixture modeling.

More specifically, Chapter 1 introduces basic concepts from structural equation modeling and possible extensions of conventional growth curve models. The subsequent chapters are structured around explaining these extensions. In addition, the data and measures used throughout this book are also introduced. Chapter 2 presents an introduction to the conventional modeling of multidimensional panel data, including confirmatory factor analysis (CFA) and growth curve modeling. Limitations of these conventional approaches are discussed. Chapter 3 introduces the logical and theoretical extension of a CFA to a second-order growth curve, known as curve-of-factors model (CFM) with an emphasis on the substantive meaning and applicability of a CFM. Chapter 4 illustrates the estimation and interpretation of unconditional and conditional CFMs. Chapter 5 presents the logical and theoretical extension of a parallel process model (PPM) to a second-order growth curve, known as a factor-of-curves model (FCM). Particular attention is focused on the substantive meanings and applicability of a FCM. Chapter 6 illustrates the estimation and interpretation of unconditional and conditional FCMs.

The second part of the book, building on the previous chapters, then shifts to growth mixture modeling. Chapter 7 is an introduction to unconditional growth mixture modeling. First, this chapter illustrates how individuals' trajectories can be heterogeneous, resulting in the presence of different subgroups (or sub-populations), each with a unique pattern of change over time. The use of a categorical latent variable that represents a mixture of sub-populations is then presented as a way to extend a growth curve to a growth mixture model when sub-population membership is not known but must be inferred from the data. Chapter 8 illustrates conditional growth mixture models; that is, how to incorporate predictors and outcomes of these various growth patterns or classes. Chapter 9 presents how to extend two previously introduced types of second-order growth curves (curve-of-factors and factor-of-curves models) to growth mixture models. The extension of a first-order parallel process growth curve to a multidimensional growth mixture model is also presented.

Acknowledgements

Our book is a collective effort that would not have materialized without the dedication of each of the authors. We also appreciate the thoughtful feedback provided by graduate students, including DaYoung Bae, Tara Sutton, Josie Kwon, and JihYoung Kim. We are thankful to the reviewers, D. Daniel Boduszek (University of Huddersfield, UK), Joel Hektner (North Dakota State University), and one anonymous reviewer, for their constructive comments and positive reviews on our initial book proposal. We also appreciate the support of Lisa L. Harlow, the Multivariate Applications Series editor, Debra Riegert, senior editor for Routledge/Taylor & Francis, and Freddie Moore, editorial assistant for Routledge/Taylor & Francis. We acknowledge the support of Dr. Rand Conger and all of those involved in the Iowa Transitions Project for allowing us to utilize data from the project for this book. Over the years of the project, support for this research has come from multiple sources, including the National Institute of Mental Health (MH00567, MH19734, MH43270, MH48165, MH51361), the National Institute on Drug Abuse (DA05347), the Bureau of Maternal and Child Health (MCJ-109572), the MacArthur Foundation Research Network on Successful Adolescent Development among Youth in High-Risk Settings, the Iowa Agriculture and Home Economics Experiment Station (Project No. 3320), and the Spencer Foundation. Last, but certainly not least, we greatly appreciate the encouragement and support of our spouses for this effort.

ABOUT THE AUTHORS

Kandauda A. S. Wickrama is a Georgia Athletic Association Endowed Professor in the Department of Human Development and Family Science at the University of Georgia. His research focuses on social determinants of health and health inequality across the life course, and the application of advanced statistical methods to social epidemiology.

Tae Kyoung Lee is a Senior Research Associate at the University of Miami in the Department of Public Health Sciences. His research interests include longitudinal models, finite mixture models, multilevel designs, generalized linear models, structural equation models, and the development of internalizing and externalizing problems and life course developmental perspective.

Catherine Walker O'Neal is an Assistant Research Scientist at the University of Georgia in the Department of Human Development and Family Science. Her major research interests include development over the life course and cumulative life experiences that are influential in health and well-being outcomes.

Frederick O. Lorenz is University Professor of Statistics and Psychology at Iowa State University. His research focuses on the relationship between questionnaire reports and observer ratings of family processes and relationship quality.

PART 1
Introduction

1

INTRODUCTION

This introductory chapter serves two purposes. First, this chapter introduces a set of incrementally-related structural equation models (SEMs) using basic concepts from structural equation modeling, including manifest variables, latent variables, causal paths, and correlated associations. These incrementally-related models include first-order SEMs as well as their second-order extensions, such as second-order growth curves and second-order growth mixture models. These models serve as an organizing guide for the remaining chapters of this book. The second purpose of this chapter is to introduce the data and measures used as examples for illustrating these models throughout the book.

A Layout of Incrementally-Related SEMs: An Organizing Guide

The organization of this book is presented in Figures 1.1, 1.2, and 1.3, which contain nine distinct longitudinal structural equation models, to be introduced shortly. Each SEM draws on the classical JKW notation (Bollen, 1989). SEMs are comprised of two components: a measurement model and a structural model. The measurement model links manifest variables to latent variables (or latent constructs). The structural model specifies the associations between latent variables. Circles represent latent variables, while squares represent manifest variables. Double-headed arrows signal correlations between variables, and single-headed arrows indicate directional paths between variables. The measurement errors (ε) of manifest variables are also shown in Figure 1.1. Thus, as shown in Figure 1.1, our latent variable at time 2 (t2) is correlated with the same latent variable at time 1 (t1) as well as at times t3, t4, and t5. This is a SEM for a longitudinal confirmatory factor analysis. There are 15 manifest variables: measures of three symptom types (depression, anxiety, and hostility) over five time points (3 x 5 = 15). Thus, for example, the first three *boxes* in Figure 1.1 could represent the t1 measures of symptoms of depression, anxiety, and hostility, each of which is a manifestation of the latent variable, internalized symptom (IS), at t1.

Illustrative Example 1.1: Examining Alternative Growth Curve Models

Figure 1.2 presents a set of alternative growth curve models that can be estimated using the same manifest variables or composites (sum or mean scores) of these manifest variables. Panels A and B in Figure 1.2 illustrate first-order growth curve models with single-level latent variables, whereas Panels C and D in Figure 1.2 illustrate second-order growth curve models with latent variables at two levels (i.e., first and second levels). The model in Panel A of Figure 1.2 uses five composite measures of IS (adding the three measures of symptoms of depression, anxiety, and hostility at each time point to create a single measure of IS at each of the five time points) as five indicators of two latent variables. The two latent variables in this model, initial level (L) and rate of change or slope (S), define individual trajectories, or the growth curve of IS over time. This is a conventional latent growth curve model (LGCM) and will be discussed in detail in Chapter 2. The model shown in Panel B of Figure 1.2 presents another alternative SEM. This model uses the 15 repeated measures as indicators to define six latent variables representing three growth curves. That is, five repeated measures of depressive symptoms define one latent growth curve (two latent variables: L-Depression and S-Depression), another five repeated measures define a latent growth curve of anxiety, and the remaining five repeated measures define a latent growth curve of hostility. This type of model is known as a parallel process model (PPM). The models shown in Figure 1.1 and Panels A and B of Figure 1.2 are all first-order SEMs because they have only one level of latent variables. The models shown in Panels C and D in Figure 1.2 contain two levels of latent variables and are known as second-order growth curve models. In these models, the first-level latent variables serve as indicators of the second-level latent variables. The second-level latent variables capture the common variances of first-level latent variables and define higher-level factors. For example, in the model shown in Panel C of Figure 1.2, five time-specific latent variables (variables measured at the same time point) are used as indicators of two second-level latent variables of IS. These two latent variables define a second-order growth curve (also known as a curve-of-factors model or CFM), to be discussed in detail in Chapters 3 and 4. The second-order model shown in Panel D of Figure 1.2 is a second-order extension of the model illustrated in Panel B of Figure 1.2. In this model, six latent variables (defined as growth factors of three growth curves in Panel B of Figure 1.2) are used to define

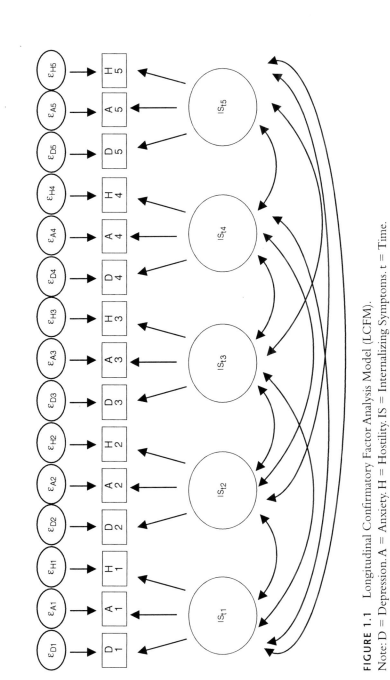

FIGURE 1.1 Longitudinal Confirmatory Factor Analysis Model (LCFM).

Note: D = Depression. A = Anxiety. H = Hostility. IS = Internalizing Symptoms. t = Time.

A. Conventional Latent Growth Curve Model   B. Parallel Process Latent Growth Curve Model (PPGCM)

C. Curve-of-Factors Model (CFM)

D. Factor-of-Curves Model (FCM)

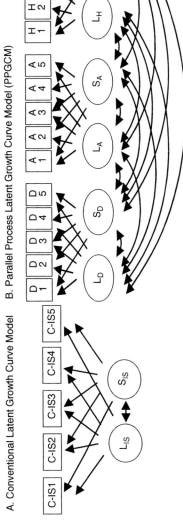
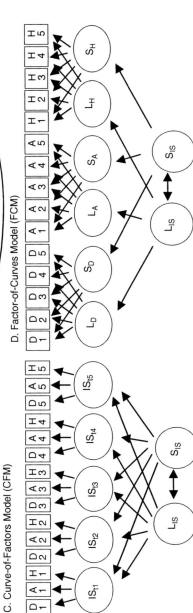

FIGURE 1.2   Variations of Latent Growth Curve Models (LGCM).
Note: D = Depression. A = Anxiety. H = Hostility. L = Level. S = Slope. t = Time. IS = Internalizing Symptoms.
C-IS = Composite Measure of Internalizing Symptoms.

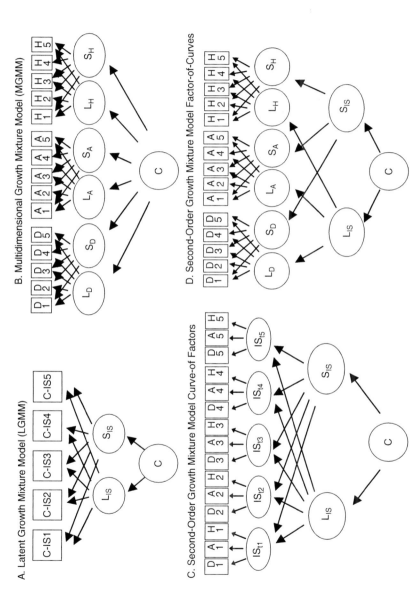

FIGURE 1.3 Variations of Growth Mixture Models (GMM).

Note: D = Depression. A = Anxiety. H = Hostility. L = Level. S = Slope. C = Class. IS = Internalizing Symptoms. C-IS = Composite Measure of Internalizing Symptoms.

two second-order latent variables. In our examples, these two second-order latent variables define a second-order growth curve of IS—that is, a more global construct captured by the three initial growth curves. This type of model is known as a factor-of-curves model or FCM, to be discussed in detail in Chapters 5 and 6.

All of the models illustrated in Figure 1.2 plot a trajectory for each individual within the sample. Yet, it is possible that all individual trajectories may not belong to the same trajectory population. That is, there can be clustering, or unobserved heterogeneity in trajectories, that can be modeled as latent classes or groups of trajectories within the same modeling framework. This clustering refers to subgroups or classes that may exist within the trajectory population; such that one "cluster" of respondents experience one average trajectory while others experience a different average trajectory. Figure 1.3 presents growth curve models with latent classes of trajectories known as growth mixture models. For example, a circle labeled "C" in the latent growth curve model illustrated in Panel A of Figure 1.3 denotes potential latent classes of IS growth curves. If these classes are incorporated into a first-order growth curve model, it is known as a first-order growth mixture model (GMM), to be discussed in detail in Chapters 7 and 8.

Panel B in Figure 1.3 illustrates another type of first-order growth mixture model known as a multidimensional growth mixture model (MGMM), to be discussed in detail in Chapter 9. A MGMM uses latent classes to capture heterogeneity in different dimensional growth curves simultaneously. This model is a direct extension of the PPM shown in Panel B of Figure 1.2. This MGMM includes classes of three first-order growth curves of symptoms of depression, anxiety, and hostility (a total of six latent variables).

Panels C and D in Figure 1.2 and in Figure 1.3 present two pairs of incremental models. The models shown in Panels C and D of Figure 1.3 are extensions of the second-order growth curves presented in Panels C and D of Figure 1.2, respectively. That is, these figures illustrate extensions of a curve-of-factors model (CFM) and a factor-of-curves model (FCM) to include classes of second-order growth curves known as second-order growth mixture models. These extensions are discussed in detail in Chapter 9.

Chapters 2 through 9 of this book instruct readers on how to analyze the models in Figures 1.1, 1.2, and 1.3 using M*plus* software and how to interpret the results. Note that we used M*plus* Version 7.4 for all models. Small differences may exist when running these models with other versions of the software. We begin with the simplest models and move to more complex models in incremental modeling steps. All of the example models use the same dataset and build on one another. As a practical guide for modeling with M*plus*, each chapter of this book presents (a) basic structural equations, (b) figures with manifest and latent variables, (c) M*plus* program commands, and (d) interpretation of the results for each model. The manifest, or observed, variables (measurement model) and latent variables (structural model) of the models are clearly shown in the figures and also presented as regression equations. M*plus* program syntax is provided in the text or in figures, and each part of the model is labeled. Also, instructions for adding predictors and outcomes to each type of model are provided, and the results are compared across different models. In the next section we discuss the study samples, data, and measures used in the examples throughout this book.

Adolescents' Internalizing Symptoms (IS) Trajectories

The example models presented are designed to capture dynamic patterns of change over time, and previous family and developmental studies have suggested that internalizing symptoms, such as anxiety (Nelemans et al., 2013), depressive symptoms (Wickrama, Conger, & Abraham, 2008), and hostility (Jester et al., 2005), are good exemplars. These variables are available in numerous previous studies, including the Iowa-based Family Transition Project (FTP) (Conger & Conger, 2002). We use data from the FTP to illustrate higher-order growth curves and higher-order growth mixture modeling. More specifically, our examples focus on modeling internalizing symptoms in adolescence along with a description of how to incorporate social and family predictors and outcomes into higher-order models, such as a CFM and FCM. First, we introduce the samples and measures for the datasets used in these illustrations.

Datasets used in Illustrations

The primary data source for illustrations in this book is the FTP which includes an "Iowa adolescent dataset" and an "Iowa parent dataset." The adolescent dataset is used for the examples throughout the chapters. Additional illustrations and exercises utilize the parent dataset as a secondary data source. These datasets come from studies conducted from 1989 to 2001. While the project is collectively known as the "Family Transitions Project," it is compiled from multiple studies including the Iowa Youth and Family Project (IYFP), the Iowa Single Parent Project (ISPP), the Family Transition Project (FTP), and the Midlife Project (MLP).

The FTP began as a longitudinal study that combined participants from two earlier research projects: the IYFP and the ISPP. These two earlier longitudinal studies were designed to study family functioning during the economic downturn that affected rural Iowa in the 1980s. The IYFP began in 1989 (Wave 1) and involved two-parent families with adolescent children (451 adolescents) from rural counties in Iowa. Families selected to participate in the study included a "target" seventh-grade child who lived with his or her two biological parents and had a sibling within four years of the target child's age (target adolescents' mean age = 12.7 years at Wave 1 in 1989, 53% female). The ISPP was initiated two years after the IYFP in 1991 and used the same procedures and measures as the IYFP. ISPP households were selected for the study because they had adolescents who were in the same grades (mean age = 14.3 years in 1991, 53% female) as those in the IYFP and were located in the same geographic location. Data came from 107 mother-only families with adolescents. In 1995 (Wave 7), the IYFP and ISPP were combined and data collection continued under the FTP name with a shift in focus to emerging adulthood and the families the target adolescents were creating. The parents of these target adolescents were followed by another project, the Midlife Project (MLP) in 2001. Given the developmental changes of internalizing symptoms during adolescence and their influence on outcomes in young adulthood, FTP data from the IYFP and ISPP are used as the main data source for the illustrations in this book in order to capture the entire period of adolescence. More specifically, our analyses begin with measures included in 1991 (Wave 3 of the initial study or t1 for the purpose of our analyses, mean age: 14.59), 1992 (Wave 4 or t2, mean age: 15.58), 1994 (Wave 6 or t3, mean age: 17.69), 1995 (Wave 7 or t4, mean age: 18.52), 1997 (Wave 9 or t5, mean age: 20.52), and 2001 (Wave 13, mean age: 24.52).

For the supplemental illustrations in this book, the parent dataset (MLP data) was used. Additional information regarding the study procedures is available from Conger and Conger (2002). Across waves of data collection, 90% of the original FTP respondents participated, on average. In total, 537 individuals provided complete information for all of the study variables.

Measures

For measuring symptoms of depression, anxiety, and hostility in adolescence, the current study used 28 items from the SCL-90-R (Derogatis & Melisaratos, 1983), which were collected in 1991 (W3 of the overall project but t1 for the purpose of our analyses), 1992 (W4; t2), 1994 (W6; t3), 1995 (W7; t4), and 1997 (W9; t5). These three symptoms are conceptualized as components of a broader construct of internalizing symptoms (IS). The questionnaire self-reported data are not available for Wave 5 (1993) and Wave 8 (1996). Throughout the book, these three symptoms are used to demonstrate most of the longitudinal models we discuss (e.g., growth curve models, second-order growth curve models, growth mixture models). For our examples of how to assess predictors in second-order growth curves and second-order growth mixtures models, we utilize five indicators of early risk factors beginning in W3 (1991). The indicators include: family economic hardship, marital conflict, hostile parenting, school context, and social isolation. Similarly, for our examples of how to assess outcomes, or consequences, in second-order models five measures (i.e., romantic violence, civic involvement, financial cutbacks, income, and educational attainment) capture the social consequences of IS in adulthood (W13, 2001). All predictor and outcome measures were created by averaging or summing items so that high scores represent high levels of each variable. More details regarding study measurements are available in Table 1.1.

Each predictor and outcome measure was selected on the basis of prior research (with consideration given to its illustrative capability, not merely the strength of correlations) and theory linking it to elevated IS. The bivariate correlation

TABLE 1.1 Measures of Adolescent Risk Factors and Social Outcomes in Young Adulthood.

Measures	# of Items	Range of Scale	α	Example Items
Internalizing Symptoms (W3–W9)				
(for both adolescent and parent datasets)				
Anxiety	10	1 (Not at all) – 5 (Extremely)	.83–.85	"Nervousness or shakiness inside"
(SCL-90-R; Derogatis & Melisaratos, 1983)				
Depressive Symptoms	13		.84–.91	"Feeling lonely" / "Blaming yourself for things"
(SCL-90-R; Derogatis & Melisaratos, 1983)				
Hostility	5		.80–.85	"Temper outbursts that you could not control"
(SCL-90-R; Derogatis & Melisaratos, 1983)				
Early Family and Psychological Factors (W2)				
Family Economic Hardship	4	1 (Never) – 5 (Always)	.73	"How often do you have enough money for things like clothes, school activities, or other things you need?"
Parents' Marital Conflict	3	1 (Never) – 4 (Always)	.75	"How often your parents argue or disagree with each other"
Hostile Parenting	15	1 (Never) – 7 (Always)	.92	"Get angry at you" / "Criticize you or your ideas"
Negative School Context	11	1 (No problems at all) – 4 (A very serious problem)	.86	"Frequent absence from classes"
(Thornberry, 1989)				
Social Isolation (Hirsch & Rapkin, 1987)	3	1 (Not at all) – 4 (Always)	.79	"I am a shy person" / "It is a hard to make a new friend"
Social Outcomes (W13)				
Romantic Violence	21	1 (Never) – 7 (Always)	.84	"Hit partner" / "Shouted or yelled at partner"
(BARS; Melby, Conger, Ge, & Warner, 1995)				
Civic Involvement	3	0 (I don't do this) – 3 (more than 30 hours)	.87	"Taking part in community activities like volunteer work, civic clubs, recreation program, and so on"
Financial Cutback	29	0=No / 1 = Yes		"Have you changed your residence to save money?"

coefficients between covariates (i.e., predictors and outcomes) and IS (i.e., depression, anxiety, and hostility) ranged from .03 to .29, $p < .05$. All growth curve and growth mixture models demonstrated throughout the book were estimated in M*plus* version 7.4 (Muthén, & Muthén, 1998–2015) using all available data with full information maximum likelihood (FIML) estimation.

References

Bollen, K. A. (1989). *Structural equations with latent variables*. New York: John Wiley & Sons.

Conger, R. D., & Conger, K. J. (2002). Resilience in Midwestern families: Selected findings from the first decade of a prospective, longitudinal study. *Journal of Marriage and the Family*, 64(2), 361–373.

Derogatis, L. R., & Melisaratos, N. (1983). The brief symptom inventory: An introductory report. *Psychological Medicine*, 13(3), 596–605.

Hirsch, B. J., & Rapkin, B. D. (1987). The transition to junior high school: A longitudinal study of self-esteem, psychological symptomology, school life, and social support. *Child Development*, 58(5), 1235–1243.

Jester, J. M., Nigg, J. T., Adams, K., Fitzgerald, H. E., Puttler, L. I., Wong, M. M., & Zucker, R. A. (2005). Inattention/hyperactivity and aggression from early childhood to adolescence: Heterogeneity of trajectories and differential influences of family environment characteristics. *Development and Psychopathology*, 17(1), 99–125.

Melby, J. N., Conger, R. D., Ge, X., & Warner, T. D. (1995). The use of structural equation modeling in assessing the quality of marital observations. *Journal of Family Psychology*, 9(3), 280–293.

Muthén, L. K., & Muthén, B. O. (1998–2015). *Mplus user's guide* (7th ed.). Los Angeles: Authors.

Nelemans, S. A., Hale, W. W., Branje, S. J., Raaijmakers, Q. A. W, Frijns, T., van Lier, P. A. C., & Meeus, W. H. J. (2013). Heterogeneity in development of adolescent anxiety disorder symptoms in an 8-year longitudinal community study. *Development and Psychopathology*, 26(1), 181–202.

Thornberry, T. (1989). Panel effects and the use of self-reported measures of delinquency in longitudinal studies. In M. W. Klein (Ed.), *Cross-national research in self-reported crime and delinquency* (pp. 347–369). Dordrecht, Netherlands: Kluwer Academic Publishers.

Wickrama, K. A. S., Conger, R. D., & Abraham, W. T. (2008). Early family adversity, youth depressive symptoms trajectories, and young adult socioeconomic attainment: A latent trajectory class analysis. *Advances in Life Course Research*, 13, 161–192.

2
LATENT GROWTH CURVES

Introduction

This chapter discusses latent growth curve models (LGCMs) in detail. First, this chapter provides an introduction to LGCMs, including the estimation of univariate growth curves, the interpretation of growth parameters, and an examination of covariates of growth curves (including predictors and distal outcomes, or consequences). For each model, figures and M*plus* syntax are also presented. This chapter also discusses the merits and limitations of growth curves and provides the modeling and conceptual foundation required for the remaining chapters.

Growth Curve Modeling

There are a number of traditional analytical approaches that allow researchers to document change across time, including regression methods, mean comparisons, and repeated measures analysis of variance (RMANOVA). In recent years, with the advent of structural equation modeling (SEM) and multi-level modeling, methods of estimation that are sensitive to within-individual change and inter-individual differences in within-individual change are gaining popularity. More specifically, the flexibility of a LGCM allows researchers to focus on within-individual changes, resulting in more accurate and nuanced conclusions concerning the outcome(s) of interest as well as insight into the role of predictor variables and their consequences (Karney & Bradbury, 1995; Rogosa, Brand, & Zimowski, 1982).

Conventional Latent Growth Curve Models (LGCM)

As a multivariate application of SEM, a LGCM analyzes change in an observed attribute over time. Repeated measurements of observed variables are used as manifest indicators of latent variables that represent individuals' initial level of an attribute and its change over time. At the conceptual level, growth curve modeling can be thought of as a two-stage process (i.e., a multilevel process). At the first stage, the goal is to describe change over time for each individual in the study (Wickrama, Lorenz, Conger, & Elder, 1997; Willet & Sayer, 1994). Conceptually, this is done by fitting a regression line (i.e., a growth curve) that plots the variable of interest over time for each individual in the study. Panel A in Figure 2.1 illustrates estimated regression lines for four individuals. At the second stage, these individual regression equations for all of the individuals in the study can then be summarized to obtain an average intercept, or initial level, (i.e., the mean of the variable of interest at the first time point) and an average slope, or rate of change, for all individuals, each with its own variance (see the bolded line in Panel A of Figure 2.1). The goal is then to use theoretically-driven covariates to explain the variation in initial levels (intercepts) and slopes across individuals (e.g., why is the initial level higher for some individuals than others?).

Linear Growth Curve Modeling

In the example we will use throughout the book, it is possible to fit a regression line linking a composite measure IS (a mean of summed scores from depressive, anxiety, and hostility symptoms) over time (t) for each person with an error term at each time point. Person i's regression can be represented by the following linear equation, where π_0 and π_1 represent person i's intercept and slope, respectively. Using our sample, this model can be written as:

$$\text{Internalizing Symptoms (IS)}_{it} = \pi_{0i} + \pi_{1i} + \varepsilon_{it}, \varepsilon_{it} \sim \text{NID}(0, \sigma_{it}^2) \tag{2.1}$$

where i = 1, 2, ..., n respondents, and ε_{it} = the error term for the ith individual at time t. For a linear change with four equally-spaced measurements t can be 0, 1, 2, and 3.

If time intervals are not equidistant, appropriate factor loadings for a linear change that are proportionate to the unequal time intervals should be specified. Thus, as shown in Panel B of Figure 2.1, in the present example, t = 0, 1, 3, 4, 6 (because Wave 5 and Wave 8 data—corresponding to t2 and t5—are not available). This numbering of the time variable indicates the availability of a measure of internalizing symptoms (IS) at an initial time (t = 0) and subsequent years 1, 3, 4, and 6. Of course, measurement intervals can vary greatly from minutes to decades. Note that this figure and others throughout the book, rely on SEM convention to name parameters and draw their shapes. See Table 2.1 for a review of the name and meaning of each Greek character according to SEM convention.

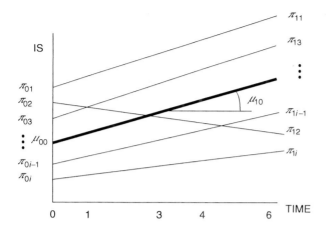

Panel A. Representation of a Collection of Linear Trajectories
(Growth Curves).

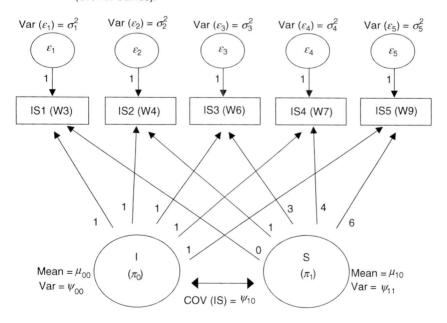

Panel B. Unconditional Linear Trajectory Model for Five Repeated
Measures Assessed at Unequal Intervals.

FIGURE 2.1 Linear Growth Curve Model (LGCM) (Unconditional Trajectory Model).

Note: IS = Internalizing Symptoms. i = Respondents. I = Initial Level. S = Slope. Var = Variance.

TABLE 2.1 Notation for SEM Parameters.

Name	Parameter	Description
Lambda	λ	Regression coefficients or factor loadings
Sigma	σ	The standard deviation of a manifest indicator; σ^2 is the (error) variance
Delta (lowercase)	δ	Residual or measurement error of an exogenous variable
Epsilon	ε	Residual or measurement error of an endogenous variable
Tau	τ	Intercept of manifest indicators
Psi	ψ	The (error) variance or covariance of a latent variable
Mu	μ	The mean of a latent variable
Eta	η	Name of a latent variable
Zeta	ζ	Residuals of a latent variable
Pi	π	Name of a latent growth variable
Gamma	γ	An estimated regression of one construct (either a latent variable or a manifest indicator) onto another. Here, endogenous variables are predicted by exogenous variables.
Beta	β	An estimated regression of one construct (either a latent variable or a manifest indicator) onto another (endogenous on endogenous)
Delta (uppercase)	Δ	Used to denote the change between any two nested estimates, such as $\Delta\chi^2$ or ΔCFI
Alpha	α	The mean of a second-order latent variable (i.e., global factor mean)

The corresponding within-individual regression equations are as follows:

$$\text{Internalizing Symptoms (IS)}_{it} = \pi_{0i} + \pi_{1i} \times t + \varepsilon_{it} \tag{2.2}$$

$$\text{For } t = 0, \text{ Internalizing Symptoms (IS)}_{i0} = \pi_{0i} + \pi_{1i} \times 0 + \varepsilon_{i0} \tag{2.3}$$

$$\text{For } t = 1, \text{ Internalizing Symptoms (IS)}_{i1} = \pi_{0i} + \pi_{1i} \times 1 + \varepsilon_{i1} \tag{2.4}$$

$$\text{For } t = 3, \text{ Internalizing Symptoms (IS)}_{i3} = \pi_{0i} + \pi_{1i} \times 3 + \varepsilon_{i3} \tag{2.5}$$

$$\text{For } t = 4, \text{ Internalizing Symptoms (IS)}_{i4} = \pi_{0i} + \pi_{1i} \times 4 + \varepsilon_{i4} \tag{2.6}$$

$$\text{For } t = 6, \text{ Internalizing Symptoms (IS)}_{i6} = \pi_{0i} + \pi_{1i} \times 6 + \varepsilon_{i6} \tag{2.7}$$

For each individual, these equations link IS measurements at the five time points using an intercept (π_{0i}) and a slope (π_{1i}). (In the above equations, residuals at each point in time are assumed to be normally distributed with means of zero and are assumed to have no correlations with other residuals (i.e., no correlated residuals).

These five equations (Equations 2.3 through 2.7) correspond to the measurement component of a structural equation model for the LGCM shown in Panel B of Figure 2.1. In this structural equation model, the intercept (π_{0i}) and slope (π_{1i}) can be considered as two latent variables measured by five manifest variables. The coefficients of π_0 in the five equations represent the factor loadings of the five indicators (i.e., measurements) for the "level" latent construct. These are almost always set to "1." The coefficients of π_1 for the five equations correspond to the factor loadings of the five measurements for the "rate of change" latent construct. Recall that in our example, these are set to 0, 1, 3, 4, and 6 respectively due to the spacing of available data time points. The level (intercept) is defined by the measurement of IS at the first time point, the point at which the factor loading is set to 0.

At the second stage, each person's intercept (π_0) and slope (π_1) can be combined with the intercepts and slopes of all sample respondents (i.e., inter-individual process) so that we have as many intercepts and slopes as we have study participants (n). Using this information, we can estimate an average intercept (μ_{00}), an average slope (μ_{10}), a correlation (ψ_{10}) between the intercept and slope, and an aggregate error variance of intercept (ψ_{00}) and slope (ψ_{11}). The univariate (i.e., unconditional) version of this model can be written as:

$$\pi_{0i} = \mu_{00} + \zeta_{0i}, \quad \text{where} \quad \zeta_{0i} \sim \text{NID}\ (0, \psi_{00}) \tag{2.8}$$

$$\pi_{1i} = \mu_{10} + \zeta_{1i}, \quad \text{where} \quad \zeta_{1i} \sim \text{NID}\ (0, \psi_{11}) \tag{2.9}$$

$$\Psi = \begin{bmatrix} \psi_{00} & \\ \psi_{10} & \psi_{11} \end{bmatrix} \tag{2.10}$$

where there are $i = 1, 2, \ldots, n$ respondents in the study, and Ψ represents the variance-covariance structure of π_{0i} and π_{1i}. The average slope (μ_{10}) describes the average overall rate of change in individuals' IS over time, or the mean trajectory of IS across the sample, while the population variance for the change parameter (ψ_{11}) reflects individual differences in the rate of change. The pattern of change over time in an attribute is determined by the means and variances of the intercepts (initial level) and slopes (rate of change) as well as the covariance between the intercepts and slopes for all the individuals. Figure 2.2 presents several possible patterns of linear change of an attribute (Lorenz, Wickrama, & Conger, 2004).

When estimating a conditional LGCM, we recommend following three steps: (1) investigate the longitudinal covariance patterns of observed variables in order to examine the feasibility of estimating growth curves (i.e., model identification), (2) estimate an unconditional LGCM to analyze change, and (3) estimate a conditional LGCM (adding covariates to the LGCM) to examine the associations between growth parameters and antecedents and/or distal outcomes. We discuss each of these steps individually.

Investigating Longitudinal Covariance Patterns

Although latent growth curves are easy to estimate, they may not always be successful. Successful growth curve modeling requires sufficient variability in the slope (i.e., significant variance of slope); without adequate variance in the slope there is no variation in the rate of change to be explained (Lorenz et al., 2004). In order to obtain some evidence for slope variance before estimating a LGCM,

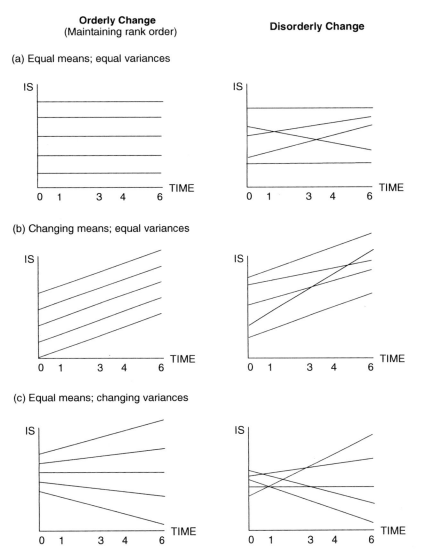

FIGURE 2.2 Patterns of Change Over Six Time Points (Adapted from Lorenz et al., 2004).
Note: IS = Internalizing Symptoms.

the longitudinal covariance patterns of the repeated measures should be examined because the growth parameters of the hypothesized model are derived from the sample correlation (covariance) matrix of the measured variables (Brown, 2006). As one example, using observed scores (y_{t1-t3}), Lorenz and colleagues

(2004) reported that, for three waves of data (assuming equal variances across the three waves), the equation for the variance of slope can be expressed as:

$$\mathrm{Var}\,(\pi_1) = \frac{(\sigma_{12} + \sigma_{23} - 2\sigma_{13})}{2} \tag{2.11}$$

where σ_{12} = observed covariance between observed y_{t1} and y_{t2}, σ_{23} = observed covariance between observed y_{t2} and y_{t3}, and σ_{13} = observed covariance between observed y_{t1} and y_{t3}. According to this formula, if the covariance between y_{t1} and y_{t3} is higher than covariance between y_{t1} and y_{t2} and between y_{t2} and y_{t3}, a negative variance of the slope is estimated. This negative slope variance leads to inadmissible solutions and the estimation of a LGCM is not possible (i.e., biased growth parameters; Hertzog, Oertzen, Ghisletta, & Lindenberger, 2008). Also, this formula indicates if there is a lack of variability in the slope (i.e., the three covariances [off-diagonal components] are equal or near equal [$\sigma_{12} = \sigma_{23} = \sigma_{13}$]). Empirical evidence of a lack of systematic change over time may be inferred when the variance of the slope is near zero.

Illustrative Example 2.1: Examining the Longitudinal Covariance Pattern of Indicators

In Table 2.2, a covariance matrix among internalizing symptoms (composite mean scores of anxiety, depression, and hostility symptoms) is provided. Observed covariances between two adjacent occasions (t and t+1) equal .148, .148, .136, and .083 (see Table 2.2). These adjacent covariances are higher than covariances between non-adjacent occasions (ranged from .057 to .127). This high covariance between two adjacent time points is preliminary evidence for the existence of a non-negative variance of slope parameter. Given this indication of a significant slope variation, a LGCM will likely fit well with the data structure. This approach can be extended to growth curve modeling with a larger number of repeated measures to examine the estimation feasibility.

TABLE 2.2 Covariance Matrix among Variables Comprising the Measurement Model (Unstandardized Coefficients).

	1	2	3	4	5
1. IS(W3)	.216				
2. IS(W4)	.148	.270			
3. IS(W6)	.127	.148	.289		
4. IS(W7)	.091	.103	.136	.222	
5. IS(W9)	.057	.075	.094	.083	.155

Note: IS = Internalizing Symptoms. All covariances were significant at $p < .001$.

Estimating an Unconditional Linear Latent Growth Curve Model (LGCM) Using Mplus

The next step is to estimate an unconditional LGCM. The following example illustrates the estimation of a linear latent growth curve model using five repeated measures.

Illustrative Example 2.2: Estimating a Linear Latent Growth Curve Model (LGCM)

M*plus* syntax and a figure representing an unconditional LGCM are shown in Panel A of Figure 2.3 (measurement errors are not shown). In the current dataset, there are seven variables, and all missing values are coded as "9." (Note that any number can be assigned to indicate missing values.) For the estimation of this LGCM, the five composite mean scores of internalizing symptoms (IS) were utilized (USEVAR = IS1−IS5). The model syntax specifies the appropriate factor loadings corresponding to the (t) time intervals, 0, 1, 3, 4, and 6, for computing the intercept (I) and slope (S). The coefficients of the intercept growth factors are fixed at one as the default. The | symbol is used in the syntax to name and define random effect variables and to specify growth models.

The M*plus* output is shown in Table 2.3. As can be seen in Table 2.3, the average intercept (I) and slope (S) were statistically significant (1.490, $p < .001$ and -.027, $p < .001$, respectively). The mean trajectory of IS declined over time. The variances of both the intercept and slope were statistically significant (.165, $p < .000$ and .004, $p < .001$, respectively). The statistically significant variance in the level suggests that some adolescents have higher initial levels of IS, while others have lower initial levels of IS, and still others have initial IS levels that are close to the mean. The statistically significant slope variance suggests that some adolescents have a higher rate of change in IS over time, compared to others with lower rate of change in IS, and still others maintain the same level of IS over time. A negative covariance (-.016, $p < .001$) was found between the intercept and slope, which indicates that, on average, individuals with a high initial level (i.e., intercept) were more likely to experience declining IS over time (i.e., a regression toward the mean trajectory) compared to other individuals. Using the drop-down menu in M*plus* (Plot → View Plots → Estimated Means) allows the user to obtain a visual depiction of the trajectory estimated by the PLOT command.

Curvilinear Growth Curve Modeling (i.e., A Quadratic Growth Curve Model)

Thus far, we have assumed that the longitudinal pattern (i.e., the trajectory) of internalizing symptoms (IS) develops linearly over time. That is, on average, a one unit change in time is associated with a μ_{10} unit change in the outcome, and the magnitude of this association is constant across all time points. Although this may

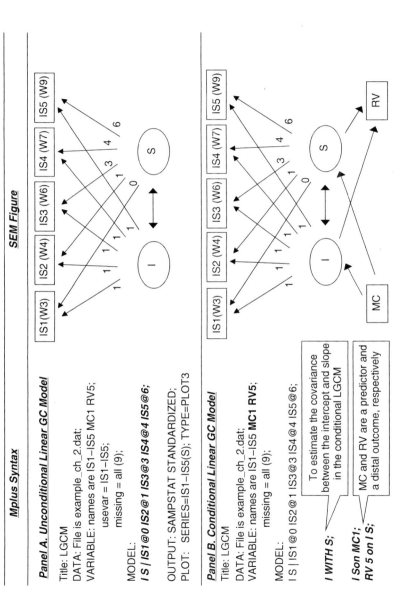

FIGURE 2.3 *Mplus* Syntaxes and SEM Figures for a Linear Latent Growth Curve Model (LGCM).
Note: IS = Internalizing Symptoms. I = Initial Level. S = Linear Slope. MC = Marital Conflict. RV = Romantic Violence.

TABLE 2.3 Unconditional Linear Growth Curve Model (See Panel A of Figure 2.3; Unstandardized Coefficients).

MODEL FIT INFORMATION

Information Criteria
Akaike (AIC)   2175.250
Bayesian (BIC)   2216.026
Sample–Size Adjusted BIC   2184.291

Chi–Square Test of Model Fit
Value   70.741
Degrees of Freedom   10
P-Value   0.0000

RMSEA (Root Mean Square Of Approximation)
Estimate   0.118
90 Percent C.I.   0.093   0.145
Probability RMSEA <= .05   0.000

CFI / TLI
CFI   0.892
TLI   0.892

SRMR (Standardized Root Square Residual)
Value   0.067

MODEL RESULTS

	Estimate	S.E.	Est. / S.E.	Two-Tailed P-Value
S WITH				
I	−0.016	0.002	−6.587	0.000
Means				
I	1.490	0.023	65.804	0.000
S	−0.027	0.004	−6.458	0.000
Variances				
I	0.165	0.015	10.771	0.000
S	0.004	0.001	6.009	0.000
Residual Variances				
IS1	0.066	0.010	6.348	0.000
IS2	0.121	0.011	10.872	0.000
IS3	0.171	0.014	12.356	0.000
IS4	0.129	0.011	12.218	0.000
IS5	0.060	0.010	5.702	0.000

accurately describe the stability or change in a behavior over time, there may be theoretical or empirical reasons to believe that the repeated measures are related to time in some nonlinear fashion, where change in an attribute is not consistent across repeated assessments. In some of these instances, growth may be described as following a quadratic form, and a quadratic growth model (i.e., a polynomial function) may provide a significant improvement over the linear equation. Unlike the linear model, which is defined by an intercept factor and a slope factor, the quadratic model includes a third latent factor to capture any curvature in the slope that might be present (see Curran & Hussong, 2003). The individual IS trajectory for a quadratic model is:

$$(IS)_{it} = \pi_{0i} + \pi_{1i} t + \pi_{2i} t^2 + \varepsilon_{it}, \quad \varepsilon_{it} \sim NID\,(0,\,\sigma_{it}^2) \tag{2.12}$$

where π_{0i} is the intercept of the trajectory, π_{1i} is the linear component of the trajectory, and π_{2i} is the quadratic component of the trajectory. Like the linear model, in a quadratic trajectory model each person's intercept (π_0), linear slope (π_1), and quadratic slope (π_2) can be combined with all other intercepts and slopes (i.e., an inter-individual process) to estimate an average intercept (μ_{00}), an average linear slope (μ_{10}), an average quadratic slope (μ_{20}), and aggregate error variances for the intercept (ψ_{00}), linear slope (ψ_{11}), and quadratic slope (ψ_{22}). The model (i.e., an unconditional quadratic LGCM) can be written as:

$$\pi_{0i} = \mu_{00} + \zeta_{0i}, \quad \text{where} \quad \zeta_{0i} \sim NID\,(0,\,\psi_{00}) \tag{2.13}$$

$$\pi_{1i} = \mu_{10} + \zeta_{1i}, \quad \text{where} \quad \zeta_{1i} \sim NID\,(0,\,\psi_{11}) \tag{2.14}$$

$$\pi_{2i} = \mu_{20} + \zeta_{2i}, \quad \text{where} \quad \zeta_{2i} \sim NID\,(0,\,\psi_{22}) \tag{2.15}$$

$$\Psi = \begin{bmatrix} \psi_{00} & & \\ \psi_{10} & \psi_{11} & \\ \psi_{20} & \psi_{21} & \psi_{22} \end{bmatrix} \tag{2.16}$$

where $i = 1, 2, \ldots, n$ respondents in the study. Ψ represents the variance-covariance structure of π_{0i}, π_{1i}, and π_{2i}. ψ_{00}, ψ_{11}, and ψ_{22} denote the variances of π_{0i}, π_{1i}, and π_{2i}, respectively. ψ_{10}, ψ_{20}, and ψ_{21} denote the covariances between π_{0i} and π_{1i}, π_{0i} and π_{2i}, and π_{1i} and π_{2i}, respectively.

The average quadratic slope (μ_{20}; rate of a change) describes the average quadratic change (i.e., the accelerating or decelerating change) in participants' IS over time, while the population variance for the quadratic change parameter (ψ_{22}) reflects differences in the quadratic change across individuals after controlling for linear changes. Factor loadings for this quadratic slope (π_2) should be fixed to correspond to the square of the linear slopes' loadings (i.e., 0, 1, 9, 16, and 36). The unconditional quadratic model is presented in Figure 2.4.

It is important to note that the number of time points available will influence how change over time can be modeled. With three time points, it is possible to

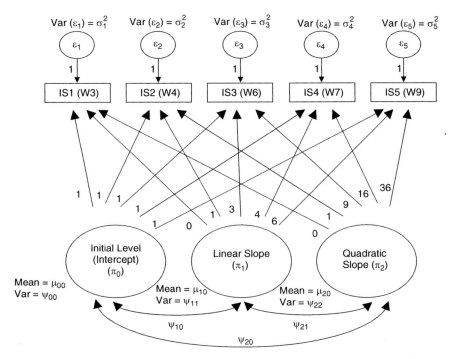

FIGURE 2.4 Quadratic Growth Curve Model (Unconditional Trajectory Model).
Note: Var = Variance.

model a linear pattern and a latent basis model; with four or more time points, linear and quadratic patterns can be modeled.

Illustrative Example 2.3: Estimating a Quadratic Latent Growth Curve Model (LGCM)

First, we estimated an unconditional quadratic LGCM using M*plus*. The M*plus* syntax and a figure of the unconditional quadratic LGCM estimated are presented in Panel A of Figure 2.5 (measurement errors are not shown).

For the estimation of a quadratic LGCM, the model syntax specifies the appropriate factor loadings corresponding to the time intervals (0, 1, 3, 4, and 6) for a linear slope (S) and the square of the linear slope loadings (0, 1, 9, 16, and 36) for a quadratic slope (Q). The remaining syntax is identical to the syntax used to estimate a linear LGCM. The coefficients of the intercept growth factors are fixed at one as the default. The M*plus* output for this unconditional quadratic LGCM is shown in Table 2.4.

As can be seen in Table 2.4, compared to the unconditional linear LGCM, overall model fit indices (except for the chi-square statistic) of the unconditional

Mplus Syntax

SEM Figure

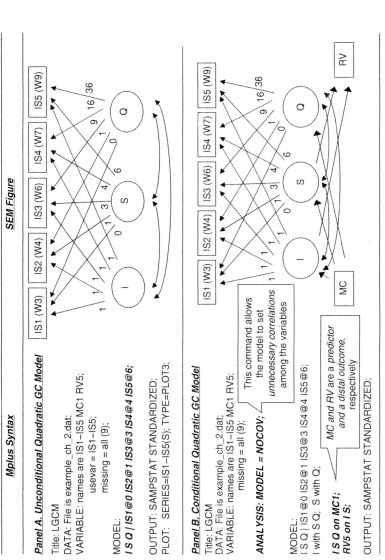

Panel A. Unconditional Quadratic GC Model

Title: LGCM
DATA: File is example_ch_2.dat;
VARIABLE: names are IS1–IS5 MC1 RV5;
    usevar = IS1–IS5;
    missing = all (9);

MODEL:
I S Q | IS1@0 IS2@1 IS3@3 IS4@4 IS5@6;

OUTPUT: SAMPSTAT STANDARDIZED;
PLOT: SERIES=IS1–IS5(S); TYPE=PLOT3;

Panel B. Conditional Quadratic GC Model

Title: LGCM
DATA: File is example_ch_2.dat;
VARIABLE: names are IS1–IS5 MC1 RV5;
    missing = all (9);

ANALYSIS: MODEL = NOCOV;

> This command allows the model to set unnecessary correlations among the variables

MODEL:
I S Q | IS1@0 IS2@1 IS3@3 IS4@4 IS5@6;
I with S Q; S with Q;

I S Q on MC1;
RV5 on I S;

> MC and RV are a predictor and a distal outcome, respectively

OUTPUT: SAMPSTAT STANDARDIZED;

FIGURE 2.5 *Mplus* Syntaxes and SEM Figures for a Quadratic Latent Growth Curve Model (LGCM).
Note: IS = Internalizing Symptoms. I = Initial Level. S = Linear Slope. Q = Quadratic Slope. MC = Marital Conflict. RV = Romantic Violence.

TABLE 2.4 Unconditional Quadratic Growth Curve Model (See Panel A of Figure 2.5; Unstandardized Coefficients).

MODEL FIT INFORMATION

Information Criteria
Akaike (AIC)   2155.159
Bayesian (BIC)   2212.246
Sample–Size Adjusted BIC   2167.817

Chi–Square Test of Model Fit
Value   42.651
Degrees of Freedom   6
P-Value   0.0004

RMSEA (Root Mean Square Of Approximation)
Estimate   0.118
90 Percent C.I.   0.086   0.153
Probability RMSEA <= .05   0.000

CFI / TLI
CFI   0.935
TLI   0.891

SRMR (Standardized Root Square Residual)
Value   0.037

MODEL RESULTS

		Estimate	S.E.	Est. / S.E.	Two-Tailed P-Value
S	WITH				
I		−0.013	0.010	−1.307	0.191
Q	WITH				
I		−0.001	0.001	0.618	0.537
S		−0.002	0.001	−1.256	0.209
Means					
I		1.467	0.023	64.651	0.000
S		0.021	0.013	1.621	0.105
Q		−0.008	0.002	−3.895	0.000
Variances					
I		0.162	0.019	8.633	0.000
S		0.014	0.009	1.878	0.060
Q		0.000	0.000	0.872	0.838
Residual Variances					
IS1		0.055	0.015	3.717	0.000
IS2		0.122	0.012	10.432	0.000
IS3		0.141	0.014	10.144	0.000
IS4		0.117	0.012	9.563	0.000
IS5		0.089	0.035	2.860	0.004

quadratic model do not show significant improvement (model fit indices will be discussed in the next section). The average intercept (I) was similar to the linear model (1.467, $p < .000$), and the model indicates that the linear trajectory of IS (that is, the slope, or S) is now stable over time (i.e., no average change; .021, $p = .105$) after controlling for the quadratic trajectory. The quadratic trajectory declined significantly over time (i.e., a decelerating change; -.008, $p < .001$). Only the variance of the intercept remained statistically significant (.162, $p < .001$). These results indicate that there were inter-individual differences in the intercept (but non-significant inter-individual differences in linear and quadratic trajectories at the $p < .05$ level). Covariances among the intercept, linear slope, and quadratic slope variances were not statistically significant.

Another alternative growth curve model is a slope-segment model (i.e., a piecewise growth model). As shown in Panel A of Figure 2.6, a linear growth curve may be modeled to have two linear slopes corresponding to different time periods. This model may be appropriate when different rates of change are expected for different stages over time. For instance, in an intervention study, a strong positive rate of change may be expected between pre- and post-intervention with a weak negative rate of change over the following six months. For the estimation of a piecewise growth model, the model syntax specifies the appropriate factor loadings for two slope segments (S1 and S2) for two adjoining time periods with a single intercept.

Model Fit Indices

One of the primary concerns researchers have when assessing models is the overall fit of the model to the data. In evaluating the goodness-of-fit of a LGCM, several model fit statistics are used. Most model fit statistics are derived from the chi-square goodness-of-fit statistic. Previous structural equation modeling textbooks often provide a detailed discussion of these indices (e.g., Brown, 2006). Thus, we only provide a short introduction to several of the most useful model fit indices.

The most often used model fit index is the chi-square statistic (χ^2), which is based on maximum likelihood model estimation. It is calculated as:

$$\chi^2 = F_{MI}(\text{sample size} - 1) \tag{2.17}$$

where F_{ML} represents the F-value reflecting minimization of the maximum likelihood criterion. The χ^2 value is associated with df (degrees of freedom) and is used to test the conventional null hypothesis for the goodness of model fit (Barrett, 2007). If the discrepancy between the covariances implied by the model and the observed sample covariances is large (as expressed by χ^2), then the model is not supported. However, the chi-square statistic is sensitive to sample size. Models based on small sample sizes are not easily rejected, whereas models based on large

Mplus Syntax

SEM Figure

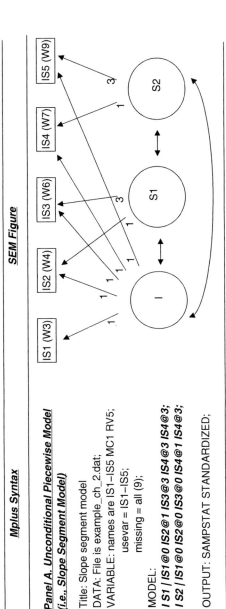

Panel A. Unconditional Piecewise Model (i.e., Slope Segment Model)

Title: Slope segment model
DATA: File is example_ch_2.dat;
VARIABLE: names are IS1–IS5 MC1 RV5;
    usevar = IS1–IS5;
    missing = all (9);

MODEL:
I S1 | IS1@0 IS2@1 IS3@3 IS4@3 IS4@3;
I S2 | IS1@0 IS2@0 IS3@0 IS4@1 IS4@3;

OUTPUT: SAMPSTAT STANDARDIZED;

Panel B. Unconditional Linear Model with Correlated Measurement Errors between Adjacent Time Points

Title: LGCM with correlated measurement errors
DATA: File is example_ch_2.dat;
VARIABLE: names are IS1–IS5 MC1 RV5;
    usevar = IS1–IS5;
    missing = all (9);

MODEL:
I S | IS1@0 IS2@1 IS3@3 IS4@4 IS5@6;

IS1 IS2 IS3 IS4 PWITH IS2 IS3 IS4 IS5;

OUTPUT: SAMPSTAT STANDARDIZED;

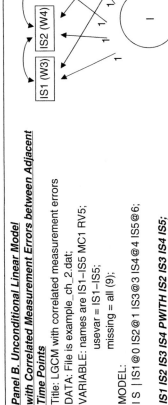

FIGURE 2.6 *Mplus* Syntaxes and SEM Figures for Other Types of Latent Growth Curve Models (LGCMs).
Note: IS = Internalizing Symptoms. I = Initial Level. S1 = Slope Segment 1. S2 = Slope Segment 2.

sample sizes are easily rejected. In order to take into account the effect of sample size as well as model complexity, several alternative model fit indices have been developed. These include the standardized root mean residual (SRMR), the root mean square error of approximation (RMSEA), the comparative fit index (CFI), and the Tucker-Lewis index (TLI). In the following paragraphs, we discuss each of these summary statistics.

The standardized root mean residual (SRMR) is calculated by dividing the sum of the squared elements of the residual correlation matrix by the number of elements in the matrix (on and below diagonal), and then taking the square root (SQRT) of this value. This value is the standardized discrepancy between observed and predicted correlations. The SRMR ranges from 0.0 to 1.0 with smaller SRMR values indicating better model fit.

The root mean square error of approximation (RMSEA) (Steiger & Lind, 1980) is another model fit index that compensates for model complexity. The RMSEA is calculated as:

$$\text{RMSEA} = \text{SQRT } (d/df) \tag{2.18}$$

where $d = \chi^2 - df$ / (sample size $- 1$) and $df =$ the degrees of freedom available in the hypothesized model. The RMSEA is relatively insensitive to sample size. RMSEA values of 0 indicate perfect model fit to the data.

The comparative fit index (CFI) evaluates the fit of a hypothesized model compared to a reduced nested model (known as the restricted model) (Hu & Bentler, 1998). The CFI is computed as follows:

$$\text{CFI} = 1 - \text{maximum}\left[(\chi_T^2 - df_T), 0\right] / \text{maximum}\left[(\chi_T^2 - df_T), (\chi_B^2 - df_B), 0\right] \tag{2.19}$$

where χ_T^2 is the χ^2 value of the hypothesized model, df_T is the df of the hypothesized model, χ_B^2 is the χ^2 value of the reduced/restricted model (i.e., the "null" model), and df_B is the df of the restricted model. The CFI has a range of possible values from 0.0 to 1.0, with values closer to 1.0 indicating a good model fit.

The Tucker-Lewis index (TLI) (Tucker & Lewis, 1973) is another model fit index that takes model complexity into account. The TLI is calculated by using the formula:

$$\text{TLI} = \left[(\chi_B^2 / df_B) - (\chi_T^2 / df_T)\right] / \left[(\chi_B^2 / df_B) - 1\right] \tag{2.20}$$

The TLI also compares the hypothesized model with a reduced/constrained model. Similar to the CFI, TLI values approaching 1.0 indicate a good model fit. Hu and Bentler (1999) suggested model fit cut-off values as follows: (1) SRMR values close to .08 or below, (2) RMSEA values close to .06 or below, and (3) CFI and TLI values close to .95 or greater.

The chi-square (χ^2), CFI, TLI, RMSEA, and SRMR values for the linear growth curve (Table 2.3) were 70.741(10 df), .892, .892, .118, and .067, respectively.

Only the SRMR indicates that the linear growth curve model was an acceptable fit to the data. The chi-square (χ^2), CFI, TLI, RMSEA, and SRMR values for the quadratic growth curve (Table 2.4) were 42.651 (6 *df*), .935, .891, .118, and .037, respectively, and only the SRMR and CFI indicate that the quadratic model was an acceptable fit with the data.

Comparing Nested Models

The nested χ^2 difference test (χ^2_{DIFF}) is widely used to compare two competing nested models where one model is a subset (or reduced form) of another model and determine which model provides a better fit to the data (e.g., the linear trajectory vs. the quadratic trajectory). A nested reduced model is obtained simply by fixing/constraining or dropping parameters from the more complex model (i.e., the reduced model is the parsimonious model). Using our example, the linear growth curve model is nested in the quadratic growth curve model; if all of the paths comprising the quadratic factor are dropped (or all factor loadings and variances are fixed to zero), the linear model is formed. To compute a χ^2_{DIFF}, the difference of the χ^2 values for the two competing models is utilized as well as the difference in the degrees of freedom (df) between the two models. The test formula can be written as:

$$\chi^2_{\text{DIFF}} = \chi^2_{\text{r}} - \chi^2_{\text{f}} \text{ and } df_{\text{diff}} = df_{\text{r}} - df_{\text{f}} \tag{2.21}$$

Here, r denotes the "reduced" model with fewer parameters and, therefore, more degrees of freedom (i.e., the linear model), whereas f denotes the "full" model with more parameters and, therefore, fewer degrees of freedom (e.g., the quadratic model).

This χ^2_{DIFF} value is distributed with df_{diff} and can be checked manually for significance using a χ^2 table. If the χ^2_{DIFF} value is statistically significant, the "full" model with more freely estimated parameters fits the data better than the "reduced" model with fewer parameters. When the χ^2_{DIFF} value is not statistically significant, both models fit equally well from a statistical perspective, so the additional free parameter can be dropped from the model (fixed to zero) and the reduced, more parsimonious, model is considered the better fitting model.

Illustrative Example 2.4: Nested Model Comparison Between Linear and Quadratic Models

In our example, considering the two competing models (linear vs. quadratic), as seen in Tables 2.3 and 2.4, the quadratic model has a $\chi^2 = 24.537$ and df = 6, whereas the linear model has a $\chi^2 = 65.592$ and df = 10. Thus, following Equation 2.21, the χ^2_{DIFF} is calculated as follows:

	Degree of Freedom (df)	χ^2
Linear Growth Model (reduced)	10	70.741
Quadratic Growth Model	6	42.651
χ^2 Difference (χ^2_{DIFF})	4	28.090

As shown above, for our example the χ^2_{DIFF} (4) = 28.090. The critical value of χ^2_{DIFF} is 9.49 (based on an alpha = .05, df = 4). Because the χ^2_{DIFF} value exceeds 9.49, it would be concluded that the quadratic model provides a significantly better fit to the data than the linear model. However, the linear growth model has a higher TLI value than the quadratic model suggesting the linear model has a better fit than the quadratic model when model complexity is taken into account. Thus, different model fit indices provide different information. We recommend looking for consistent evidence across multiple tests/indices when making conclusions about the model fit.

Illustrative Example 2.5: Nested Model Comparison Between Models with and Without Correlated Errors

Similarly, nested model comparisons can be used to compare two competing models between a model with correlated measurement errors and one with uncorrelated measurement errors because an uncorrelated error model is a reduced form of a correlated error model. It is a common phenomenon with longitudinal data to have high correlations between adjacent time points due to reporter bias (Campbell & Reichardt, 1991). In order to model these correlated error structures, measurement errors (see Panel B of Figure 2.6) can be correlated in a growth curve model (i.e., autocorrelated errors). The M*plus* output for this model is shown in Table 2.5. Using the unconditional linear model shown in Panel A of Figure 2.3 (i.e., a reduced linear model of the quadratic model), the chi-square difference test indicates that incorporating correlated measurement errors in the linear growth model significantly improved the model fit (χ^2_{DIFF} (df) = 17.043(4), critical value of χ^2 (4) = 9.49). However, all model indices (except for the CFI) indicate that both the uncorrelated error model (see Table 2.3; CFI/TLI = .892/.892, RMSEA = .118, and SRMR = .067) and the correlated error measurement model (CFI/TLI = .915/.858, RMSEA = .135, and SRMR = .074) were not acceptable.

Illustrative Example 2.6: Non-nested Model Comparison Between Linear and Piecewise Models

Another alternative growth curve model is a slope-segment model (i.e., a piecewise growth model). As shown in Panel A of Figure 2.6, a linear growth

TABLE 2.5 Results from the Unconditional Linear Growth Curve Model with Correlated Measurement Errors between Two Adjacent Time Points (See Panel B of Figure 2.6; Unstandardized Coefficients).

MODEL FIT INFORMATION

Information Criteria	
Akaike (AIC)	2166.206
Bayesian (BIC)	2223.293
Sample–Size Adjusted BIC	2178.864

Chi–Square Test of Model Fit	
Value	53.698
Degrees of Freedom	6
P-Value	0.0000

RMSEA (Root Mean Square Of Approximation)	
Estimate	0.135
90 Percent C.I.	0.103 0.169
Probability RMSEA <= .05	0.000

CFI / TLI	
CFI	0.915
TLI	0.858

SRMR (Standardized Root Square Residual)	
Value	0.074

MODEL RESULTS

		Estimate	S.E.	Est. / S.E.	Two-Tailed P-Value
S	WITH				
I		−0.020	0.004	−4.500	0.000
IS1	WITH				
IS2		−0.010	0.018	−0.578	0.563
IS2	WITH				
IS3		0.028	0.010	2.689	0.007
IS3	WITH				
IS4		0.028	0.011	2.496	0.013
IS4	WITH				
IS5		−0.019	0.013	−1.448	0.148
Means					
I		1.479	0.023	64.153	0.000
S		−0.027	0.004	−6.288	0.000
Variances					
I		0.177	0.025	7.130	0.000
S		0.005	0.001	4.575	0.000
Residual Variances					
IS1		0.043	0.022	1.943	0.052
IS2		0.121	0.019	6.419	0.000
IS3		0.194	0.016	11.824	0.000
IS4		0.124	0.014	8.567	0.000
IS5		0.028	0.020	1.381	0.167

curve may be modeled to have two linear slopes corresponding to different time periods. The M*plus* output for a piecewise growth model is shown in Table 2.6. The model fit indices indicate this type of model is a good fit to the data (χ^2(df) = 31.362, p < .05, CFI/TLI = .955/.925, RMSEA = .098, and SRMR = .031). However, this is a non-nested model because this model is not nested within a linear growth model with a single slope. Thus, a nested chi-square difference test ($\Delta\chi^2$) cannot be utilized for model comparison.

In order to compare non-nested models (e.g., a linear growth curve model with two linear slope segments compared to a linear model with one slope), the Akaike information criteria (AIC; Brown, 2006) can be used. The model with the lower AIC value is preferred. In our example, the piecewise model with two slope segments has an AIC value of 2143.870 whereas the linear growth model has an AIC value of 2175.250. However, for the purpose of our illustrations, we will use a conventional linear growth curve model (LGCM) for all subsequent analyses.

Adding Covariates to an Unconditional Model

Statistically significant variances in the growth parameters (intercept and slope) show heterogeneity among individual-specific trajectories. This variance is the variability that predictors should explain in the analysis of change. Thus, in addition to the simple description of change, growth curve modeling allows us to explain systematic inter-individual differences in both the level and rate of change for internalizing symptoms (IS). When the change parameter (e.g., rate of change, or slope, for IS) covaries significantly with a predictor variable (e.g., parents' marital conflict at Time 1 [MC1]), differences in the rate of change in symptoms across individuals are said to be systematic (Willet & Sayer, 1994).

Thus, predictors of a LGCM can be written as the structural part of a SEM in equation form:

$$\pi_{0i} = \mu_{00} + \gamma_0\, MC1_i + \zeta_{0i}, \quad \text{where} \quad \zeta_{0i} \sim NID(0, \psi_{00}) \tag{2.22}$$

$$\pi_{1i} = \mu_{10} + \gamma_1\, MC1_i + \zeta_{1i}, \quad \text{where} \quad \zeta_{1i} \sim NID(0, \psi_{11}) \tag{2.23}$$

For our illustrative purpose, we have used parents' marital conflict (MC1) as a predictor. γ_0 is the magnitude of the coefficient linking the covariate (MC1) to the intercept, and γ_1 is the magnitude of the coefficient linking MC1 to the slope of internalizing symptoms. A growth curve model, like that presented in the above two-level equations, can also be considered a multilevel model. Hence, regression coefficients and variance components can also be estimated as a multilevel model using multilevel software (e.g., SAS Procmixed, HLM).

One advantage of utilizing a SEM approach instead of a multilevel approach is that a SEM approach allows for the inclusion of subsequent outcomes (e.g., young adults' romantic relationship violence at Time 5 [RV5]) predicted by growth parameters within the same analytical framework. In a SEM, the level of

TABLE 2.6 Unconditional Piecewise Growth Curve Model (See Panel A of Figure 2.6; Unstandardized Coefficients).

MODEL FIT INFORMATION

Information Criteria	
Akaike (AIC)	2143.870
Bayesian (BIC)	2200.957
Sample–Size Adjusted BIC	2156.528
Chi–Square Test of Model Fit	
Value	31.362
Degrees of Freedom	6
P-Value	0.0153
RMSEA (Root Mean Square Of Approximation)	
Estimate	0.098
90 Percent C.I.	0.066 0.134
Probability RMSEA <= .05	0.008
CFI / TLI	
CFI	0.955
TLI	0.925
SRMR (Standardized Root Square Residual)	
Value	0.031

MODEL RESULTS

		Estimate	S.E.	Est. / S.E.	Two-Tailed P-Value
S1	WITH				
I		−0.013	0.005	−2.369	0.018
S2	WITH				
I		−0.022	0.004	−5.302	0.000
S1		0.000	0.002	0.015	0.988
Means					
I		1.464	0.023	65.015	0.000
S1		0.008	0.008	1.039	0.299
S2		−0.063	0.008	−7.642	0.000
Variances					
I		0.159	0.017	9.408	0.000
S1		0.008	0.003	3.132	0.002
S2		0.005	0.004	1.241	0.215
Residual Variances					
IS1		0.059	0.012	4.729	0.000
IS2		0.122	0.011	10.937	0.000
IS3		0.124	0.016	7.852	0.000
IS4		0.120	0.011	10.635	0.000
IS5		0.080	0.026	3.112	0.002

RV can be expressed as a function of the level and slope of internalizing symptoms, and written as:

$$RV_i = \tau + \beta_1 \pi_{0i} + \beta_2 \pi_{1i} + \varepsilon_i \tag{2.24}$$

τ is the intercept for the regression of RV. β_1 is the magnitude of the coefficient linking the initial level parameter (π_{0i}) of IS to the outcome (RV). Likewise, β_2 is the magnitude of the coefficient linking the slope parameter (π_{1i}) of IS to the outcome (i.e., a conditional LGCM).

Using these formulas, we can investigate whether covariates are associated with inter-individual differences in the intercept and slope of IS.

Illustrative Example 2.7: Adding a Predictor and Outcome to a Linear LGCM

The M*plus* syntax and a figure of this conditional LGCM are provided in Panel B of Figure 2.3. In this model, we examine the effects of the covariate, parents' marital conflict (MC1), as a predictor of the intercept and slope of IS. We also simultaneously examine romantic violence of young adults at Time 5 (RV5) as an outcome of the intercept and slope of IS. The bold portions of the syntax add the covariates into the LGCM. Seven variables are used in this example (USEVAR= IS1–IS5 MC1 RV5). Note that when estimating the effects of both a predictor and outcome simultaneously using a conditional LGCM, the covariance between the intercept (I) and slope (S) is not estimated if the WITH statement (i.e., I WITH S) is not specified in the conditional LGCM. Thus, we specified I WITH S under the MODEL command. The first ON statement regresses the intercept and slope growth factors on the covariate (MC1). The second ON statement estimates the influence of the intercept and slope on an outcome variable of interest (RV5). This LGCM analysis can be extended to include multiple predictors (e.g., family economic hardship, hostile parenting, etc.) and outcomes (e.g., romantic violence, financial cutbacks, etc.). The M*plus* output for this conditional linear LGCM is shown in the dotted box in Table 2.7.

For interpreting the coefficients, marital conflict (MC1) positively influenced the initial level of adolescents' IS ($b = 0.153$, $p < .001$), whereas MC1 did not influence the slope of IS over time ($b = -0.011$, $p = .106$). That is, on average, adolescents experiencing higher levels of parents' MC exhibited higher initial levels of IS. Both the intercept and slope positively influenced romantic violence ($b = 1.153$, $p < .001$ and $b = 4.288$, $p < .01$, respectively). That is, both adolescents with a higher initial level of IS and those with a steeper rate of change in IS encountered more romantic violence compared to adolescents with a lower initial level and a less steep rate of change in IS.

TABLE 2.7 Conditional Linear Growth Curve Model (See Panel B of Figure 2.3; Unstandardized Coefficients).

MODEL FIT INFORMATION

Information Criteria	
Akaike (AIC)	2505.382
Bayesian (BIC)	2569.718
Sample–Size Adjusted BIC	2518.947

Chi–Square Test of Model Fit	
Value	85.784
Degrees of Freedom	17
P-Value	0.0000

RMSEA (Root Mean Square Of Approximation)	
Estimate	0.099
90 Percent C.I.	0.079 0.120
Probability RMSEA <= .05	0.000

CFI / TLI	
CFI	0.887
TLI	0.860

SRMR (Standardized Root Square Residual)	
Value	0.065

MODEL RESULTS

		Estimate	S.E.	Est. / S.E.	Two-Tailed P-Value
I	ON				
MC		0.153	0.036	4.242	0.000
S	ON				
MC		−0.011	0.007	−1.618	0.106
RV	ON				
I		1.153	0.213	5.424	0.000
S		4.588	1.777	2.582	0.010

Illustrative Example 2.8: Adding a Predictor and Outcome to a Quadratic LGCM

Like estimating a linear LGCM in a SEM framework, estimating a quadratic LGCM in a SEM framework also allows for the addition of multiple covariates (i.e., predictors and outcomes; a conditional quadratic LGCM). Thus, we can investigate whether covariates are systematically associated with the intercept, linear slope, and quadratic slope of a quadratic growth curve model (however, recall that in our example [see Table 2.4], variance of the quadratic slope was not significant at .05 level).

As with the linear model, we included parents' marital conflict as a predictor and romantic violence of young adults as an outcome. The M*plus* syntax and a figure of the conditional quadratic LGCM are shown in Panel B of Figure 2.5. Similar to the model specification of the conditional LGCM (see Panel B of Figure 2.3), the WITH statement should be specified under the MODEL command to estimate the covariances among the intercept (I), linear slope (S), and quadratic slope (Q) (i.e., I WITH S Q; S WITH Q). The effects of *both* a predictor and outcome are estimated by specifying the bold portions of the syntax in the MODEL

TABLE 2.8 Conditional Quadratic Growth Curve Model (See Panel B of Figure 2.5; Unstandardized Coefficients).

MODEL FIT INFORMATION

Information Criteria	
Akaike (AIC)	2477.007
Bayesian (BIC)	2561.448
Sample–Size Adjusted BIC	2494.811
Chi–Square Test of Model Fit	
Value	47.409
Degrees of Freedom	12
P-Value	0.0000
RMSEA (Root Mean Square Of Approximation)	
Estimate	0.085
90 Percent C.I.	0.060 0.111
Probability RMSEA <= .05	0.011
CFI / TLI	
CFI	0.942
TLI	0.898
SRMR (Standardized Root Square Residual)	
Value	0.031

MODEL RESULTS

		Estimate	S.E.	Est. / S.E.	Two-Tailed P-Value
I	ON				
MC		0.133	0.037	3.631	0.000
S	ON				
MC		0.025	0.019	1.333	0.183
Q	ON				
MC		−0.006	0.003	−2.060	0.039
RV	ON				
I		0.843	0.151	5.578	0.000
S		2.900	0.773	3.751	0.000

command. We have constrained some of the parameters in this example model to be zero in order to achieve model identification. The M*plus* output for the conditional quadratic LGCM is provided in Table 2.8.

The model fit indices suggest that the quadratic conditional model has an acceptable fit (CFI = .942 and SRMR = .031). In terms of path coefficients for the predictor (MC1), results indicated that MC1 was positively associated with the initial level of youth's internalizing symptoms (unstandardized b = .133, $p < .001$). Also, like the results of the conditional linear LGCM (see Table 2.7), MC1 was not associated with the linear slope (unstandardized b = .025, $p = .183$). Instead, MC1 was negatively associated with a quadratic slope (b = -.006, $p < .05$), indicating that parents' marital conflict was associated with deceleration (slowing down the escalation) of IS trajectories over time. For the distal outcome coefficients, the initial level (I) and linear slope (S) of IS predicted romantic violence (intercept: b = .843, $p < .001$, linear slope: b = 2.900, $p < .001$) indicating similar results as the conditional linear LGCM (see Table 2.7). That is, both adolescents with a higher initial level of IS and those with a steeper rate of linear change in IS encountered more romantic violence compared to adolescents with a lower initial level and a less steep rate of change in IS after taking the quadratic trend into account.

Methodological Concerns in Longitudinal Analysis: Why Growth Curves?

Latent growth curve modeling addresses important methodological concerns for analyzing panel data. In the next section, we discuss these concerns and the specifics of how they are addressed by a LGCM.

The Need to Preserve the Continuity of Change

Traditional regression methods and mean comparisons are not capable of preserving the continuity of change. Instead, regression techniques examine change over time using only two repeated measurements of a variable, incorporating very little information about change over time (Rogosa et al., 1982). Taking only two time points into consideration provides only two discrete snapshots linked together by the assumption of a continuous process (Karney & Bradbury, 1995). The traditional regression methods for analyzing change in individual outcomes typically include regressing the final measurement of an attribute on predictors after controlling for the initial level of the attribute. This is known as "taking residualized scores," "autoregression," or "cross-lagged models" because covariates are used to predict residualized scores of the final measurement after removing the "effect" (or association) of the initial measurement. This method entails several limitations. First, the non-dynamic nature of these models views the repeated measure as a status at two distinct time points rather than a process that unfolds over time (Coyne & Downey, 1991). Second, when individual change follows a non-linear trajectory, regression methods are unlikely to reveal intricacies of such change (Willet & Sayer, 1994).

Moreover, "ignoring the continuous nature of change process, traditional methods prevent empirical researchers from entertaining a richer, broader spectrum of research questions, questions that deal with the nature of individual development" (Willet, 1988, p. 347). For example, questions regarding the nature of change in developmental research—such as the prodromal development of symptoms—require researchers to view change as a continuous process using more than two time points, because developmental processes may be non-linear, which can only be estimated with more than two time points. Moreover, this process may be systematically associated with sociocontextual and developmental processes.

The Need to Investigate Different Growth Parameters

Individual-specific growth parameters (e.g., the initial level and rate of change) can capture different facets of a multifaceted process of development. These parameters include not only the intensity or severity (i.e., the initial level) but also the amount of growth or decline (i.e., the rate of change or slope) in an attribute (e.g., internalizing symptoms) over time. For example, the developmental course of an already depressed individual who has experienced a sharp increase in symptom level from "moderate" to "very high" is qualitatively different from the developmental course of a mentally healthy individual who has experienced the same amount of increase in depressive symptoms from "zero" to "moderate" over the same period of time. Although the aforementioned two trajectories have the same amount of change, they have important differences: specifically, different symptom levels (or intensities). Traditional methods are not sensitive enough to capture this difference and distinguish between these two courses of development, although these courses may have different antecedents and/or consequences (Karney & Bradbury, 1995; Rueter, Scaramella, Wallace, & Conger, 1999). That is, to understand and investigate the full course of an attribute's development (e.g., psychological symptoms or a behavior), multiple facets of change (e.g., both the intensity and the shape/rate of change) have to be taken into account. Furthermore, it is important to understand the relative contributions of different growth parameters of an attribute (e.g., level and rate of change) to an outcome (e.g., a specific disease of interest) (Eaton, Badawa, & Melton, 1995; Judd, 1994; Rueter et al., 1999). Similarly, the intensity level and rate of change may be independently affected by different factors, including sociocontextual factors.

The Need to Incorporate Growth Parameters as Either Predictors or Outcomes in the Same Model

It is important to test simultaneously the influence of antecedents on the developmental course of an individual attribute (e.g., symptom) and the relationship between the developmental course and the onset of subsequent outcomes (e.g., disorders). Testing antecedents and outcomes in this manner enables researchers to disentangle direct and indirect effects among the variables. Traditional regression and multivariate analysis of variance (MANOVA) methods are unable to include all growth parameters (e.g., the level and change) in the model simultaneously, both as independent outcomes of antecedents (e.g., sociocontextual factors) and predictors of later outcomes.

The Need to Incorporate Time-Varying Predictors

All aspects of change in a *predictor* (i.e., the level and rate of change) are important for understanding the associations with various aspects of change in another *study attribute* (i.e., its level and rate of change). For example, within-individual change in sociocontextual predictors may result in within-individual change in IS. Thus, a proper analysis should relate variations in growth parameters of an individual attribute to variations in sociocultural variables over time. This type of parallel-process, or associated growth curve, model (PPM) will be discussed in Chapter 5. Such an analysis provides a richer and deeper understanding of the dynamic relationship between time-varying sociocontextual factors and individual attributes than traditional analytical methods. Traditional regression and MANOVA methods are not flexible enough to entertain time-varying predictors while preserving their continuous nature. This limitation is an important drawback for researchers in various fields exploring sociocontextual factors, such as family, work, marriage, and other life experiences, because their impacts on individual psychological, health, and behavioral outcomes are constantly changing. A LGCM within a SEM framework fulfills the above need for analyzing change and allows for an investigation of the dynamic relationship between time-varying sociocontextual factors and time-varying individual outcomes.

Limitations

Although a LGCM approach within a SEM framework addresses several methodological issues related to the proper analysis of change, several potential limitations remain. First, although a LGCM estimates individual-specific growth parameters and examines systematic associations between growth parameters and other correlates, SEM is similar to traditional regression analysis, in that, the strength and nature of such associations are assumed to be the same for all individuals. Similarly, although the magnitude of growth parameters can differ across individuals, growth shapes (e.g., linear) are assumed to be the same for all individuals. It is possible that individual differences in growth shape and associations between growth parameters and covariates exist. Second, unlike traditional regression analysis, regression parameters from a SEM should be interpreted after acknowledging that some of the error terms in the model are allowed to correlate. Extreme caution should be used when error terms are correlated in a manner that is not theoretically or methodologically meaningful. Third, the statistical assumptions in SEM regarding distributional characteristics of the variables are more restrictive. Specifically, a SEM assumes univariate and multivariate normality of the variables. Last, there are various fit indices available to evaluate a SEM, and the appropriate indices for any given model must be identified depending on factors such as sample size, number of variables, and deviation from distributional assumptions.

Beyond Latent Growth Curve Modeling

Researchers in a variety of disciplines have applied LGCM to describe and analyze change in personal characteristics, behaviors, and health outcomes as well as relationship attributes over time (Liu, Rovine, & Molenaar, 2012). However, the basic model has limitations and the proper analysis of change may warrant extensions of a conventional LGCM for several reasons. First, a conventional LGCM does not allow for higher-order factor structures. For instance, a complex multidimensional data structure may exist (e.g., anxiety, depression, and hostility symptoms as different dimensions of internalizing symptoms). To properly investigate this possibility, a LGCM can be extended to a second-order latent growth model, which captures potential factor structures in a longitudinal context. Second, a conventional LGCM does not take into account potential heterogeneity, or clustering, that may exist among individual trajectories. Instead, it assumes that all individuals come from a single population and the same pattern of growth can be applied to the entire population (homogeneity). A growth mixture model (GMM), an extension of a LGCM, can be used to address this by identifying potential unobserved heterogeneity in individual trajectories. A GMM, as an emerging statistical approach, models heterogeneity by classifying individuals into groups with similar patterns, or latent classes of trajectories (Muthén, 2003).

More importantly, it is possible to combine these two LGCM extensions, incorporating both higher-order factor structures and growth mixture modeling (GMM) into various types of higher-order GMMs. Although these extensions of a LGCM are increasingly being used in research spanning a wide variety of fields, including psychological, epidemiological, socioeconomic, political, and developmental studies, a practical guide for applying these advanced statistical analyses is not currently available. Thus, one important contribution of this volume is to provide an introduction and practical guide to the cutting-edge application of various LGCM extensions (rather than a conceptual and statistical presentation of LGCMs) for social science and related fields (e.g., psychological, epidemiological, socioeconomic, political, and developmental studies).

Revisiting the Layout of Models: Figures 1.1, 1.2, and 1.3

Having discussed latent growth curve modeling in detail, we revisit Figures 1.1—1.3 from Chapter 1 in order to illustrate our arrangement of models throughout the book with an emphasis on their sequential incremental relationships and also to introduce acronyms for these models. Following this arrangement of models, this book provides an overview of how to model multidimensional data structures using first-order models, second-order growth curve models, and growth mixture models.

First-Order Structural Equation Models

As shown in Figures 1.1–1.3, three basic structural equation models are used to develop second-order models. We discuss these three first-order SEMs: (1) latent growth curve models, LGCM; (2) longitudinal confirmatory factor models, LCFM, and (3) parallel process growth curve models, PPM. We have already discussed latent growth curve models in this chapter. The remaining first-order SEMs will be discussed in subsequent chapters.

Second-Order Growth Curve Modeling

Next, we introduce second-order growth curve models (SOGCMs). As shown in Figures 1.1–1.3, a first-order model can be extended to a second-order LGCM in two alternative modeling pathways. A latent confirmatory factor model (LCFM) can be extended to a curve-of-factors growth curve model (CFM) (see Figure 1.1 and Panel C of Figure 1.2, respectively) and a parallel process latent growth curve model (PPM) can be extended to a factor-of-curves growth curve model (FCM) (see Panels B and D of Figure 1.2, respectively) (Duncan, Duncan, & Strycker, 2006). Like first-order growth model parameters, these two second-order growth models have a mean and variance for both the intercept and slope factors as well as a factor covariance.

SOGCMs have the advantage of creating theoretically error-free latent constructs to be used as growth indicators (i.e., time-varying indicators) for higher-order growth modeling rather than using observed scores as the growth indicators. In the basic LGCM example (see Figure 2.1), we used composite scores (mean scores constructed from scores of the three symptoms at each time point) without considering measurement errors when estimating the growth curve. Failure to account for measurement errors in the model can lead to specification error, misfit, and biased parameter estimates. Creating each latent variable using multiple indictors (i.e., symptoms of anxiety, depression, and hostility) of internalizing symptoms for each time point (measurement model: LCFM; see Figure 1.1; Little, 2013) accounts for measurement error separately for each symptom. These latent factors are then used as indicators of a second-order growth curve (i.e., CFM). Alternatively, measurement errors can be accounted for in first-order growth curve models (PPM), which can serve as a measurement model for a FCM.

Growth Mixture Modeling

Third, we introduce growth mixture models (GMMs) to take the heterogeneity of trajectories into account. We use step-by-step examples to illustrate three prominent types of GMMs. As shown in Figure 1.3, this includes two second-order growth mixture models (SOGMMs), including a factor-of-curves model (FCM) and a curve-of-factors model (CFM) as well as a GMM incorporating parallel process growth curves (PPM) (Grimm & Ram, 2009). GMMs can also be extended to include predictors and subsequent outcomes (i.e., a conditional model). Recent studies have suggested a 3-step approach or specific M*plus* syntax to estimate unbiased class proportions when adding covariates in a GMM (Asparouhov & Muthén, 2014). Thus, we will introduce M*plus* procedures for taking uncertainty in class proportions into account when incorporating covariates in a GMM.

Chapter 2 Exercises

Note that all chapter exercises and examples were created using M*plus* Version 7.4 (with Windows Version 7). When using other versions, results may vary somewhat.

This sample dataset includes composite mean scores for three indicators of internalizing symptoms (depression, anxiety, and hostility symptoms) for mothers at T1~T5. The M*plus* syntax for the data is as follows:

DATA: File is exercise_Ch.2.dat;

VARIABLE: names are IS1M IS2M IS3M IS4M IS5M IS5T MU1 DEP5T;

    MISSSING ARE ALL (9);

Note that IS1M~IS5M indicate mothers' internalizing symptoms at T1~T5. MU1 indicates marital unhappiness at T1. DEP5T indicates target adolescents' depressive symptoms at T5.

1. Researchers commonly encounter several convergence issues leading to the inability to produce standard errors when estimating an unconditional latent growth curve model (LGCM). One of the main reasons for these convergence issues is the negative variance of slope factors, which is not admissible for estimation.

   By specifying the *SAMPSTAT* option in the *OUTPUT* command of the M*plus* syntax, a covariance matrix among all repeated measures can be estimated. Using this command option, investigate whether the example dataset is appropriate for estimating parameters of a LGCM.

2. In this chapter, we introduced growth curves with several functions of time (or age change). The optimal time function could be linear or non-linear, such as a quadratic function. In order to detect the optimal trajectory,

   A. Fit models assessing three possible time functions: linear, quadratic, and piecewise (using T3 as a distinct time point), and complete Exercise Table 2.1 on the following page by entering the results.

EXERCISE TABLE 2.1 Model Comparisons among Various Unconditional LGCMs.

Growth curve models with different time functions	Model Fit Indices							
	χ^2	df	$\Delta\chi^2(df)$, p-value	Model Comparison	CFI / TLI	SRMR	RMSEA	AIC
Linear Time Function (M1)								
Quadratic Time Function (M2)				M1 vs. M2				
Piecewise Time Function (M3, use T3 as a distinct time point)				M3 vs. M2				

B. Check all available model fit indices for each model. Do you think that each of the models has an acceptable fit?

C. Analyze competing models. If competing models are nested, use the nested chi-square test, $\Delta\chi^2$. If the models are not nested, use the AIC index for comparison purposes. Which model is the best fitting model? Why?

D. Once you have selected the optimal model, use the PLOT option to estimate a mean plot of the trajectory.

E. Interpret all growth mean parameters for the optimal model.

F. Next, covariates can be inserted into the optimal model. Using marital unhappiness (MU1) as a covariate, investigate how MU1 influences growth parameters (initial level and slopes) in the piecewise model. Insert the bold M*plus* syntax below into the unconditional model, and interpret all statistically significant coefficients (p-value < .05) related to the predictor.

VARIABLE: NAMES = IS1-IS5 MU1 DEP5T;
                    USEV = IS1 IS2 IS3 IS4 IS5 **MU1**;
                    ⋮

MODEL:
... ON MU1;

G. Last, insert target adolescents' depressive symptoms at T5 as a distal outcome in the optimal model by adding the M*plus* syntax in bold below to the unconditional model, and interpret all statistically significant coefficients (p-value < .05) related to the distal outcome.

VARIABLE: NAMES = IS1-IS5 MU1 DEP5T;
                    USEV = IS1 IS2 IS3 IS4 IS5 **DEP5T;**
                    ⋮

MODEL:
DEP5T ON ...;

References

Asparouhov, T., & Muthén, B. (2014). Auxiliary variables in mixture modeling: Three-step approaches using Mplus. *Structural Equation Modeling*, 21(3), 155–167.

Barrett, P. (2007). Structural equation modeling: Adjusting model fit. *Personality and Individual Difference*, 42(5), 815–824.

Brown, T. A. (2006). *Confirmatory factor analysis for applied research.* New York, NY: Guilford.

Campbell, D. T., & Reichardt, C. S. (1991). Problems in assuming the comparability of pretest and posttest in autoregressive and growth models. In R. E. Snow & E. Wiley (Eds.), *Improving inquiry in social science: A volume in honor of Lee J. Cronbach* (pp. 201–219). Mahwah, NJ: Erlbaum.

Coyne, J. C., & Downey, G. (1991). Social factors in psychopathology: Stress, social support, and coping processes. *Annual Review of Psychology*, 43(1), 401–425.

Curran, P. J., & Hussong, A. M. (2003). The use of latent trajectory models in psychopathology research. *Journal of Abnormal Psychology*, 112(4), 526–544.

Duncan, T. E., Duncan, S. C., & Strycker, L. A. (2006). *An introduction to latent variable growth curve modeling. Concepts, issues and applications*. (2nd ed.). Mahwah, NJ: Erlbaum.

Eaton, W. W., Badawa, M., & Melton, B. (1995). Prodromes and precursors: Epidemiological data for primary prevention of disorders with slow onset. *American Journal of Psychiatry*, 152(7), 967–972.

Grimm, K. J., & Ram, N. (2009). A second order growth mixture model for developmental research. *Research in Human Development*, 6 (2&3), 121–143.

Hertzog, C., Oertzen, T., Ghisletta, P., & Lindenberger, U. (2008). Evaluating the power of latent growth curve models to detect individual differences in change. *Structural Equation Modeling*, 15(5), 541–563.

Hu, L., & Bentler, P. M. (1998). Fit indices in covariance structure modeling: Sensitivity to underparameterized model misspecification. *Psychological Methods*, 3(4), 424–453.

Hu, L., & Bentler, P. M. (1999). Cutoff criteria for fit indexes in covariance structure analysis: Conventional criteria versus new alternatives. *Structural Equation Modeling*, 6(1), 1–55.

Judd, L. L. (1994). Clinical characteristics of subsyndromal symptomatic depression. *International Academy of Biomedical Drug Research*, 9(6), 67–74.

Karney, B. R., & Bradbury, T. N. (1995). Assessing longitudinal change in marriage: An introduction to the analysis of growth curves. *Journal of Marriage and the Family*, 57(4), 1091–108.

Little, T. D. (2013). *Longitudinal structural equation modeling*. New York, NY: Guilford.

Liu, S., Rovine, M. J., & Molenaar, P. C. M. (2012). Selecting a linear mixed model for longitudinal data: Repeated measures analysis of variance, covariance pattern model, and growth curve approaches. *Psychological Methods*, 17(1), 15–30.

Lorenz, F. O., Wickrama, K. A. S., & Conger, R. D. (2004). Modeling continuity and change in family relations with panel data. In R. D. Conger, F. O. Lorenz, & K. A. S. Wickrama (Eds.), *Continuity and change in family relations: Theory, methods, and empirical findings* (pp. 15–62). Mahwah, NJ: Erlbaum.

Muthén, B. (2003). Statistical and substantive checking in growth mixture modeling: Comment on Bauer and Curran (2003). *Psychological Methods*, 8(3), 369–377.

Rogosa, D. R., Brand, D., & Zimowski, M. (1982). A growth curve approach to the measurement of change. *Psychological Bulletin*, 90(6), 726–748.

Rueter, M. A., Scaramella, L., Wallace, L. E., & Conger, R. D. (1999). First onset of depressive or anxiety disorders predicted by the longitudinal course of internalizing symptoms and parent-adolescent conflict. *Archives of General Psychiatry*, 56(8), 726–739.

Steiger, J. H., & Lind, J. M. (1980). *Statistically based tests for the number of common factors*. Paper presented at the meeting of the Psychometric Society, Iowa City, IA.

Tucker, L. R., & Lewis, C. (1973). A reliability coefficient for maximum likelihood factor analysis. *Psychometrika*, 38(1), 1–10.

Wickrama, K. A. S., Lorenz, F. O., Conger, R. D., & Elder, G. H. (1997). Marital quality and physical illness: A latent growth curve analysis. *Journal of Marriage and Family*, 59(1), 143–155.

Willet, J. B. (1988). Questions and answers in the measurement of change. *Review of Research in Education*, 15, 345–422.

Willet, J. B., & Sayer, A. G. (1994). Using covariance structure analysis to detect correlates and predictors of individual change over time. *Psychology Bulletin*, 116(2), 363–381.

3

LONGITUDINAL CONFIRMATORY FACTOR ANALYSIS AND CURVE-OF-FACTORS GROWTH CURVE MODELS

Introduction

This chapter serves as a conceptual orientation to confirmatory factor analysis (CFA) in a longitudinal context (longitudinal confirmatory factor analysis, LCFA) and how a LCFA can be extended to a second-order growth curve model (also known as a curve-of-factors model, or CFM). This CFA-CFM is presented in an incremental manner with several steps: (1) an introduction to confirmatory factor analysis (CFA), (2) a discussion of longitudinal confirmatory factor analysis (LCFA), and (3) an explanation of how a LCFA can be conceptualized as the measurement model of a second-order growth curve model (a curve-of-factors model, CFM). Steps 1 and 2 refer to the confirmatory factor model with first-order primary latent factors. Step 3 introduces the addition of second-order latent growth factors to the LCFA model to form a second-order latent growth curve model (or curve-of-factors model, CFM). The chapter also discusses the useful features of a CFM over a conventional latent growth curve model (LGCM).

Confirmatory Factor Analysis (CFA) (Step One)

Factor analysis is a data-reduction technique that reduces many items to a single factor or a few factors. For example, depressive symptoms, anxiety symptoms, and hostility symptoms can be reduced to an underlying global factor representing internalizing symptoms (IS). There are two general approaches to factor analysis: the classic exploratory factor analysis (EFA) and the confirmatory factor analysis (CFA) in a structural equation framework. We usually think about CFA in the context of constrained factor estimation, meaning that we impose a hypothesized factor structure to be tested, whereas an EFA is an unconstrained

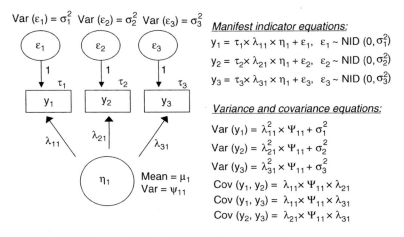

Var $(\varepsilon_1) = \sigma_1^2$ Var $(\varepsilon_2) = \sigma_2^2$ Var $(\varepsilon_3) = \sigma_3^2$

Manifest indicator equations:

$$y_1 = \tau_1 \times \lambda_{11} \times \eta_1 + \varepsilon_1, \quad \varepsilon_1 \sim NID\ (0, \sigma_1^2)$$
$$y_2 = \tau_2 \times \lambda_{21} \times \eta_1 + \varepsilon_2, \quad \varepsilon_2 \sim NID\ (0, \sigma_2^2)$$
$$y_3 = \tau_3 \times \lambda_{31} \times \eta_1 + \varepsilon_3, \quad \varepsilon_3 \sim NID\ (0, \sigma_3^2)$$

Variance and covariance equations:

$$Var\ (y_1) = \lambda_{11}^2 \times \Psi_{11} + \sigma_1^2$$
$$Var\ (y_2) = \lambda_{21}^2 \times \Psi_{11} + \sigma_2^2$$
$$Var\ (y_3) = \lambda_{31}^2 \times \Psi_{11} + \sigma_3^2$$
$$Cov\ (y_1, y_2) = \lambda_{11} \times \Psi_{11} \times \lambda_{21}$$
$$Cov\ (y_1, y_3) = \lambda_{11} \times \Psi_{11} \times \lambda_{31}$$
$$Cov\ (y_2, y_3) = \lambda_{21} \times \Psi_{11} \times \lambda_{31}$$

Mean $= \mu_1$
Var $= \psi_{11}$

FIGURE 3.1 Confirmatory Factor Analysis (CFA) Model.
Note: This CFA distinguishes unobserved latent factors (η) from observed indicator variables (y) using ellipses and boxes.

factor estimation that provides an empirical solution for a *potential* factor structure. CFA is used to *affirm* a pre-existing opinion about the structure of a global factor reflecting a domain or phenomenon of interest. It shows how different indicators, or specific items assessing multiple subdomains, contribute to (or "load onto") a latent global factor (see Figure 3.1). A CFA model can be estimated using repeated measures in a longitudinal context (LCFA) to examine the structure of a latent factor over multiple measurement occasions. Estimating a LCFA tests the factor invariance over time; that is, it tests whether items contribute to factors of the same phenomenon of interest in a similar manner over different time points. In this chapter, we use the term "subdomain" to describe indicator, or manifest, variables, which are hypothesized to contribute to a broader global factor. We use the term "global domain" to refer to the latent global factor comprised of multiple indicator variables, or subdomains, in a CFA model.

Specification of a Simple CFA

Suppose we assume that a global domain of internalizing symptoms (IS) is comprised of three subdomains—symptoms of depression (y1), anxiety (y2), and hostility (y3) (see Figure 3.1)—all of which are obtained by measures asked in a questionnaire. Arrows in Figure 3.1 connect the global latent variable to these subdomain responses (manifest indicators), which are typically considered consequences of the latent global factor. That is, manifest variables are predicted by latent factors. Residuals are measurement errors, which are also conceptualized as latent variables. Accordingly, in Figure 3.1, circles are used to represent latent variables including the global domain and the error terms (η and ε), squares represent manifest indicators (y_{1-3}), and single-headed arrows represent regression weights.

As depicted in Figure 3.1, each manifest indicator variable, y, is regressed on a latent factor, η, producing: (1) an intercept, τ, (2) a regression coefficient, λ, and (3) a residual, ε, assuming the observations are such that the residuals are normally and independently distributed (NID) with a mean of zero and variance, σ^2.

When summing question responses from all respondents in the study (i = 1, 2, …, n), we obtain a mean and a variance for each of the three subdomain indicators as well as a covariance between each of the subdomain indicators. The variances for each of the three subdomain indicators can be partitioned into two parts, the common variance due to the underlying global factor (η_1 = IS) and a variance that is unique to each of the three subdomain indicators. The unique variances for symptoms of depression, anxiety, and hostility are labeled var(ε_1), var(ε_2), and var(ε_3), respectively in Figure 3.1. The relative strength of the relationship between the latent variable IS (η_1) and its first subdomain (depressive symptoms) is expressed as a regression coefficient (λ_{11}) and is often referred to as a "factor loading." The relative strength of the second indicator (anxiety symptoms) is designated as λ_{21}, and so on. Notice that we are assuming the three subdomains are correlated with each other, as indicated by "Cov(y_1, y_2)" etc. We assume that the error terms are uncorrelated (e.g., cov(ε_1, ε_2) = 0). This is because once the common variance between any two subdomains is removed by the common factor (η_1) the remaining subdomain variance is unique (i.e., uncorrelated with the other subdomains). Using linear regression, the three sub-domain manifest indicators (i.e., y_1, y_2, and y_3) for individual i (= 1, 2, …, n) can be expressed as:

$$y_{1i} = \tau_{1i} + \lambda_{11} \times \eta_{1i} + \varepsilon_{1i}, \qquad \varepsilon_{1i} \sim \text{NID}\,(0, \sigma_1^2) \tag{3.1}$$

$$y_{2i} = \tau_{2i} + \lambda_{21} \times \eta_{1i} + \varepsilon_{2i}, \qquad \varepsilon_{2i} \sim \text{NID}\,(0, \sigma_2^2) \tag{3.2}$$

$$y_{3i} = \tau_{3i} + \lambda_{31} \times \eta_{1i} + \varepsilon_{3i}, \qquad \varepsilon_{3i} \sim \text{NID}\,(0, \sigma_3^2) \tag{3.3}$$

CFA reproduces the variance and covariance by modeling parameter estimates of the measurement solution (Bollen, 1989). Thus, most parameters in a CFA, including the variances of latent variables, factor loadings, and residual variances, can be used to reproduce the variance of each subdomain indicator (y) and the covariance between indicators. As shown in Figure 3.1, the variance of each indicator and covariance between indicators can be estimated as:

$$\text{Var}(y_1) = \lambda_{11}^2 \times \Psi_{11} + \sigma_1^2 \tag{3.4}$$

$$\text{Var}(y_2) = \lambda_{21}^2 \times \Psi_{11} + \sigma_2^2 \tag{3.5}$$

$$\text{Var}(y_3) = \lambda_{31}^2 \times \Psi_{11} + \sigma_3^2 \tag{3.6}$$

$$\text{Cov}(y_1, y_2) = \lambda_{11} \times \Psi_{11} \times \lambda_{21} \tag{3.7}$$

$$\text{Cov}(y_1, y_3) = \lambda_{11} \times \Psi_{11} \times \lambda_{31} \tag{3.8}$$

$$\text{Cov}(y_2, y_3) = \lambda_{21} \times \Psi_{11} \times \lambda_{31} \tag{3.9}$$

where Var $(y_{1\sim3})$ are variances of the three indicators. $\lambda_{1\sim3}$ are factor loadings of the three indicators. Ψ_{11} is the variance of the latent variable. $\sigma^2_{1\sim3}$ are residual variances of the three indicators. $\text{Cov}(y_i, y_j)$ are covariances between any two indicators (for $i, j = 1, 2, 3, i \neq j$).

CFA Model Identification

We can obtain estimates of each parameter in a CFA model, if the model is "empirically identified." For a CFA model to be empirically identified, the number of unknowns in the model, t, (that is, the number of freely estimated model parameters, such as factor loadings, factor variance and covariances, and error variances) must be less than the number of elements in the observed variance-covariance matrix of k indicators. Thus, for a unique solution of a CFA:

$$t = \frac{[k \times (k + 1)]}{2} \tag{3.10}$$

Accordingly, a minimum of three indicators is preferred for a CFA model with one latent factor. If the number of parameters, t, is more than $k(k+1)$, the model is "under identified" and there is not sufficient information for model estimation. In such a case, some parameters in the model have to be "fixed" to zero, fixed to a constant value, or constrained to equal each other. For example, in order to estimate a model with only two indicators ($k = 2$), the variance of the latent factor is constrained to 1.0, the indicator loadings are constrained to equal each other, and the error variances are also fixed to equal each other.

Scale Setting in a CFA

The unstandardized coefficients of a CFA model is influenced by how parameter scales are set (scale setting). Several scale setting approaches have been suggested within the SEM literature, including "marker variable," "fixed effect," and "effect coding" approaches. We will discuss CFA scale setting and the implication of these settings for the unstandardized solution of second-order growth curves in Chapter 4. For now, it is sufficient to be aware that scale setting has important implications.

Scale setting in a CFA is important because the unstandardized solutions of second-order growth parameters are influenced by how the parameter scales are set for the CFA. Because the primary goal of latent growth curve models is to examine longitudinal trajectories and developmental change over time, estimates of these growth parameters in unstandardized units are meaningful and easy to interpret as they retain the scale's metric. Moreover, growth curve models are, fundamentally, models about the structure of means (Little, 2013), which utilizes the unstandardized solution of longitudinal growth parameters (Bollen & Curran, 2006).

Longitudinal Confirmatory Factor Analysis (LCFA): Model Specification (Step Two)

So far, Figure 3.1 assumes that the data were collected at one point in time. Suppose we want to investigate the hypothesis that internalizing symptoms (IS) change over time, and our goal is to track change in the global domain of internalizing symptoms (IS) over time. This can be done by extending the model in Figure 3.1 to the model shown in Figure 3.2, where η_1, η_2, and η_3 represent repeated measures of the same construct (IS) at three different time points.

Within the model specification for a CFM shown in Figure 3.2 an LCFA is illustrated in the top portion of the figure. The first-order growth factor model (measurement model) at each time point, t, is defined by: (1) measurement intercept values (τ_{jt}) for each observed indicator (y_{jit}), (2) factor loadings (λ_{jt}), (3) a latent factor variable (η_{it}), and (4) a residual for each indicator variable (ε_{jit}), where the subscript j refers to particular indicator j and the subscript i refers to individuals. Thus, the regression equation of y_j for individual i (= 1, 2, ... n) at the time point t (= 0, 1, 2, 3, 4, 5, ... t) can be written as:

$$y_{jit} = \tau_{jt} + \lambda_{jt} \times \eta_{it} + \varepsilon_{jit,} \qquad \varepsilon_{jit} \sim NID\,(0,\, \sigma^2_{jt}) \qquad (3.11)$$

For each participant (i = 1, 2, ... n), we have a replication of the model shown in Figure 3.2.

In the model in Figure 3.2, the factor loadings $\lambda_{11}, \lambda_{21}$, and λ_{31} represent the strength of relationships between IS at Time 1 and the three manifest variables at Time 1 (symptoms of depression, anxiety, and hostility, respectively), while $\lambda_{12}, \lambda_{22}$, and λ_{32} reflect the strength of the relationship between IS at Time 2 and symptoms of depression, anxiety, and hostility, respectively, at Time 2; similar factor loadings represent the strength of relationships among manifest variables and IS at Time 3. When combining observations from all respondents (i = 1, 2, 3, ... n), it is possible to obtain estimates of the means and variances for IS at Times 2 and 3, just as we did at Time 1 in the CFA. These three measures of IS at different time points are likely to be highly correlated if IS is relatively stable over time.

A Second-Order Growth Curve: A Curve-of-Factors Model (Step Three)

Specification of a Curve-of-Factors Model (CFM)

We are now in familiar territory because the three repeated measures of IS can be coalesced into an aggregate intercept (π_0) and an aggregate slope (π_1), just as

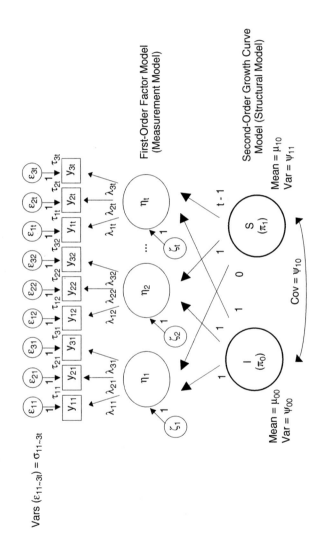

FIGURE 3.2 A Curve-of-Factors Model (CFM).

Note: I = Intercept. S = Slope (rate of change).

we explained in Chapter 2. These aggregated variables (*latent variables,* η_{it}) can be specified as the second-order factors of the CFM. This can be expressed as the following equations: (Note that normally the factor loadings (λs) are set to 1.0 for the latent intercept factor (π_{0i}), and appropriate loadings for the latent slope factor (π_{1i}) are based on the length of time between measurement occasions.)

$$\eta_{it} = \pi_{0i} + \lambda_t \times \pi_{1i} + \varsigma_{it}, \tag{3.12}$$

$$\pi_{0i} = \mu_{00} + \varsigma_{0i}, \qquad\qquad \varsigma_{0i} \sim \text{NID}\,(0, \Psi_{00}) \tag{3.13}$$

$$\pi_{1i} = \mu_{10} + \varsigma_{1i}, \qquad\qquad \varsigma_{1i} \sim \text{NID}\,(0, \Psi_{11}) \tag{3.14}$$

$$\Psi = \begin{bmatrix} \Psi_{00} & \\ \Psi_{10} & \Psi_{11} \end{bmatrix} \tag{3.15}$$

As in the case of a LGCM, second-order *intercepts* and *slopes* in a CFM also have a mean (μ) (μ_{00} for the intercept and μ_{10} for the slope) and residuals (ςs), and these residuals are assumed to be NID with a mean of zero and a variance-covariance structure (ψ).

The variance for the latent factor IS at Time 1 can be partitioned into two parts, one part is due to the common intercept and slope, and the second part is due to error (e.g., ς). The unique errors (e.g., ς_1, ς_2, and ς_3) are assumed to be uncorrelated once we partition out the common variance due to the intercept and slope. The model shown in Figure 3.2 and the related equations represent linear change over time; more elaborate models, such as quadratic models, are possible if data are available from four or more points in time.

A curve-of-factors model (CFM) is a method for multivariate analysis of the relations among different subdomains over time, also known as a multiple-indicator growth curve model. As we have discussed earlier, the structure of this second-order growth curve is the same as that of a conventional LGCM; that is, repeated measures define the intercept and slope growth factors. In a CFM, the indicator variables representing different subdomains are factor analyzed (LCFA) to produce factor scores of a global domain (e.g., internalizing symptoms, or IS), which is represented as a latent factor. Essentially, factor scores of the global domain over time (CFA latent factors) are then used as indicators of a second-order growth curve. In other words, the primary global latent factors (e.g., IS) defined in the LCFA model are used as indicators (or repeated measures) of the second-order growth factors (e.g., level and slope). *These CFA primary latent factors operate as refined repeated indicators of second-order growth curves because they are free from measurement error (at the population level).* A CFM fits a growth curve to factor scores representing what the study subdomains (e.g., depression, anxiety, and hostility symptoms) have in common at each time point (Duncan, Duncan, & Strycker, 2006). As with a conventional LGCM, this second-order growth curve includes growth factors, and in this case, these growth factors pertain to change in a global domain over time, including the initial level of IS and its slope (or rate of change over time). Before estimating a CFM, we have to ensure that several subdomain indicators (e.g., depression, anxiety, and hostility symptoms, in our example) contribute to repeated latent factors reflecting the same global factor (e.g., IS) with a common structure (same meaning) over time.

In conceptualizing a CFM, we also assume that repeated latent factors reflecting the same global factor (or global domain) at different time points have a time trend (or trajectory) with inter-individual variation. As in the case of a conven- · tional LGCM, it is expected that there is variation in the trajectory of the latent global factor (e.g., IS) across individuals. That is, the trajectories of the global factor over time are also thought to be specific to the individual. Some individuals may experience an increase in the global factor over time, whereas others may experience a decrease in the global factor. Still others may not experience any change in the global factor. The growth factors of these second-order individual trajectories (i.e., the level and slope of IS) have a mean (representing the average trend) and variance (representing the amount of inter-individual variation around the mean). This approach is known as a curve-of-factors model because it first conceptualizes global factors and then estimates a growth curve of those global factors (Duncan et al., 2006).

POINT TO REMEMBER…

In summary, a CFM can be estimated using the latent global factors as "refined" repeated indicators (see Panel C in Figure 1.2 from Chapter 1). That is, a CFM is estimated by incorporating a CFA as the measurement model within the latent growth curve modeling framework. A CFM is considered a second-order growth curve because it contains primary latent variables (latent global factors defined by subdomain indicators) and secondary latent variables (latent growth factors defined by first-order global factors as indicators).

Why Analyze a Curve-of-Factors Model? Improvements Over a Conventional LGCM

A CFM is an improvement over a conventional growth curve model (see Panels A and C of Figure 1.2 in Chapter 1), which is often estimated using composite measures as indicators (e.g., summing scores of depression, anxiety, and hostility symptoms to create a composite measure of IS at each time point). A CFM avoids several potential problems that are common in conventional growth curve models utilizing composite measures as indicators (Hancock, Kuo, & Lawrence, 2001). Three specific improvements and advantages of a CFM compared to a conventional LGCM are now presented in detail.

Equal Contribution of Items to the Composite Measure

First, unlike using composite measures, which assume that each subdomain measure contributes equally to the global domain (e.g., assuming equal weights for symptoms of depression, anxiety, and hostility when computing a composite measure of IS by summing the three symptom values), a CFM allows subdomain measures to have different contributions to the global domain because the global domain is assessed as a latent factor (e.g., IS). The unequal weighting may be theoretically meaningful because some subdomains defining a latent global factor may be more salient than other subdomains. In a conventional LGCM, if the assumption of equal contribution of subdomain measures is violated, the results may include biased conclusions about any observed change in the global domain over time. Using composite measures in a longitudinal context is also problematic because this approach assumes a constant (i.e., invariant) relationship between the subdomain indicators and the latent global factor across time points. If these relationships vary across time points, the ability to measure real change in the latent factors over time is compromised and biases may result (Bollen & Curran, 2006). However, the use of multiple subdomain indicators as observed indicators, rather than a composite measure that collapses across the subdomains, increases the number of parameters to be estimated. Thus, for relatively small samples, a CFM may not be appropriate because it could lead to convergence problems and unreliable solutions.

Longitudinal Measurement Invariance (Factorial Invariance)

Second, a CFM allows researchers to examine the quality of latent factors used to assess a global domain (e.g., IS) by estimating a CFA *before* modeling these factors as indicators of higher-order growth factors. That is, researchers can test whether the factor structure (the factor loadings of subdomain indicators on a latent global domain factor) is invariant over time. If factor variance exists, which indicates a lack of stability in the factor structure over time, the quality and meaning of the latent global factor also varies. Consequently, this non-invariance can result in the misinterpretation of any observed change in the study phenomenon. Particularly, there can be changes in how respondents answer survey items over time as they develop and mature.

Variance Components of Indicators: Measurement Error and Time-Specific Variance

Third, using multiple indicators to define latent factors separates the variance of each global indicator into an item-specific variance component and a time-specific variance component. The item-specific variance component is conceptualized as the measurement error of the manifest indicator, whereas the time-specific variance component is captured by the latent factor. That is, "refined" repeated indicators are used to define growth factors in a CFM. Finally, this CFA-CFM approach allows for autocorrelations among manifest indicators. These autocorrelations take into account the potential influence of conditions at one point in time on the conditions at subsequent time points. In this CFA-CFM approach, secondary growth factors reflect the temporal comorbidity of different subdomains. This reflection suggests a latent global factor explanation for the co-occurrence of subdomain outcomes. For example, temporal comorbidity among symptom subdomains of depression, anxiety, and hostility may be attributed to a global domain of internalizing symptoms. The temporal comorbidity may suggest a "common syndrome." Symptom comorbidity across different subdomains has also been explained as the undifferentiated accumulation of distress (Krueger, 1999; Lilienfeld, 2003). Others have indicated that a global factor might suggest the presence of common risks, such as a genetic risk that is shared across symptoms (Silberg, Rutter, D'Onofrio, & Eaves, 2003) or a common vulnerability (Krueger, Caspi, Moffitt, & Silva, 1998).

Thus, the CFA-CFM approach may be more applicable than a conventional LGCM when investigating temporal comorbidity, including addressing change in a common syndrome or change in a global domain comprised of different subdomains after taking measurement errors into account. Also, this approach allows for the investigation of common risks and consequences associated with the temporal comorbidity. For example, the CFA-CFM approach can be used to investigate temporal comorbidity among various types of substance use, as a global factor comprised of alcohol use, drug use, and tobacco use. Change in the global factor of substance use over time can also be assessed. In marital research, a CFA-CFM can be used to investigate the global domain, or temporal comorbidity, underlying marital happiness, marital satisfaction, and marital integration and to analyze change in the global factor (e.g., marital well-being) over time. In the area of work, a CFA-CFM can be used to investigate the global domain comprising work satisfaction, work autonomy, and work security and to analyze change in this global factor of work quality over time.

It is important to note that although a CFM allows for the investigation of change in global domains, it does not allow for the investigation of change in individual subdomains (within-subdomain change) over time. That is, with a CFM the researcher is unable to investigate differential growth of related subdomains and their associations. Particularly, a CFM does not reveal any potential convergence or divergence of subdomain-specific trajectories over time. Instead, it explicitly models the trajectories of the global domain.

Chapter 3 Exercises

NOTE: We recommend completing these exercises after reading Chapter 4 because the content is closely related.

You are provided with a panel dataset consisting of the following variables from mothers:

- DEP1M~DEP4M indicate mothers' depressive symptoms at T1~T4.
- ANX1M~ANX4M indicate mothers' anxiety symptoms at T1~T4.
- HOS1M~HOS4M indicate mothers' symptoms of hostility at T1~T4.

1. Draw a CFA model using three manifest indicators (i.e., DEP1M, ANX1M, and HOS1M) of a latent factor (internalizing symptoms, or IS). Use the marker variable approach for scale setting (use ANX1 as the marker variable).

2. Draw a longitudinal confirmatory factor analysis (LCFA) using ANX1M~ANX4M, DEP1M~DEP4M, and HOS1M~HOS4M. Use ANX1M~ANX4M as the marker variables for scale setting. Label all of the parameters using conventional symbols. Do not specify autocorrelated errors in the LCFA.

3. Indicate equality constraints to impose in the LCFA to test the assumptions of measurement invariance.

4. Draw a curve-of-factors model (CFM) based on your LCFA from Question 2 and label all of the parameters. Specify factor loadings for linear change (t = 0, 1, 2, and 3).

References

Bollen, K. A. (1989). *Structural equations with latent variables.* New York: John Wiley & Sons, Inc.

Bollen, K. A., & Curran, P. J. (2006). *Latent curve models: A structural equation approach.* Wiley Series on Probability and Mathematical Statistics. New Jersey: John Wiley & Sons.

Duncan, T. E., Duncan, S. C., & Strycker, L. A. (2006). *An introduction to latent variable growth curve modeling: Concepts, issues, and applications* (2nd ed.). Mahwah, NJ: Lawrence Erlbaum Associates.

Hancock, G. R., Kuo, W., & Lawrence, F. R. (2001). An illustration of second-order latent growth models. *Structural Equation Modeling,* 8(3), 470–489.

Krueger, R. F. (1999). The structure of common mental disorders. *Archives of General Psychiatry,* 56(10), 921–926.

Krueger, R. F., Caspi, A., Moffitt, T. E., & Silva, P. A. (1998). The structure and stability of common mental disorders (DSM-III-R): A longitudinal epidemiological study. *Journal of Abnormal Psychology,* 107(2), 216–227.

Lilienfeld, S. O. (2003). Comorbidity between and within childhood externalizing and internalizing disorders: Reflections and directions. *Journal of Abnormal Child Psychology*, 31(3), 285–291.

Little, T. D. (2013). *Longitudinal structural equation modeling*. New York, NY: Guilford.

Silberg, J., Rutter, M., D'Onofrio, B., & Eaves, L. (2003). Genetic and environmental risk factors in adolescent substance use. *Journal of Child Psychology and Psychiatry*, 44(5), 664–676.

4

ESTIMATING CURVE-OF-FACTORS GROWTH CURVE MODELS

Introduction

This chapter serves as a guide for the practical application of confirmatory factor analysis (CFA) in a longitudinal context (longitudinal confirmatory factor analysis, LCFA) and how a LCFA can be extended to a second-order growth curve model known as a curve-of-factors model, or CFM. The chapter provides a practical guide on how to: (1) incorporate autocorrelations, (2) impose factor invariance over time, and (3) extend a LCFA to a second-order growth curve (a curve-of-factors model, or CFM) using M*plus* statistical software. Steps 1, 2, and 3 refer to the confirmatory factor model (LCFA) with first-order, or primary, latent factors, which represent the measurement model of a curve-of-factors model (CFM). Step 3 addresses the estimation of a second-order latent growth curve model by adding secondary latent growth factors to a LCFA.

This chapter also explains different scale-setting approaches for the measurement model and their implication for second-order growth curve modeling. Finally, guidance is provided for incorporating time-invariant and time-varying covariates into a CFM to test various hypotheses related to antecedents and outcomes of change in a global latent factor. For each model, figures and M*plus* syntax are presented.

Steps for Estimating a Curve-of-Factors Model (CFM)

In order to estimate a CFM, we recommend the following four steps:

Step One
> Investigate the longitudinal correlation patterns among indicators (subdomain manifest variables).

Step Two

> Estimate an unconstrained longitudinal confirmatory factor model (LCFA) using subdomain indicators and incorporating autocorrelations among the errors.

Step Three

> Establish measurement invariance for the factor loadings, mean parameters, and error variances of indicators in the LCFA model across time points.

Step Four

> Estimate a second-order latent growth curve model using latent factors of the identified LCFA as the measurement model.

Steps 1, 2, and 3 refer to the estimation of a confirmatory factor model with primary latent factors. These steps represent the measurement model of the CFM. Step 4 (the estimation of the second-order latent growth curve) adds the secondary latent growth factors to the model.

Investigating the Longitudinal Correlation Patterns of Subdomain Indicators (Step One)

Before performing a LCFA, we recommend examining the observed correlation matrix of the subdomain indicators in order to obtain evidence as to whether, in general, the model fits the data structure. Little (2013) suggested that a LCFA model is appropriate when the correlation coefficients among subdomain indicators of the global latent domain at the same time point are higher than the correlation coefficients among the same subdomain indicators at different time points, or autocorrelations.

Illustrative Example 4.1: Examining the Longitudinal Correlation Patterns Among Indicators

In Table 4.1, an observed correlation matrix of indicators is shown for three subdomains over time. Correlations among the three subdomain indicators (symptoms of depression, anxiety, and hostility) at the same occasion are highly correlated. More specifically, these correlation coefficients range from .71 to .78 for anxiety and depressive symptoms over time, .61 to .72 for depressive and hostility symptoms over time, and .60 to .74 for hostility and anxiety symptoms over time (see the bold coefficients on the diagonal). The correlations for the same subdomain at the different time points (autocorrelations) are lower and range from .30 to .57 for anxiety symptoms, .31 to .61 for depressive symptoms, and .23 to .57 for hostility symptoms (see the coefficients within boxes). The large correlation coefficients among the three indicators (i.e., symptoms of depression, anxiety, and hostility) at the same time point provide evidence for the existence of a global latent factor, or

TABLE 4.1 Correlation Matrix among Internalizing Symptoms (n = 436).

	Correlation matrix among internalizing symptoms														
	Anxiety					Depression					Hostility				
	T1	T2	T3	T4	T5	T1	T2	T3	T4	T5	T1	T2	T3	T4	T5
Anxiety															
ANX1 (T1)	—														
ANX2 (T2)	.57	—													
ANX3 (T3)	.49	.55	—												
ANX4 (T4)	.34	.38	.48	—											
ANX5 (T5)	.30	.39	.45	.51	—										
Depression															
DEP1 (T1)	*.75*	.50	.45	.32	.29	—									
DEP2 (T2)	.50	*.77*	.46	.31	.32	.61	—								
DEP3 (T3)	.40	.43	*.77*	.39	.40	.49	.49	—							
DEP4 (T4)	.33	.38	.48	*.78*	.44	.44	.42	.53	—						
DEP5 (T5)	.21	.29	.35	.34	*.71*	.31	.35	.44	.40	—					
Hostility															
HOS1 (T1)	.67	.39	.35	.25	.23	*.61*	.37	.27	.26	.18	—				
HOS2 (T2)	.47	*.70*	.43	.30	.29	.46	*.71*	.35	.34	.24	.57	—			
HOS3 (T3)	.40	.35	*.67*	.32	.26	.39	.37	*.65*	.37	.23	.39	.44	—		
HOS4 (T4)	.36	.32	.46	*.74*	.35	.38	.29	.37	*.70*	.26	.34	.35	.51	—	
HOS5 (T5)	.15	.24	.30	.31	*.60*	.23	.25	.37	.32	*.66*	.23	.22	.31	.33	—
Mean	1.40	1.36	1.40	1.23	1.21	1.51	1.56	1.64	1.49	1.43	1.47	1.49	1.54	1.34	1.31
Variance	.21	.26	.27	.19	.14	.31	.36	.42	.35	.28	.31	.38	.36	.26	.19

Notes: ANX = Anxiety Symptoms. DEP = Depressive Symptoms. HOS = Hostility Symptoms. T = Time. Boxes indicate correlations among the same symptoms over time. Bold-italic numbers indicate correlations among different symptoms at the same measurement occasion.

global domain, for each of the five time points (a LCFA model, see Step 2 below). We refer to this global latent domain as internalizing symptoms (IS).

Performing an Unconstrained Longitudinal Confirmatory Factor Analysis (LCFA) (Step Two)

After investigating the longitudinal correlation pattern among indicators, the next step is to examine the longitudinal associations among the global latent factor domain using an unconstrained LCFA model (a configural model).

Illustrative Example 4.2: Longitudinal Confirmatory Factor Analysis (LCFA) Using Mplus

M*plus* syntax and an illustration of an unconstrained LCFA model using IS as the global latent factor with autocorrelated errors are shown in Figures 4.1 and 4.2, respectively. Notice that we set the factor loading (λ) of one indicator to 1 for each time point (DEP1~5), and its intercept is set to zero for model identification purposes (this is known as "marker variable" scale setting for the CFA parameters). However, other scale setting approaches can also be used, and we elaborate on this possibility later in the chapter.

A CFA defines unobserved global domains (identified as latent factors) using subdomain measures as manifest variables. Recall that for our example, the global domain is internalizing symptoms (IS) and the subdomain measures assess symptoms of depression, anxiety, and hostility. The latent factors (IS at multiple time points) are allowed to correlate. M*plus* syntax for this specification is shown below, and the complete syntax is provided in Figure 4.1. The BY statement specifies the estimation of the latent factors (i.e., IS) utilizing the three subdomain symptoms (symptoms of depression, anxiety, and hostility) as the manifest variables.

```
IS1 by DEP1 ANX1 HOS1;
IS2 by DEP2 ANX2 HOS2;
IS3 by DEP3 ANX3 HOS3;
IS4 by DEP4 ANX4 HOS4;
IS5 by DEP5 ANX5 HOS5;
```

IS1-IS5 correspond to internalizing symptoms for the five measurement occasions (IS1, IS2, IS3, IS4, and IS5). In M*plus*, unless otherwise specified, the metric of global factors of the three IS symptoms are set automatically by fixing the first specified symptom loading in each BY statement to 1 (i.e., the first manifest variable listed is used as the marker variable indicator). In our LCFA model (see Figure 4.1), we set the depressive symptoms manifest variables to be

Title: LCFA with auto correlated errors
DATA: File is C:\example_ch_4.dat;
VARIABLE: names are ANX1-ANX5 DEP1-DEP5 HOS1-HOS5;
        usevar = ANX1-ANX5 DEP1-DEP5 HOS1-HOS5;
        missing = all (9);

MODEL:
  IS1 by DEP1
      ANX1 HOS1;
  IS2 by DEP2
      ANX2 HOS2;
  IS3 by DEP3
      ANX3 HOS3;
  IS4 by DEP4
      ANX4 HOS4;
  IS5 by DEP5
      ANX5 HOS5;

> By specifying the DEP variables as the first indicator of IS, all DEP variables are automatically fixed to 1 (*marker variable approach*)

  [DEP1-DEP5@0];
  [ANX1-ANX5];
  [HOS1-HOS5];
  DEP1-DEP5;
  ANX1-ANX5;
  HOS1-HOS5;

> DEP1-DEP5 should be fixed at zero across time for model identification (*marker variable approach*)

[IS1-IS5];

> Estimating means of latent global factors

DEP1 WITH DEP2 DEP3 DEP4 DEP5;
DEP2 WITH DEP3 DEP4 DEP5;
DEP3 WITH DEP4 DEP5; DEP4 WITH DEP5;
ANX1 WITH ANX2 ANX3 ANX4 ANX5;
ANX2 WITH ANX3 ANX4 ANX5;
ANX3 WITH ANX4 ANX5; ANX4 WITH ANX5;
HOS1 WITH HOS2 HOS3 HOS4 HOS5;
HOS2 WITH HOS3 HOS4 HOS5;
HOS3 WITH HOS4 HOS5; HOS4 WITH HOS5;

> Specifying **autocorrelated errors** to avoid model misspecification and investigate the existence of indicator-specific variance in the same symptoms over time (IT model)

OUTPUT: STANDARDIZED MOD (3.84);

FIGURE 4.1 *Mplus* Syntax for a Configural Latent Confirmatory Factor Analysis Model (LCFA) with Autocorrelated Errors.
Note: DEP = Depressive Symptoms. ANX = Anxiety Symptoms. HOS = Hostility Symptoms. IS = Internalizing Symptoms. The marker variable scale setting approach was used.

used as the marker variables by fixing the factor loadings at one and the intercept at zero for the five measurement occasions using the syntax:

[DEP1–DEP5@0]

[ ] Brackets in the syntax are used to estimate variable means.

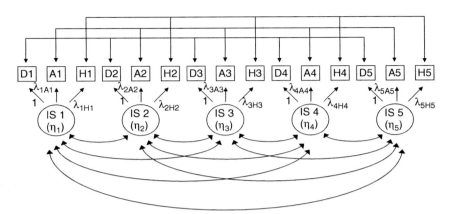

FIGURE 4.2 SEM of a Configural Latent Confirmatory Factor Analysis Model (LCFA) with Auto correlated Errors.
Note: D = Depressive Symptoms. A = Anxiety Symptoms. H = Hostility Symptoms. IS = Internalizing Symptoms. The marker variable scale setting approach was used. Intercepts (τs) and error variances (Var (εs)) are not shown in the figure for simplicity.

However, as we previously mentioned in Chapter 3, this marker variable approach has implications for the estimates of latent factors because the variance of each latent factor and the covariance among the latent factors are defined depending on the metric of the marker variable (e.g., DEP1) (Brown, 2006). For example, the means and variances of the five latent constructs (IS at five time points) are defined by the means and variances of the manifest variables identified as the marker variables (DEP1-DEP5). The associations among the other indicators are estimated relative to their association with the marker variable. Thus, when estimating a second-order growth model (CFM) (which will be discussed in Step 4), the metric used to assess the secondary growth factors' variance and covariance can change arbitrarily depending on which indicator has a fixed intercept of zero and factor loadings of 1 (i.e., depending on which variable is used as the marker variable). These scaling issues for latent factors will be discussed in detail when estimating a second-order growth curve model later in this chapter. The M*plus* syntax to estimate the variance of each subdomain indicator is as follows:

DEP1–DEP5;
ANX1–ANX5;
HOS1–HOS5;

Residual correlations (autocorrelated errors) are specified within the model in order to investigate correlations among the same symptoms (i.e., subdomains or manifest variables) across different time points (see Figure 4.2).

The *Mplus* syntax for autocorrelations is as follows (see Figure 4.1 for the complete LCFA syntax):

DEP1 with DEP2 DEP3 DEP4 DEP5;
DEP2 with DEP3 DEP4 DEP5;
DEP3 with DEP4 DEP5;
DEP4 with DEP5;
ANX1 with ANX2 ANX3 ANX4 ANX5;
ANX2 with ANX3 ANX4 ANX5;
ANX3 with ANX4 ANX5;
ANX4 with ANX5;
HOS1 with HOS2 HOS3 HOS4 HOS5;
HOS2 with HOS3 HOS4 HOS5;
HOS3 with HOS4 HOS5;
HOS4 with HOS5;

Such autocorrelated errors allow researchers to investigate the error variances and covariances. Also, even if these autocorrelation coefficients are not statistically significant, Little (2013) recommends retaining autocorrelated errors in the LCFA model to avoid model misspecification, which often leads to biased model parameter estimates. It should be mentioned that error variances not only contain random measurement error but also contain reliable components that are specific to the indicator (i.e., "trait" effects) and generalize across time (Geiser, Eid, Nussbeck, Courvoisier, & Cole, 2010). We will come back to these "trait" effects later.

The results of the initial LCFA model indicate that the model fit the data structure well (χ^2(df) = 75.506(50), $p < .05$; CFI/TLI = .994/.987; RMSEA = .034; SRMR = .031; see Panel A of Table 4.2).

POINT TO REMEMBER...

Modification indices can be used to investigate the possibility of improving the model fit. Corresponding to any fixed path in the model, modification indices represent the change in chi-square (with df = 1) when the path is freed (that is, when the path is estimated within the model). A statistically significant reduction in chi-square suggests that when the path or correlation of interest is freed the model achieves a better fit. Modification indices are specified as options in the output command (MOD (3.84)). Note that 3.84 is the chi-square critical value for one degree of freedom.

TABLE 4.2 M*plus* Output: Model Fit Indices.

a. *Configural model*

Chi-Square Test of Model Fit

Value	75.506
Degrees of Freedom	50
P-Value	0.0114

RMSEA (Root Mean Square Error Of Approximation)

Estimate	0.034	
90 Percent C.I.	0.017	0.049
Probability RMSEA <= .05	0.958	

CFI/TLI

CFI	0.994
TLI	0.987

Chi-Square Test of Model Fit for the Baseline Model

Value	4293.898
Degrees of Freedom	105
P-Value	0.0000

SRMR (Standardized Root Mean Square Residual)

Value	0.031

b. *Modified configural model*

Chi-Square Test of Model Fit

Value	48.933
Degrees of Freedom	46
P-Value	0.3562

RMSEA (Root Mean Square Error Of Approximation)

Estimate	0.012	
90 Percent C.I.	0.000	0.035
Probability RMSEA <= .05	0.999	

CFI/TLI

CFI	0.999
TLI	0.998

Chi-Square Test of Model Fit for the Baseline Model

Value	4293.898
Degrees of Freedom	105
P-Value	0.0000

SRMR (Standardized Root Mean Square Residual)

Value	0.027

Modification indices (see Table 4.3) revealed that estimating four additional error correlations would improve the model fit indices (DEP1 with HOS4 and HOS5 with DEP3, ANX1, and ANX5). The decision to incorporate additional correlations suggested by the modification indices should be grounded in existing theoretical and empirical research. Given that existing research has illustrated a high level of comorbidity among each of these symptoms in adolescence, it is not surprising that some comorbidity exists among symptoms across time. Consequently, the LCFA model was re-specified to include the four correlations indicated. Results of the reanalysis yielded a statistically and substantially better fitting model (χ^2(df) = 48.933(46), p = .356; CFI/TLI = .999/.998; RMSEA = .013; SRMR = .027; see Panel B of Table 4.2). The standardized coefficients of parameters from the reanalyzed model are shown in Table 4.4.

As shown in Table 4.4, the standardized factor loadings for each subdomain manifest variable (e.g., DEP1, ANX1, and HOS1) comprising the global latent domain (i.e., internalizing symptoms, or IS) ranged from .73 to .91 [T1], .81 to .88 [T2], .76 to .89 [T3], .83 to .90 [T4], and .80 to .89 [T5]. Conventionally, according to factor analysis literature, factor loadings $\geq$ .60 are deemed acceptable (Matsunaga, 2010). Thus, these factor loadings appear acceptable and suggest that together these three specific symptoms are indicators of latent factors of IS. Also, the correlations

TABLE 4.3 M*plus* Output: Modification Indices (MIs).

MODEL MODIFICATION
INDICES

		M.I.	E.P.C.	Std E.P.C.	StdYX E.P.C.
BY Statements					
IS1	BY ANX4	7.373	−0.095	−0.044	−0.103
IS1	BY ANX5	4.780	0.074	0.035	0.091
IS1	BY HOS4	4.829	0.089	0.042	0.080
IS2	BY DEP4	3.917	0.082	0.043	0.072
IS4	BY ANX5	8.429	0.097	0.050	0.133
IS4	BY DEP1	5.284	0.096	0.050	0.089
IS4	BY DEP5	4.330	−0.094	−0.048	−0.092
IS5	BY ANX1	4.435	−0.083	−0.037	−0.081
IS5	BY DEP1	3.845	0.092	0.041	0.074
WITH Statements					
HOS2	WITH DEP3	3.986	−0.015	−0.015	−0.128
HOS4	WITH DEP1	7.262	0.015	0.015	0.161
HOS5	WITH ANX1	4.164	−0.010	−0.010	−0.168
HOS5	WITH ANX5	5.012	−0.016	−0.016	−0.253
HOS5	WITH DEP3	4.350	0.013	0.013	0.141
HOS5	WITH DEP5	8.341	0.031	0.031	0.407

TABLE 4.4 M*plus* Output: Standardized Parameter Estimates of a Configural LCFA Model.

	Estimate	S.E.	Est./S.E.	Two-Tailed P-Value
IS1 BY				
DEP1	0.830	0.021	38.772	0.000
ANX1	0.907	0.020	46.078	0.000
HOS1	0.732	0.026	27.964	0.000
IS2 BY				
DEP2	0.880	0.017	51.619	0.000
ANX2	0.865	0.019	46.584	0.000
HOS2	0.810	0.021	39.416	0.000
IS3 BY				
DEP3	0.855	0.020	41.766	0.000
ANX3	0.892	0.020	45.664	0.000
HOS3	0.758	0.025	30.624	0.000
IS4 BY				
DEP4	0.863	0.017	51.178	0.000
ANX4	0.898	0.016	55.128	0.000
HOS4	0.825	0.019	43.881	0.000
IS5 BY				
DEP5	0.798	0.034	23.643	0.000
ANX5	0.890	0.036	24.617	0.000
HOS5	0.809	0.035	22.959	0.000
IS2 WITH				
IS1	0.638	0.035	18.154	0.000
IS3 WITH				
IS1	0.549	0.046	12.070	0.000
IS2	0.565	0.041	13.731	0.000
IS4 WITH				
IS1	0.428	0.047	9.040	0.000
IS2	0.437	0.046	9.438	0.000
IS3	0.556	0.043	12.838	0.000
IS5 WITH				
IS1	0.344	0.053	6.459	0.000
IS2	0.396	0.050	7.907	0.000
IS3	0.470	0.046	10.140	0.000
IS4	0.489	0.045	10.893	0.000
DEP1 WITH				
DEP2	0.493	0.059	8.383	0.000
DEP3	0.357	0.069	5.187	0.000
DEP4	0.483	0.070	6.939	0.000
DEP5	0.265	0.068	3.880	0.000

TABLE 4.4 (*cont.*)

	Estimate	S.E.	Est./S.E.	Two-Tailed P-Value
DEP2 WITH				
DEP3	0.311	0.074	4.204	0.000
DEP4	0.315	0.070	4.489	0.000
DEP5	0.260	0.066	3.934	0.000
DEP3 WITH				
DEP4	0.464	0.061	7.644	0.000
DEP5	0.333	0.067	4.933	0.000
DEP4 WITH				
DEP5	0.184	0.064	2.889	0.004
ANX1 WITH				
ANX2	0.274	0.085	3.226	0.001
ANX3	0.114	0.108	1.062	0.288
ANX4	−0.029	0.110	−0.261	0.794
ANX5	0.099	0.121	0.820	0.412
ANX2 WITH				
ANX3	0.360	0.078	4.617	0.000
ANX4	0.155	0.079	1.965	0.049
ANX5	0.300	0.090	3.325	0.001
ANX3 WITH				
ANX4	0.089	0.100	0.884	0.377
ANX5	0.266	0.104	2.559	0.011
ANX4 WITH				
ANX5	0.448	0.093	4.816	0.000
HOS1 WITH				
HOS2	0.491	0.047	10.440	0.000
HOS3	0.235	0.060	3.941	0.000
HOS4	0.184	0.062	2.977	0.003
HOS5	0.144	0.080	1.789	0.074
HOS2 WITH				
HOS3	0.305	0.059	5.177	0.000
HOS4	0.166	0.063	2.646	0.008
HOS5	0.011	0.079	0.141	0.888
HOS3 WITH				
HOS4	0.484	0.053	9.190	0.000
HOS5	0.238	0.070	3.411	0.001
HOS4 WITH				
HOS5	0.246	0.068	3.626	0.000
DEP1	0.193	0.062	3.119	0.002
HOS5 WITH				
DEP3	0.188	0.077	2.430	0.015
ANX1	−0.293	0.110	−2.661	0.008
ANX5	−0.440	0.263	−1.672	0.095

among the latent factors (i.e., IS over time) are in the moderate range (ranged from .34 to .64, $p < .001$), which indicates modest correlations among the latent global factors (or acceptable discriminant validity of IS over time) (see Table 4.4). Most of the autocorrelated errors among specific symptoms were statistically significant and in the expected direction (ranged from .18 to .49 for depressive symptoms, -.03 to .44 for anxiety symptoms, -.44 to .49 for hostility symptoms) even after controlling for the correlations between the latent factors of IS at different time points. These statistically significant autocorrelated errors among within-subdomain indicators imply the existence of certain trait-specific factors, which will be explained in more detail with an extended LCFA model later in this chapter.

Measurement Invariance of the LCFA Model (Step Three)

POINT TO REMEMBER...

When a measure has been shown to be consistent over multiple occasions, it is assumed that the measure has assessed the same construct (in other words, a construct with the same meaning over time) (Meredith & Horn, 2001). This consistency is referred to as measurement, or factor, invariance. More specifically, longitudinal measurement invariance refers to consistency in a measure over time, and group invariance refers to consistency in a measure across multiple groups of people. Thus, measurement invariance indicates that any observed change in the construct over time is true construct difference (Little, 2013).

If the assumption of longitudinal measurement invariance (also referred to as factorial invariance) is not met, any conclusions regarding change in a construct of interest over time are questionable or, perhaps, invalid due to measurement artifacts or item bias (Meredith, 1964). For longitudinal models, such as a LCFA, the assumption of measurement invariance can be tested by systematically constraining parameters and comparing the fit of competing models (Kim & Willson, 2014). The formulas for three potential constraints that build on each other are:

Weak invariance (constraining factor loadings, λs);

$$\lambda_t = \lambda_{t+1} = \lambda_{t+2} = \lambda_{t+3} \ldots$$

Strong invariance (constraining both factor loadings, λs, and mean parameters, τs);

$$\tau_t = \tau_{t+1} = \tau_{t+2} = \tau_{t+3} \ldots$$

Strict invariance (constraining factor loadings, λs, mean parameters, τs, and residual variances, σs);

$$\sigma_t^2 = \sigma_{t+1}^2 = \sigma_{t+2}^2 = \sigma_{t+3}^2 \ldots$$

Regarding these incremental testing sequences, the highest (strict) level of invariance would be ideal. However, this level may be implausible in practice because this level of restriction tests whether or not the residual variances consisting of both systematic errors and random errors are exactly the same over time. Although systematic errors may be stable across time, it is essentially impossible for random errors to be consistent at each time point. Thus, it has been suggested that meeting the strong invariance assumption is sufficient for ensuring the meaningful interpretation of model parameters (Thompson & Green, 2006).

Illustrative Example 4.3: Systematic Incremental Testing Sequences for Assessing Measurement Invariance

Under maximum likelihood (ML) estimation, measurement invariance is typically tested using a nested chi-square difference test, $\Delta\chi^2$, between the unconstrained model and the model(s) enforcing equality constraints (Ferrer, Balluerka, & Widaman, 2008; Harring, 2009). However, in a Monte-Carlo simulation study, Meade, Johnson, and Braddy (2008) found that the χ^2 statistic is highly sensitive to sample size. Thus, for model comparison, the use of an alternative fit index was recommended, such as the comparative fit index (CFI), which is less sensitive to sample size and more sensitive to a lack of invariance than the χ^2 statistic. Cheung and Rensvold (2002) suggested that the assumption of measurement invariance is met if the change in the CFI (ΔCFI) between the unconstrained model and the constrained model is less than .01. However, Little (2013) suggests that the judgment should not be based on any single statistic. Instead, multiple indicators of the change in model fit should be considered simultaneously.

Thus, we consider several model fit indices (discussed in Chapter 2) when assessing measurement invariance. First, it is recommended to fit a configural factorial invariance model; that is, a model with no constraints placed on any model parameters. Using this *unconstrained model,* check whether each latent construct can be formed in a similar manner (comparable factor loading patterns are expected over time). Second, a weak invariance model should be estimated. This model freely estimates the mean parameters but constrains the factor loadings for the same manifest variables (or indicators) at different time points. If fit indices comparing the weak invariance model to the unconstrained model indicate that constraining the factor loadings to be equal does not significantly worsen the model fit, it can be concluded that there is evidence for equivalent relationships among the manifest variables and the latent construct over time (i.e. the assumption of weak invariance is met).

Next, a strong invariance model should be estimated; that is, a model that constrains both the indicator means and the factor loadings to be equal across different time points. If the fit indices suggest the strong invariance model is *not* similar to the weak invariance model, we *cannot* assume that the LCFA captures the longitudinal changes of "true means" (i.e., latent factor means) because it is likely that change in the mean of the latent factor over time is due to variability in the means of the observed indicators over time. However, if the strong invariance model constraining both the indicator means and factor loadings fits as well as the weak invariance model based on indicators of difference in model fit, the indicator variables are thought to have a stable relationship with the latent construct (i.e., a one unit increase in the latent factor is associated with a similar amount of change in an indicator variable at each time point – that is, the strength of the association between the indicator and the latent factor is constant across time). Furthermore, strong measurement invariance also suggests that changes in the latent factor mean over time · can be attributed to true change in the global domain construct (Brown, 2006).

To test for each type of measurement invariance, we examine overall model fit indices, including $\Delta\chi^2$ and ΔCFI. In our example, in order to test measurement invariance in the previously specified LCFA model, the bold numbers in parentheses can be added to the LCFA model in the M*plus* syntax as follows.

For testing weak invariance, use the following syntax (see Figures 4.3 and 4.4):

IS1 by DEP1
    ANX1 HOS1 *(1-2)*;
IS2 by DEP2
    ANX2 HOS2 *(1-2)*;
IS3 by DEP3
    ANX3 HOS3 *(1-2)*;
IS4 by DEP4
    ANX4 HOS4 *(1-2)*;
IS5 by DEP5
    ANX5 HOS5 *(1-2)*;

For testing strong invariance, add the following syntax (see Figures 4.5 and 4.6):

[DEP1-DEP5@0];
[ANX1-ANX5] *(3)*;
[HOS1-HOS5] *(4)*;

For testing strict invariance, add the following syntax (see Figures 4.7 and 4.8):

DEP1-DEP5 *(5)*;
ANX1-ANX5 *(6)*;
HOS1-HOS5 *(7)*;

Nested Model Comparison for Measurement Invariance

The results of the nested model comparisons are shown in Table 4.5. The analyses revealed that the configural model, or the unconstrained model (M2), has a good overall fit. Next, we created M3 by constraining the factor loadings to be equal. We tested the assumption of weak invariance by comparing M3 with M2. The results indicated that the constraints incorporated in M3 do not significantly

MODEL:
  IS 1 by DEP1
          ANX1 HOS1 (1 - 2);
  IS 2 by DEP2
          ANX2 HOS2 (1 - 2);
  IS 3 by DEP3
          ANX3 HOS3 (1 - 2);   ◁── Testing weak invariance
  IS 4 by DEP4
          ANX4 HOS4 (1 - 2);
  IS 5 by DEP5
          ANX5 HOS5 (1 - 2);

  [DEP1-DEP5@0];
  [ANX1-ANX5];
  [HOS1-HOS5];
  DEP1-DEP5;
  ANX1-ANX5;
  HOS1-HOS5;

  [IS1-IS5];

  DEP1 WITH DEP2 DEP3 DEP4 DEP5;
  DEP2 WITH DEP3 DEP4 DEP5;
  DEP3 WITH DEP4 DEP5; DEP4 WITH DEP5;

  ANX1 WITH ANX2 ANX3 ANX4 ANX5;
  ANX2 WITH ANX3 ANX4 ANX5;
  ANX3 WITH ANX4 ANX5; ANX4 WITH ANX5;

  HOS1 WITH HOS2 HOS3 HOS4 HOS5;
  HOS2 WITH HOS3 HOS4 HOS5;
  HOS3 WITH HOS4 HOS5; HOS4 WITH HOS5;
  HOS4 WITH DEP1; HOS5 WITH DEP3 ANX1 ANX5;
OUTPUT: STANDARDIZED MOD (3.84);

FIGURE 4.3 M*plus* Syntax Testing Weak Invariance for a Latent Confirmatory Factor Analysis Model (LCFA).
Note: DEP = Depressive Symptoms. ANX = Anxiety Symptoms. HOS = Hostility Symptoms. IS = Internalizing Symptoms. The marker variable scale setting approach was used.

reduce the model fit compared to M2 (the configural model) ($\Delta\chi^2$(df) = 14.50(8), p = .08; ΔCFI = .001). Thus, the assumption of weak invariance is met. However, in comparing M3 (the weak invariance model) to M4 (the strong invariance model), which adds constraints to make the manifest variable means equal across time, the constraints incorporated in M4 significantly reduced the model fit (significant increase in chi–square) ($\Delta\chi^2$(df) = 59.806(8), p < .001; ΔCFI = .014). This finding may imply the existence of partial strong invariance (not strong invariance), which occurs when one or more of the loadings and/or one or more of the intercepts cannot be constrained to be equal across time (Brown, 2006).

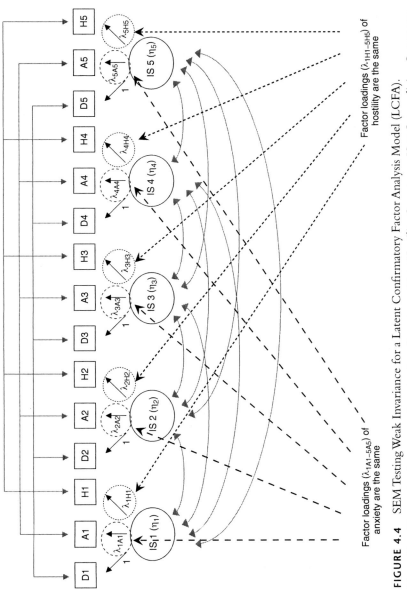

FIGURE 4.4 SEM Testing Weak Invariance for a Latent Confirmatory Factor Analysis Model (LCFA).

Note: D = Depressive Symptoms. A = Anxiety Symptoms. H = Hostility Symptoms. IS = Internalizing Symptoms. The marker variable scale setting approach was used. Intercepts (τs) and error variances (Var (εs)) are not shown in the figure for simplicity.

TROUBLESHOOTING

If the model shows evidence of partial invariance, Little (2013) suggests finding the "offending" manifest variable(s) that are not invariant over time using model fit diagnostics (e.g., modification indices [MOD]). Following this suggestion, in our example, we found that constraining the anxiety mean at T1 [ANX1] to be equal across time worsens model fit. Thus, we assumed that the measure of anxiety symptoms is not invariant over time. Most SEM literature suggests retaining the offending manifest variable(s) in the model but releasing the constraint of invariance. Releasing this constraint generally does not lead to serious problems with model specification unless there are too many loadings or intercepts that change over time (Brown, 2006; Little, 2013; Muthén & Muthén, 2007–2010).

Based on the finding that our anxiety measure is not invariant over time, we created a partial strong invariance model (M5; strong invariance model constraining the indicator means to be equal for depressive symptoms and hostility but not anxiety at T1). The $\Delta\chi^2$ statistic indicated a statistically significant reduction in model fit compared to M3 ($\Delta\chi^2(7) = 29.520, p < .001$), which suggests M3 is the best fitting model, but the ΔCFI was less than .01. Because Cheung and Rensvold (2002) suggested that the set of constrained parameters is fundamentally the same across time when the ΔCFI is less than or equal to .01, we can proceed with the understanding that the assumption of partial strong measurement invariance is met.

Last, we also tested a strict invariance model (with the exclusion of mean constraints for indicators of anxiety symptoms across time). As expected based on the previous models, the fit statistics indicated that the assumption of strict invariance was not met ($\Delta\chi^2(df) = 108.291(12), p < .001; \Delta$CFI $= .018$). Others have suggested that partial strong invariance, like we found support for in M5, is the minimum level of measurement invariance required to proceed to second-order modeling. *That is, second-order modeling requires: (a) stable relationships between the factor loadings and the latent construct over time and (b) the ability to attribute "true mean" changes in the constructs over time to true construct differences and not to indicator mean changes.*

Taking Autocorrelations Among Indicators in a LCFA into Account as a Trait Factor

After ensuring measurement invariance in the LCFA, it is then important to examine indicator-specific (j) autocorrelations among errors across different measurement points (t). Depending on the magnitude of these correlated errors, an alternative LCFA model specification can be utilized to account for effects

```
MODEL:
    IS1 by DEP1
            ANX1 HOS1 (1 - 2);
    IS2 by DEP2
            ANX2 HOS2 (1 - 2);
    IS3 by DEP3
            ANX3 HOS3 (1 - 2);
    IS4 by DEP4
            ANX4 HOS4 (1 - 2);
    IS5 by DEP5
            ANX5 HOS5 (1 - 2);
```

[DEP1-DEP5@0];

[ANX1-ANX5] (3);　　⎤

[HOS1-HOS5] (4);　　⎥———— Testing strong invariance

DEP1-DEP5;　　　　　⎦

ANX1-ANX5;

HOS1-HOS5;

[IS1 - IS5];

DEP1 WITH DEP2 DEP3 DEP4 DEP5;
DEP2 WITH DEP3 DEP4 DEP5;
DEP3 WITH DEP4 DEP5; DEP4 WITH DEP5;

ANX1 WITH ANX2 ANX3 ANX4 ANX5;
ANX2 WITH ANX3 ANX4 ANX5;
ANX3 WITH ANX4 ANX5; ANX4 WITH ANX5;

HOS1 WITH HOS2 HOS3 HOS4 HOS5;
HOS2 WITH HOS3 HOS4 HOS5;
HOS3 WITH HOS4 HOS5; HOS4 WITH HOS5;
HOS4 WITH DEP1; HOS5 WITH DEP3 ANX1 ANX5;

OUTPUT: STANDARDIZED MOD (3.84);

FIGURE 4.5 M*plus* Syntax Testing Strong Invariance for a Latent Confirmatory Factor Analysis Model (LCFA).
Note: DEP = Depressive Symptoms. ANX = Anxiety Symptoms. HOS = Hostility Symptoms. IS = Internalizing Symptoms. The marker variable scale setting approach was used.

that are specific to individual subdomains. In the alternative model specification, known as an indicator-specific trait factor model (or IT model), autocorrelation is represented by a latent "trait factor" rather than correlated errors (Bollen & Curran, 2006; Geiser & Lockhart, 2012; Grilli & Varriale, 2014). These subdomain factors are generally known as time-invariant trait (TIT) factors. This alternative approach investigates whether each specific subdomain indicator has a unique variance after controlling for the global domain factors (IS).

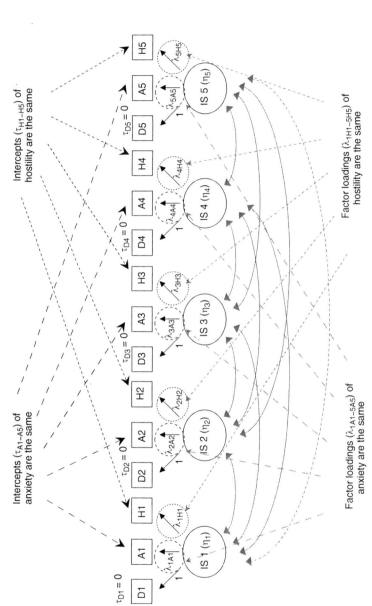

FIGURE 4.6 SEM Testing Strong Invariance for a Latent Confirmatory Factor Analysis Model (LCFA).

Note: D = Depressive Symptoms. A = Anxiety Symptoms. H = Hostility Symptoms. IS = Internalizing Symptoms. The marker variable scale setting approach was used. Error variances (Var (ε_s)) and autocorrelated errors of observed indicators are not shown in the figure for simplicity. τ = Intercept of an observed manifest variable.

```
MODEL:
   IS1 by DEP1
          ANX1 HOS1 (1 - 2);
   IS2 by DEP2
          ANX2 HOS2 (1 - 2);
   IS3 by DEP3
          ANX3 HOS3 (1 - 2);
   IS4 by DEP4
          ANX4 HOS4 (1 - 2);
   IS5 by DEP5
          ANX5 HOS5 (1 - 2);

   [DEP1-DEP5 @ 0];
   [ANX2-ANX5] (3);
   [HOS1-HOS5] (4);

   DEP1-DEP5       (5);
   ANX1-ANX5       (6);      Testing strict
   HOS1-HOS5       (7);      invariance

   [IS1-IS5];

   DEP1 WITH DEP2 DEP3 DEP4 DEP5;
   DEP2 WITH DEP3 DEP4 DEP5;
   DEP3 WITH DEP4 DEP5; DEP4 WITH DEP5;

   ANX1 WITH ANX2 ANX3 ANX4 ANX5;
   ANX2 WITH ANX3 ANX4 ANX5;
   ANX3 WITH ANX4 ANX5; ANX4 WITH ANX5;

   HOS1 WITH HOS2 HOS3 HOS4 HOS5;
   HOS2 WITH HOS3 HOS4 HOS5;
   HOS3 WITH HOS4 HOS5; HOS4 WITH HOS5;
   HOS4 WITH DEP1; HOS5 WITH DEP3 ANX1 ANX5;

   OUTPUT: STANDARDIZED MOD (3.84);
```

FIGURE 4.7 M*plus* Syntax Testing Strict Invariance for a Latent Confirmatory Factor Analysis Model (LCFA).
Note: DEP = Depressive Symptoms. ANX = Anxiety Symptoms. HOS = Hostility Symptoms. IS = Internalizing Symptoms. The marker variable scale setting approach was used.

Although this model can shed light on interesting and theoretically relevant research questions, there are several considerations that must be addressed before analyzing an IT model. First, we strongly recommend testing this model *after* confirming longitudinal factorial invariance because the autocorrelated errors can be influenced by the level of invariance tested. In order to estimate unbiased model parameters, we recommend analyzing the IT model for the LCFA model assuming at least partial strong invariance (i.e., constraining factor loadings and manifest variable means over time) because the model has unbiased autocorrelated errors. Second, in order to estimate a reliable variance of the latent trait factor, effect sizes

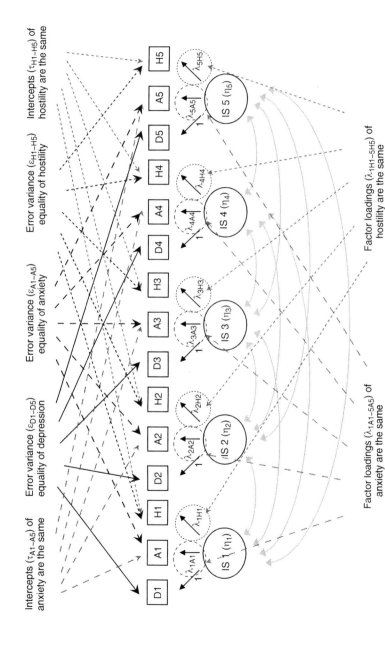

FIGURE 4.8 SEM Testing Strict Invariance for a Latent Confirmatory Factor Analysis Model (LCFA).

Note: D = Depressive Symptoms. A = Anxiety Symptoms. H = Hostility Symptoms. IS = Internalizing Symptoms. The marker variable scale setting approach was used. Intercepts (τs), error variances (Var (εs)), and autocorrelated errors of observed indicators are not shown in the figure for simplicity.

TABLE 4.5 Results from Models Testing Measurement Invariance in a Longitudinal CFA Model.

	χ^2 (df)	Model Comparison	$\Delta\chi^2$ (Δdf)	CFI	ΔCFI	RMSEA (90% CI)	SRMR	BIC
Null model (M1)	4293.898 (105)							
Configural LCFA model (with autocorrelations) (M2)	48.933 (46)			.999		.012 (.000, .035)	.027	5552.400
LCFA with weak invariance (M3)	63.437 (54)	M3 vs. M2	14.504 (8)	.998	.001	.020 (.000, .038)	.034	5518.283
LCFA with strong invariance (M4)	123.243 (62)	M4 vs. M3	59.806 (8)***	.985	.014	.048 (.035, .060)	.035	5529.468
LCFA with strong invariance (M5) except for anxiety 1	92.957 (61)	M5 vs. M3	29.520 (7)***	.992	.006	.035 (.019, .048)	.037	5505.260
LCFA with strict invariance (M6) except for anxiety 1	201.248 (73)	M6 vs. M5	108.291 (12)***	.969	.018	.063 (.053, .074)	.062	5540.619
Configural IT model (M7)	173.181 (80)	M7 vs. M5		.960		.052 (.041, .062)	.044	5517.431

Note: M = Model. LCFA = Longitudinal Confirmatory Factor Analysis. IT = Indicator-Specific Trait Factor. All of these LCFA models included autocorrelated errors. ***p < .001.

of the autocorrelated errors should be examined before fitting this model because the "trait"-specificity may be reflected in the magnitude of the autocorrelated errors over time (Geiser & Lockhart, 2012). That is, autocorrelated errors among indicators should be statistically significant and have correlation coefficients of at least moderate effect sizes in order to produce reliable factor variance. If autocorrelated errors among the same subdomain indicators over time are not statistically significant or do not have at least moderate effect sizes, a LCFA model with autocorrelated errors is more appropriate than an IT model.

Regarding the model comparison between two competing models (IT model and LCFA model), the IT model (without autocorrelated errors) is not nested within the LCFA model (with autocorrelated errors). Thus, nested model comparison tests ($\Delta\chi^2$ and ΔCFI) cannot be used to assess the IT model. Instead, the overall model fit indices (e.g., χ^2(df), CFI, RMSEA, SRMR, and BIC [a lower BIC value indicates the preferred model]) should be compared when judging which model is preferable.

Illustrative Example 4.4: Longitudinal Confirmatory Factor Analysis (LCFA) with "Trait" Factors (IT model)

In Figures 4.9 and 4.10, M*plus* syntax and a corresponding model for a configural IT model without constraints (that is, defining trait factors for each individual subdomain: D, A, and H) is shown. The specific syntax that defines these trait factors is:

D by DEP1-DEP5;
A by ANX1-ANX5;
H by HOS1-HOS5;

Similar to the process of assessing measurement invariance within the LCFA model, the IT model also can be systematically tested for measurement invariance by constraining factor loadings (λ) (loading equality).

In our example, most of the autocorrelated errors were statistically significant in the partial strong invariance model, ranging from .186 to .501 for depressive symptoms (DEP1-DEP5), -.038 to .417 for anxiety symptoms (ANX1-ANX5), and .003 to .488 for hostility symptoms (HOS1-HOS5). However, the results of the configural IT model (the unconstrained IT model) showed that all of the standardized loadings for trait factors were small (ranging from .216 to .372 for depressive symptoms, .173 to .360 for anxiety symptoms, and .166 to .450 for hostility symptoms; see Table 4.6). Furthermore, most loadings for anxiety and hostility were not statistically significant.

Consequently, these results do not suggest the existence of time-invariant trait (TIT) factors within our LCFA model. However, for the purpose of this example, we compared model fit indices between the LCFA model with autocorrelated errors and the configural IT model. As shown in Table 4.5, although each model has an adequate fit, the partial strong invariance LCFA model with autocorrelated errors (M5; χ^2(df) = 92.957(61), CFI = .985, RMSEA = .006, BIC = 5505.260) demonstrated a slightly better fit with the data structure than the configural IT model (M7; χ^2(df) = 173.181(80), CFI = .916, RMSEA = .052, BIC = 5517.431). *Thus, it was appropriate to proceed with a second-order growth curve model using the LCFA model with autocorrelated errors assuming partial strong invariance.*

Title: IT MODEL
DATA: File is C:\example_ch_4.dat;
VARIABLE: names are ANX1-ANX5 DEP1-DEP5 HOS1-HOS5;
          usevar = ANX1-ANX5 DEP1-DEP5 HOS1-HOS5;
          missing = all (9);

ANALYSIS: MODEL = NOCOV; ⟵ This command sets all *correlations* among latent variables to zero.

MODEL:
    IS1 by DEP1
        ANX1 HOS1 (1 - 2);
    IS2 by DEP2
        ANX2 HOS2 (1 - 2);
    IS3 by DEP3
        ANX3 HOS3 (1 - 2);
    IS4 by DEP4
        ANX4 HOS4 (1 - 2);
    IS5 by DEP5
        ANX5 HOS5 (1 - 2);

    [DEP1-DEP5@0];
    [ANX2-ANX5](3);
    [HOS1-HOS5](4);

    [IS1-IS5];

    IS1 WITH IS2 IS3 IS4 IS5;
    IS2 WITH IS3 IS4 IS5;
    IS3 WITH IS4 IS5; IS4 WITH IS5;

    D by DEP1-DEP5;
    A by ANX1-ANX5; ⟵ Specifying indicator-specific trait effect factors
    H by HOS1-HOS5;

OUTPUT: STANDARDIZED MOD (3.84);

FIGURE 4.9 *Mplus* Syntax for an Indicator-Specific Trait Factor (IT) Model. Note: DEP = Depressive Symptoms. ANX = Anxiety Symptoms. HOS = Hostility Symptoms. IS = Internalizing Symptoms. The marker variable scale setting approach was used. Estimator = MLR was used.

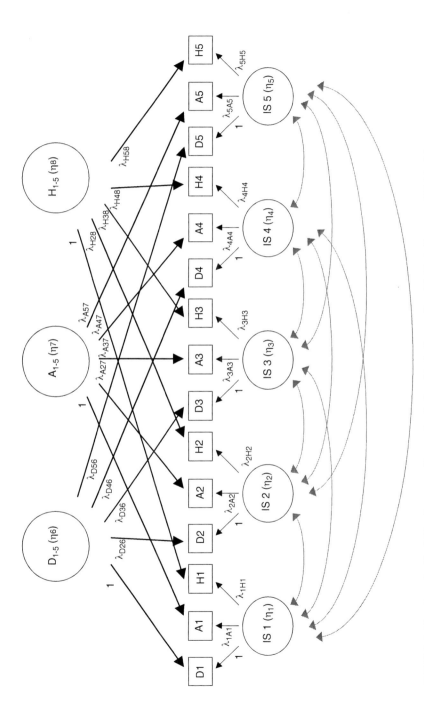

FIGURE 4.10 Longitudinal Confirmatory Factor Analysis with Indicator-Specific Trait Factors – IT Model.

Note: D = Depressive Symptoms. A = Anxiety Symptoms. H = Hostility Symptoms. IS = Internalizing Symptoms. The marker variable scale setting approach was used.

TABLE 4.6 M*plus* Output: Standardized Parameter Estimates for a Configural IT model.

		Estimate	S.E.	Est./S.E.	Two-Tailed P-Value
D	BY				
	DEP1	0.372	0.076	4.884	0.000
	DEP2	0.313	0.087	3.609	0.000
	DEP3	0.290	0.102	2.827	0.005
	DEP4	0.260	0.107	2.417	0.016
	DEP5	0.216	0.103	2.092	0.036
A	BY				
	ANX1	0.173	0.156	1.109	0.267
	ANX2	0.237	0.214	1.107	0.268
	ANX3	0.238	0.108	2.201	0.028
	ANX4	0.193	0.137	1.408	0.159
	ANX5	0.360	0.298	1.208	0.227
H	BY				
	HOS1	0.277	0.231	1.201	0.230
	HOS2	0.286	0.221	1.295	0.195
	HOS3	0.450	0.161	2.802	0.005
	HOS4	0.328	0.159	2.058	0.040
	HOS5	0.166	0.108	1.541	0.123

Estimating a Second-Order Growth Curve: A Curve-of-Factors Model (CFM) (Step Four)

If the assumption of measurement invariance is met for the LCFA model, a second-order growth curve model, such as a curve-of-factors model (CFM), can be estimated. Autocorrelations (or IT constructs) among indicators can also be included in a CFM. The covariances among global IS latent factors provide meaningful information regarding second-order growth parameters. As introduced in Equation 2.11 in Chapter 2,

$$\text{Var}\,(\pi_{1i}) = \frac{\left(\sigma_{12} + \sigma_{23} - 2\sigma_{13}\right)}{2} \tag{4.1}$$

The variance of the slope factor (π_{1i}) for the second-order growth curve model can likely be estimated when diagonal covariances among indicators (in this case, adjacent covariances between IS_t and IS_{t+1}) are higher than off-diagonal covariances. In our LCFA model with partial strong invariance, we found that the covariances between adjacent latent constructs were higher (ranged from .096

to .163, $p < .001$) than the off-diagonal covariances (ranged from .067 to .141, $p < .001$). These model covariances imply that a second-order slope variance can probably be estimated in a CFM without resulting in a solution that contains a negative variance.

Illustrative Example 4.5: Estimating a Curve-of-Factors Model (CFM)

As shown in Figures 4.11 and 4.12, a curve-of-factors model (CFM) for symptoms of depression, anxiety, and hostility can be specified in M*plus*.

```
MODEL:
    IS1 by DEP1
          ANX1 HOS1(1-2);
    IS2 by DEP2
          ANX2 HOS2(1-2);
    IS3 by DEP3
          ANX3 HOS3(1-2);
    IS4 by DEP4
          ANX4 HOS4(1-2);
    IS5 by DEP5
          ANX5 HOS5(1-2);

    [DEP1-DEP5@0];
    [ANX2-ANX5](3);
    [HOS1-HOS5](4);
```

Second-order growth model specification

I_IS S_IS | IS1@0 IS2@1 IS3@3 IS4@4 IS5@6;
[I_IS S_IS]; I_IS WITH S_IS;

```
    DEP1 WITH DEP2 DEP3 DEP4 DEP5;
    DEP2 WITH DEP3 DEP4 DEP5;
    DEP3 WITH DEP4 DEP5; DEP4 WITH DEP5;

    ANX1 WITH ANX2 ANX3 ANX4 ANX5;
    ANX2 WITH ANX3 ANX4 ANX5;
    ANX3 WITH ANX4 ANX5; ANX4 WITH ANX5;

    HOS1 WITH HOS2 HOS3 HOS4 HOS5;
    HOS2 WITH HOS3 HOS4 HOS5;
    HOS3 WITH HOS4 HOS5; HOS4 WITH HOS5;
    HOS4 WITH DEP1;
    HOS5 WITH DEP3 ANX1 ANX5;

OUTPUT: SAMPSTAT STANDARDIZED;
PLOT:   SERIES=IS1-IS5 (S_IS); TYPE=PLOT3;
```

FIGURE 4.11 M*plus* Syntax for a Curve-of-Factors Model (CFM).
Note: DEP = Depressive Symptoms. ANX = Anxiety Symptoms. HOS = Hostility Symptoms. IS = Internalizing Symptoms. I = Initial Level. S = Slope.

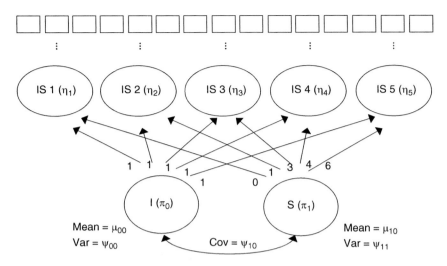

FIGURE 4.12 SEM for a Curve-of-Factors Model (CFM).
Note: IS = Internalizing Symptoms. I = Initial Level. S = Slope. Boxes represent indicators. Intercepts (τs), error variances (Var (εs)), and autocorrelated errors of observed indicators are not shown in the figure for simplicity.

The CFM results are shown in Figure 4.13. Based on the model fit indices, the specified CFM had a good fit with the data structure (χ^2(df) = 166.598(71), $p < .001$; CFI/TLI = .977/.966; RMSEA = .056, SRMR = .062). Statistically significant mean levels existed for the intercept and slope of the second-order model, indicating an initial level of IS that is greater than zero and an increasing trend in IS over time (level: 1.589, $p < .001$, linear slope: -.026, $p < .001$, respectively). Inter-individual differences in the second-order growth factors were also statistically significant with estimated variances suggesting the existence of inter-individual variation within both the second-order intercept (initial level) and the rate of change over time (slope) (level: .178, $p < .001$, slope: .004, $p < .001$). The statistically significant slope variance suggests that some individuals have a higher rate of change in IS compared to others with a lower rate of change in IS over time, and still others maintain the same level of IS over time.

The covariance (the unstandardized coefficient) between the intercept and slope was -.017, $p < .001$. Regardless of the small magnitude, the interpretation of this negative coefficient with a decreasing mean trajectory is particularly important. Figure 4.14 illustrates how to interpret this negative covariance between the intercept factor and slope factor variances with a decreasing mean trajectory. As can be seen, the negative covariance between the intercept and slope in the current example suggests that individuals with higher initial levels of IS had a steeper decrease in IS over time compared to individuals experiencing lower initial levels

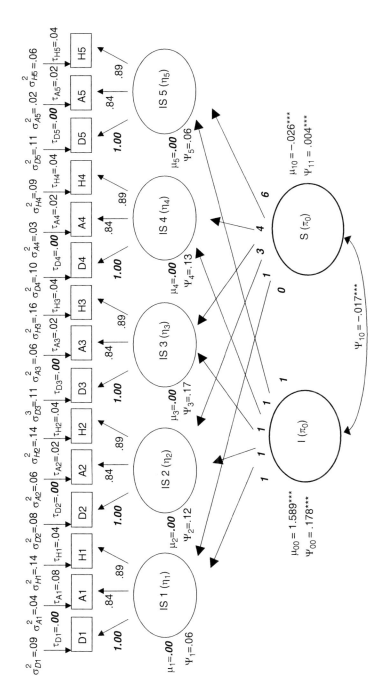

FIGURE 4.13 Results of a Second-Order Growth Curve Model (CFM) Using the Marker Variable Scale Setting Approach.

Note: Unstandardized coefficients are shown. DEP = Depressive Symptoms. ANX = Anxiety Symptoms. HOS = Hostility Symptoms. I = Intercept. S = Slope. Partial strong invariance was specified using the marker variable approach. The residual variances among the same indicators over time were correlated but are not shown in the figure for simplicity. Values in italics and bold are fixed parameters. $\chi^2(\text{df}) = 166.598(71)$, $p < .001$; CFI/TLI $= .977/.966$; RMSEA $= .062$; SRMR $= .056$, $p < .001$.

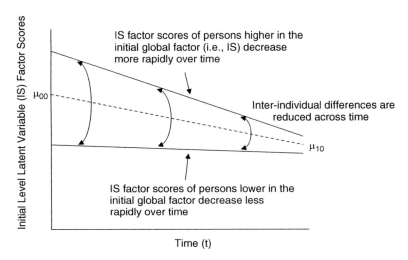

FIGURE 4.14 Description of the Negative Association Between Intercept and Slope Variances in a Model With a Negative Trajectory Mean.

Note: μ_{00} = the mean of the intercept factor. μ_{10} = the mean slope factor (trajectory). These means are indicated by the dotted line.

of IS. That is, factor scores of persons higher in the initial global factor (i.e., IS) decreased more rapidly across time than factor scores for individuals with lower values of the global factor initially.

In the current example, models with linear trajectories are utilized for clear interpretation. However, as shown in Chapter 2, this model can be extended to non-linear models, such as quadratic and exponential models, as well as piecewise or slope-segment models. In the same manner that we tested measurement invariance, we can also systematically determine the best fitting model (e.g., linear or quadratic), according to the model fit indices ($\Delta\chi^2$ and ΔCFI tests for nested model comparisons and AIC and BIC [low values of AIC and BIC are preferred] for non-nested model comparisons).

Using the drop-down menu in M*plus* (Plot →View Plots → Estimated Means) allows the user to obtain the trajectory estimated by the PLOT command (see Figure 4.15 for a sample plot output).

As we discussed, in this example, DEP1, DEP2, DEP3, DEP4, and DEP5 were used as "marker variables" for scale setting because this is the default scale setting approach in M*plus*. However, there are several approaches for scale setting.

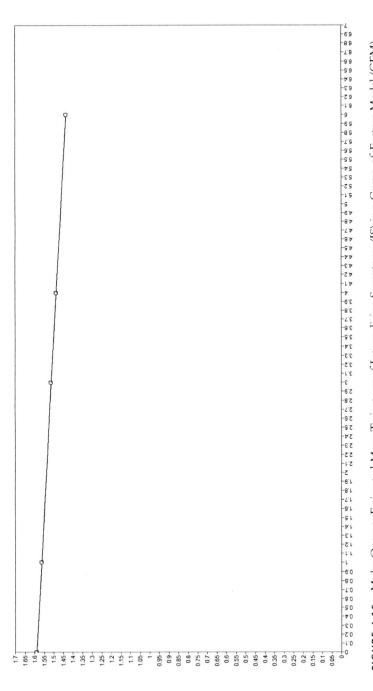

FIGURE 4.15 M*plus* Output: Estimated Mean Trajectory of Internalizing Symptoms (IS) in a Curve-of-Factors Model (CFM).

Scale Setting Approaches and Second-Order Growth Model Parameters (Curve-of-Factors Model, CFM)

In a CFM, the latent factors that account for the variance and covariance among multiple indicators of different subdomains (e.g., DEP1, ANX1, and HOS1) are considered as global factors (e.g., IS). Unstandardized solutions for these latent global factors are strongly influenced by the type of scale setting used, and these factors determine the metric of the second-order growth parameters in a CFM. Thus, the scale setting of latent global factors (which occurs in the CFA step of the CFM) is crucial when estimating second-order growth parameters in a CFM.

There have been two traditional approaches to scale setting: the marker variable approach and the fixed factor approach. A third approach, known as the effect coding approach has more recently been used by Little, Siegers, and Card (2006). This approach is only appropriate when each indicator has the same range of possible outcomes. The effect coding approach avoids several of the undesirable consequences of the marker variable and fixed factor approaches. Because these scaling issues have implications for second-order growth when modeling, we first discuss these three approaches briefly using our CFA model examples in order to provide a clear understanding of these different scale setting approaches. Each approach is then applied within a second-order growth curve model, or, more specifically, a CFM.

Marker Variable Approach

Using the marker variable approach, one indicator is selected to be the "marker" variable in the traditional SEM specification. With this method, the intercept of one of the indicators of the latent factor is fixed to be zero (usually the first indicator is selected) and the loading of the chosen indicator is fixed to 1. Recall that this is the approach used earlier in the chapter to examine a LCFA (see Step Two). The marker variable approach has been widely used and accepted within the SEM literature. *In Mplus, the marker variable approach is applied by default.* However, the marker variable scale setting approach may cause undesirable consequences within the unstandardized solution of the model.

Illustrative Example 4.6: Using the Marker Variable Approach for CFA Scale Setting

To illustrate the implications of the marker variable approach, we apply this approach to a simple CFA model with one latent factor and three manifest indicators. The manifest variable representing depressive symptoms (DEP1) at Time 1 (T_1) is selected as the marker indicator for IS (η_1). By fixing the factor loading of the first indicator (DEP1) to 1, the reliable variance component (equal to squaring the standardized factor loading [λ]) of DEP1 is passed on to the factor variance of the latent construct (IS [η_1]) (see Equations 3.1, 3.2, and 3.3 in Chapter 3). Now, the associations among the latent factor and the other two indicators (ANX1 and HOS1) are estimated relative to their association with the marker variable (DEP1). The associations between the latent factor and the other two indicators produce weighted factor loadings for these two indicators. This means that the unstandardized coefficients, which are, more specifically, the unstandardized factor loadings, can change arbitrarily depending on the variable selected as the marker variable.

As shown in Figure 4.16, we evaluate the commonly used marker variable approach with three different one-factor CFA models. These subjective factor loadings influence the unstandardized solution for the variances of the initial level and slope factors of the second-order growth model (which we elaborate on in the next section).

M*plus* syntax for a CFA model utilizing the marker variable approach is:

MODEL:
IS1 by DEP1 ANX1 HOS1;
[DEP1@0]; [IS1];

In the M*plus* syntax shown above, the *factor loadings* of DEP1 are automatically fixed to one because this variable was specified first. The third line of syntax ([DEP1@0]) fixes the *intercept* of DEP1 at zero. Consequently, this syntax utilizes DEP1 as the marker variable. [IS1] is the syntax to estimate the mean parameter of the latent factor, IS1.

Regarding the mean structures for the unstandardized solution, Little (2013) showed how to reproduce the means of manifest indicators from CFA model parameters using a simple equation:

$$E(y) = \tau_y + \lambda_y \times \mu_\eta \tag{4.2}$$

where E(y) is the mean of a manifest indicator, τ_y is the intercept of the indicator (y), λ_y is the indicator (y)'s factor loading, and μ_η is the mean of the latent global factor. This equation applies to each of the indicators. According to this equation, for example, the mean of DEP1, E(y), (as the marker variable) automatically becomes the mean of the latent factor because the factor loading, λ, equals 1 and the indicator intercept, τ, equals 0. Similar to the estimation of factor loadings, the intercepts (τ) of the other two indicators (ANX1 and HOS1) are now weighted because these two intercepts are estimated from the original indicator mean (E(y)), factor loadings (λ), and mean of the latent factor (μ_η, which is identical to the marker variable mean). *Thus, the mean of the latent factor can cause scaling problems for the second-order growth model mean parameters (both the intercept, μ_0, and slope, μ_1) because the means can change arbitrarily depending on which variable is selected to serve as the marker variable.* Panel A of Figure 4.16 illustrates a CFA model using

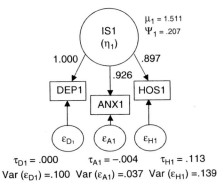

Marker variable: DEP1

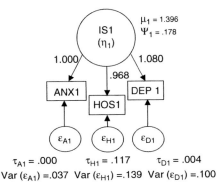

Marker variable: ANX1

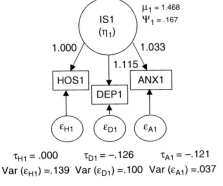

Marker variable: HOS1

FIGURE 4.16 Model Parameters Using Different Variables as the Marker Variable for CFA Model Estimation.

Note: Unstandardized coefficients are shown. DEP1 = Depressive Symptoms at T1. ANX1 = Anxiety Symptoms at T1. HOS1 = Hostility Symptoms at T1. IS = Internalizing Symptoms. All three models have the same model fit (χ^2 (df) = .000(0), p = .000; CFI/TLI = 1.00/1.00; RMSEA = .000; SRMR = .000).

depressive symptoms at T_1 (DEP1) as the marker variable (by fixing the factor loading to 1 and intercept to 0). As can be seen, the reliable variance of DEP1 (.207) is passed on to the variance of IS1 (η_1). The other two factor loadings (ANX1 and HOS1) are estimated based on the associations between the marker variable, DEP1, and the other two manifest variables, or subdomain symptoms. It is important to note that by fixing the intercept to 0, the mean of IS1 is now identical to the mean of DEP1 (see the mean of DEP1 in Table 4.1).

Previous literature has suggested that this approach is less arbitrary when the manifest indicator chosen as the marker variable is: (1) reliable, (2) well-known, and (3) measured in a clear metric (Bollen & Curran, 2006; Brown, 2006; Chen, Sousa, & West, 2005), but these rules are still subjective and unclear at times. Thus, depending on which indicator is chosen as the marker variable, this approach may lead to different unstandardized solutions. This can cause biased growth parameters, especially for the initial level and slope factor variances and means (unstandardized solution) in the second-order growth model (more specifically, a CFM). In the current CFM example (see Figure 4.13), we applied this marker variable approach. The notable feature of this CFM is that the means of the lower-order latent factors (IS_{1-5}) are fixed to 0. This feature allows the marker variable means (DEP_{1-5}) to transfer directly to the second-order intercept and slope factors. Thus, the means of the second-order intercept and slope factors (I and S) rely heavily on the means of the five indicators serving as the marker variables (DEP_{1-5}). Consequently, the mean parameters of the second-order latent factors (μ_I and μ_S) can easily change depending on which variable is selected as the marker variable.

M*plus* syntax for estimating an unconditional CFM using the marker variable approach is as follows: (Note that autocorrelations among manifest indicators are not specified.)

MODEL:
IS1 by DEP1
    ANX1 HOS1 (1-2);
IS2 by DEP2
    ANX2 HOS2 (1-2);
IS3 by DEP3
    ANX3 HOS3 (1-2);
IS4 by DEP4
    ANX4 HOS4 (1-2);
IS5 by DEP5
    ANX5 HOS5 (1-2);
[DEP1-DEP5@0];
[ANX2-ANX5](3);
[HOS1-HOS5](4);
I_IS S_IS | IS1@0 IS2@1 IS3@3 IS4@4 IS5@6;
[I_IS S_IS];

As can be seen in Figure 4.13, by constraining the factor loadings (λs) and intercepts (τs) of the marker variables (DEP1 to DEP5) to 1 and 0 respectively (see italic bold numbers), reliable variances and intercepts of the marker variables transfer directly to the variances and intercepts of the latent factors (η_1 to η_5), which, in turn, influence the estimation of growth parameters (μs and ψs) in a CFM.

Fixed Factor Approach

Using the fixed factor approach, the variance of the latent factor is fixed to a specific value (most commonly, 1.00) and the means of the latent constructs are fixed to 0 (Brown, 2006). Consequently, this approach produces a completely standardized solution; that is, there is a standardized variance for the latent constructs and their correlations with each other (standardized covariances). Compared to the marker variable approach, the fixed factor approach has several advantages for scale setting. First, this approach is less arbitrary than the marker variable approach because the researcher does not select one variable (the marker variable) to serve as the metric for the latent construct. Second, unlike the marker variable approach where only the variance of the marker variable is passed on to the variance of the latent factor, by fixing the variance of the latent factor to 1 this approach estimates factor loadings that are optimally balanced among indicators (similar to loading effect sizes) and reproduces the reliable variance component of each indicator (see Equations 3.1, 3.2, and 3.3). This means that the factor loadings are not weighted by the variance of any certain indicator (i.e., marker variable), which produces reliability for each indicator. That is, the reliability of each manifest indicator is easily reproduced by squaring the estimated factor loading (λ). Notice that the loadings are not standardized (i.e., unstandardized factor loadings) because these loadings are estimated in an unstandardized matrix of manifest indicators, where the total variances ($=\lambda^2 \times \psi + \sigma^2$) are obtained from the raw data (see each variance in Table 4.1). Regarding mean structures, by fixing the mean of the latent factor to 0, the original means of the manifest indicators are reproduced (E(y) $=\tau_y$, see Equation 4.2).

This approach is an improvement over the marker variable approach because it reproduces each indicator's mean without depending on the quality of one indicator in the CFA portion of the model. However, in this fixed factor approach, because the mean of the latent factor is equal to zero, in a second-order growth model the initial level means also become zero. Thus, the analysis of change in the global domain (IS) does not involve the mean of its initial level, which can make describing the second-order trajectory problematic.

Illustrative Example 4.7: Using the Fixed Factor Scale Setting Approach in a CFA

An example of the fixed factor approach using a simple CFA model is shown in Panel A of Figure 4.17.

Panel A: Fixed factor approach

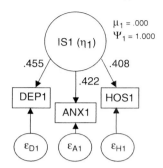

$\mu_1 = .000$
$\Psi_1 = 1.000$

$\tau_{D1} = 1.511$ $\tau_{A1} = 1.396$ $\tau_{H1} = 1.468$
Var $(\varepsilon_{D1}) = .100$ Var $(\varepsilon_{A1}) = .037$ Var $(\varepsilon_{H1}) = .139$

Panel B: Effect coding approach

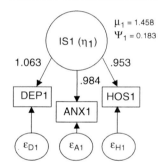

$\mu_1 = 1.458$
$\Psi_1 = 0.183$

$\tau_{D1} = -.039$ $\tau_{A1} = -.040$ $\tau_{H1} = .078$
Var $(\varepsilon_{D1}) = .100$ Var $(\varepsilon_{A1}) = .037$ Var $(\varepsilon_{H1}) = .139$

FIGURE 4.17   Comparison Between the Fixed Factor and Effect Coding Approaches. Note: Unstandardized coefficients are shown. These two models have the same model fit (χ^2 (df) = .000 (0), $p = .000$; CFI/TLI = 1.00/1.00; RMSEA = .000; SRMR = .000).

M*plus* syntax for a CFA model utilizing the fixed factor approach to scale setting is as follows:

MODEL:
IS1 by DEP1* ANX1 HOS1;
IS1@1; [IS1@0];
[DEP1 ANX1 HOS1];

In M*plus*, the asterisk (*) frees the first factor loading, which is fixed at one as the default in M*plus*, in order to define the metric of the factor. By using IS1@1, the variance of the latent factor (IS at T1) is now fixed to 1. Brackets in the syntax [ ] are used to estimate means of the three manifest variables (i.e., symptoms of depression, anxiety, and hostility).

The results are shown in Panel A of Figure 4.17. Unlike the marker variable approach, all factor loadings have similar effects (range from .408 to .455), which represents optimally balanced factor loadings. Also, all factor loadings now indicate true reliability of the indicators. For example, the reliability of ANX1 is .178 ($= .422^2$), which indicates the proportion of true variance in the observed · variable ANX1 that is explained by the latent factor (IS1, η). This variance is not influenced by the indicator selected as the marker variable (DEP1), as is the case when the marker variable approach is utilized. Another benefit to this approach is that the intercept of each indicator is identical to the observed mean scores (see the means of DEP1, ANX1, and HOS1 in Table 4.1).

Results from the CFM with this fixed factor approach are shown in Figure 4.18.

Effect Coding Approach

Little et al. (2006) have used a third approach known as "effect coding." This approach is similar to effect coding in an analysis of variance (ANOVA) and utilizes non-arbitrary scale setting to estimate the mean structures of both the manifest indicators and latent factor. To apply this approach, each indicator should have the same range of possible values. This approach constrains the sum of the number of indicator intercepts to "0," and the set of loadings are fixed to average "1.0," which is the same as having the loadings sum to the number of unique indicators. Thus, the factor loadings and intercepts are estimated as the value that would solve the constraint: $\lambda_{11} = 3 - \lambda_{21} - \lambda_{31}$ and $\tau_1 = 0 - \tau_2 - \tau_3$ (assuming there are three indicators of the latent factor). It does not matter what parameter is estimated; once any two of the three parameters are estimated, the third is determined. Thus, the model estimates the same parameters by: $\lambda_{21} = 3 - \lambda_{11} - \lambda_{31}$ or: $\lambda_{31} = 3 - \lambda_{11} - \lambda_{21}$ for loadings and by: $\tau_2 = 0 - \tau_1 - \tau_3$ or: $\tau_3 = 0 - \tau_1 - \tau_2$ for intercepts. The equation changes in an orderly fashion depending on the number of indicators. For instance, if the latent factor has four indicators, one loading would be constrained to be 4 minus the other 3 estimated loadings.

Compared to the previous two traditional scale setting approaches, the effect coding approach has two major advantages for the estimation of latent factor mean and variance parameters. First, like the fixed factor approach, the effect coding approach provides an optimal balance of factor loadings across the indicators, but the loading effects average around 1.00 with the effect coding approach. These loadings result in an estimation of the variance (ψ) of the latent factor (η) that is equivalent to the average of the indicators' reliable variances (reliability).

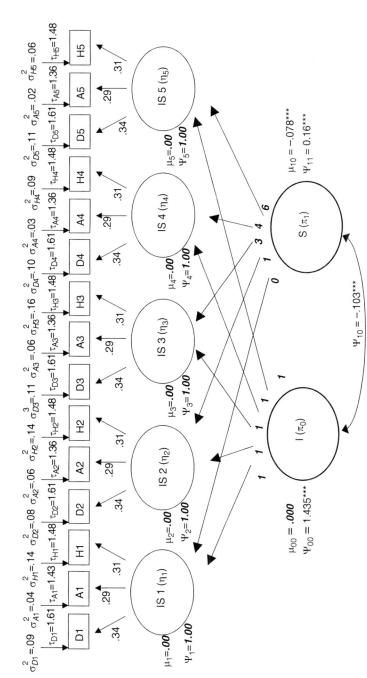

FIGURE 4.18 Results of a Curve-of-Factors Model (CFM) Using the Fixed Factor Scale Setting Approach.

Note: Unstandardized coefficients are shown. DEP = Depressive Symptoms. ANX = Anxiety Symptoms. HOS = Hostility Symptoms. I = Intercept. S = Slope. Partial strong invariance was specified using the fixed factor scale setting approach. Also, the residual variances among the same indicators over time were correlated but are not shown in the figure for simplicity. Values in italics and bold were fixed parameters. χ^2(df) = 207.352(75), $p < .001$; CFI/TLI = .968/.956; RMSEA = .064; SRMR = .079. *** $p < .001$.

Second, by constraining the sum of all intercepts to zero (e.g., $\tau_1 = 0 - \tau_2 - \tau_3$), the latent factor's mean is estimated as the average of the indicator means. That is, the mean of latent factor (η_μ) is equal to the average of the set of indicators' means $(= \sum_{i=1}^{1} E(y_i)$ / number of indicators). The average of the indicators' means is a more accurate estimate of the population value than any one indicator arbitrarily selected from the set (i.e., the marker variable approach; Little et al., 2006). More importantly, researchers can now estimate a more stable mean for the latent factor compared to previous approaches. This feature is attractive for the purpose of estimating and interpreting second-order growth mean parameters. Similar to the marker variable approach, this approach also produces a weighted intercept (τ_y) for each indicator. However, as shown in Equation 4.2, the weighted intercept can be easily converted to the observed mean score, $E(y)$, of each indicator.

Illustrative Example 4.8: Using the Effect Coding Scale Setting Approach in a CFA

An example of the effect coding approach is shown in Panel B of Figure 4.17.

M*plus* syntax for a CFA model utilizing the effect coding approach to scale setting is as follows:

```
MODEL:
IS1 by DEP1*(L1) ANX1 (L2) HOS1 (L3);
[IS1](M); [DEP1](T1); [ANX1](T2); [HOS1](T3);
MODEL CONSTRAINT: NEW (DEP_M ANX_M HOS_M);
L1 = 3 – L2 – L3;
T1 = 0 – T2 – T3;
DEP_M = T1 + (L1*M);
ANX_M = T2 + (L2*M);
HOS_M = T3 + (L3*M);
```

In the above syntax, labels are assigned to the factor loadings of the three indicators by adding L1, L2, and L3 in the MODEL command. Also, labels are assigned to the mean parameters of the three indicators and the latent factor by adding T1, T2, T3, and M, respectively. Any combination of letters and numbers can be used to assign parameter labels.

The MODEL CONSTRAINT command specifies parameter constraints using the labels assigned in the MODEL command. In our model, L1 (the factor loading of DEP1) is now estimated using $\lambda_{DEP1} = 3 - \lambda_{ANX1} - \lambda_{HOS1}$ (which is equivalent to L1 = 3 – L2 – L3 in the M*plus* syntax) and produces .183 of the latent variance in IS1. This estimation indicates that .183 is the average of the reliable variance components (= $(.455^2 + .422^2 + .408^2)$ / 3) (see the factor loadings in Panel A in Figure 4.17) across the three indicators of IS1. In the same manner, the intercept of DEP1 is now estimated by $\tau_{DEP1} = 0 - \tau_{ANX1} - \tau_{HOS1}$ (which is equivalent to T1 = 0 – T2 – T3 in the M*plus* syntax) and equals 1.458, the mean of the latent factor IS1. This latent factor mean now represents the average of the indicators' means (with some rounding error). However, as can be seen in Panel B of Figure 4.17, some of the indicators' intercepts are now negative because these are weighted intercepts, which can be difficult to interpret. These intercepts can easily be converted to mean scores using Equation 4.2. For example, the intercept of ANX1 (-.040) can be converted to the mean score of 1.40 (1.40 = -.040 + .984 × 1.458). The converted mean scores can be estimated by specifying "NEW (DEP_M ANX_M HOS_M)" along with the three equations in the MODEL CONSTRAINT command. For example, mean scores of anxiety (ANX_M) can be estimated by ANX_M = T2 + (L2*M), which is equivalent to Equation 4.2.

Next, we illustrate the use of the effect coding approach to scale setting for estimating an unconditional CFM. M*plus* syntax with the effect coding approach in the MODEL statement is as follows:

```
MODEL:
IS1 by DEP1*(L1)
        ANX1 HOS1 (L2-L3);
IS2 by DEP2*(L1)
        ANX2 HOS2 (L2-L3);
IS3 by DEP3*(L1)
        ANX3 HOS3 (L2-L3);
IS4 by DEP4*(L1)
        ANX4 HOS4 (L2-L3);
IS5 by DEP5*(L1)
        ANX5 HOS5 (L2-L3);
[DEP1-DEP5](T1); [ANX1](T2);
[ANX2-ANX5](T3); [HOS1-HOS5](T4);
I_IS S_IS | IS1@0 IS2@1 IS3@3 IS4@4 IS5@6;
[I_IS](M1); [S_IS];
MODEL CONSTRAINT: NEW (DEP_M ANX1_M ANX2_M HOS_M);
L1 = 3 – L2 – L3;
T1 = 0 – T2 – T3;
DEP_M = T1 + (L1*M1);
ANX1_M = T2 + (L2*M1);
ANX2_M = T3 + (L2*M1);
HOS_M = T4 + (L3*M1);
```

Note that autocorrelations among manifest indicators are not specified in the above MODEL command. All commands have been previously described in the simple

CFA models already introduced, with the exception of assigning labels to the four new parameters (DEP_M ANX1_M ANX2_M HOS_M). These four new parameters are used to estimate constrained mean scores of the four indicators (i.e., DEP, ANX1, ANX2, and HOS). Recall that our LCFA model met the assumption of partial longitudinal invariance rather than strong invariance because mean equality was not applicable for ANX at Time 1. Thus, the current model estimates two mean scores for anxiety symptoms: a mean score of ANX at Time 1 and a constrained mean score of ANX at Time 2 to Time 5. Results from the CFM when utilizing the effect coding approach to scale setting are shown in Figure 4.19.

Notice that the model fit indices for the CFM utilizing the effect coding approach are the same as the model fit indices for the CFM utilizing the marker variable approach (χ^2(df) = 166.598(49), $p < .001$; CFI/TLI = .977/.966; RMSEA = .056, SRMR = .062). That is, the two models are statistically identical. However, as can be seen in Figure 4.19, the mean (μ_{00}) of the second-order intercept factor in the effect coding approach model is now estimated as an average of the three indicators (τ_{D-H}). This aggregated mean provides an average initial level for the second-order growth curve that is more reliable and more easily interpreted than the mean provided by the other approaches.

Adding Covariates to a Curve-of-Factors Model (CFM)

In order to test hypotheses involving antecedents and consequences, similar to a first-order LGCM, covariates can also be incorporated into a CFM (creating a conditional CFM). In Chapter 2, we introduced a conditional first-order LGCM with a single covariate (parents' marital conflict, or MC). Covariates can be time-invariant (TIC; not changing over time) or time-varying (TVC; changing over time). Also, covariates can be manifest variables (a single indicator) or latent factors with multiple indicators. First, we will discuss time-invariant covariates (including both single-indicator and multiple-indicator covariates as well as antecedent and outcome covariates) followed by a discussion of time-varying covariates.

Time-Invariant Covariate (TIC) Model

Incorporating a single indicator variable (W) as a predictor

Recall the following CFM equations and the CFM illustrated in Figure 3.2 from Chapter 3:

For latent global factors:

$$y_{jit} = \tau_{jt} + \lambda_{jt} \times \eta_{it} + \varepsilon_{jit}, \quad \varepsilon_{jit} \sim \text{NID}\left(0, \sigma^2_{jt}\right) \tag{4.3}$$

$$\eta_{it} = \pi_{0i} + \lambda \times \pi_{1i} + \zeta_{it}, \quad \zeta_{it} \sim \text{NID}\left(0, \Psi_t\right) \tag{4.4}$$

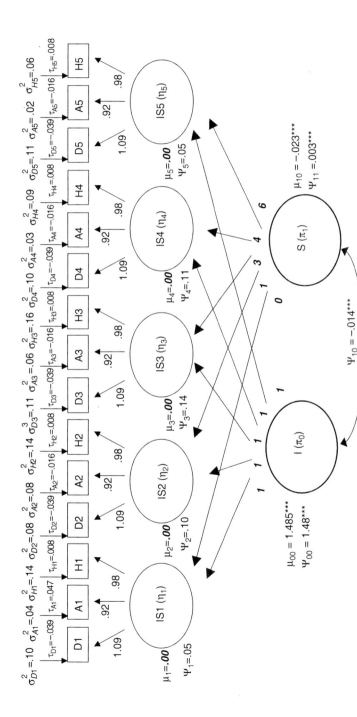

FIGURE 4.19 Results of a Curve-of-Factors Model (CFM) Using the Effect Coding Scale Setting Approach.

Note: Unstandardized coefficients are shown. DEP = Depressive Symptoms. ANX = Anxiety Symptoms. HOS = Hostility Symptoms. I = Intercept. S = Slope. Partial strong invariance was specified using the effect coding scale setting approach. Also, the residual variances among the same indicators over time were correlated but not shown in the figure for simplicity. Values in italics and bold were fixed parameters. χ^2(df) = 166.598(71), $p < .001$; CFI/TLI = .977/.966; RMSEA = .056; SRMR = .062. *** $p < .001$.

For second-order growth factors:

$$\pi_{0i} = \mu_{00} + \zeta_{0i}, \qquad \zeta_{0i} \sim \text{NID}\left(0, \Psi_{00}\right) \tag{4.5}$$

$$\pi_{1i} = \mu_{10} + \zeta_{1i}, \qquad \zeta_{1i} \sim \text{NID}\left(0, \Psi_{11}\right) \tag{4.6}$$

As shown by Equations 4.5 and 4.6, no predictors of the second-order latent factor (π) were specified as this is the unconditional model and the right-hand side of the equations only include the means and individual deviation from the means. Now, we can add time-invariant covariates (W) to the equations for the second-order growth factors.

$$\pi_{0i} = \mu_{00} + \gamma_{01} \times W_{i1} + \gamma_{02} \times W_{i2} + \ldots + \gamma_{0n} \times W_{in} + \zeta_{0i} \tag{4.7}$$

$$\pi_{1i} = \mu_{10} + \gamma_{11} \times W_{i1} + \gamma_{12} \times W_{i2} + \ldots + \gamma_{1n} \times W_{in} + \zeta_{1i} \tag{4.8}$$

where $W_{i1}, W_{i2}, \ldots, W_{in}$, are time-invariant covariates (TIC) that are uncorrelated with the variances of all other covariates and residuals.

Illustrative Example 4.9: Adding a Time-Invariant Covariate (TIC) as a CFM Predictor

For an example, we used MC (parents' marital conflict) at Time 1 as a TIC. In order to specify MC as a predictor (W), the syntax below can be added to the MODEL command of the unconditional CFM (with the marker variable coding approach).

I_IS S_IS ON MC;

I_IS and S_IS represent the intercept and slope variance for a curve-of-factors model (CFM) of IS.

The results are shown in Figure 4.20. Note that we have used the marker variable approach for scale setting of all of the time-invariant covariates (TIC). The results regarding the influence of marital conflict (MC) on growth factors are similar to those of the first-order LGCM.

The regression coefficients in Figure 4.20 can be interpreted as follows:

.516 = the expected change in the average adolescent's initial level of IS for each one unit increase in parents' marital conflict (MC).

-.020 = the expected change in the average adolescent's rate of change (slope) of IS over time for each one unit increase in parents' marital conflict (MC). Because the coefficient is negative, this coefficient indicates a decline over time. However, this path is not statistically significant.

Incorporating a Multiple-Indicator Latent Variable (P) as a Predictor

The main advantage of fitting growth curves in a SEM framework is that latent time-invariant variables (P) consisting of multiple indicators (x's) can be used as

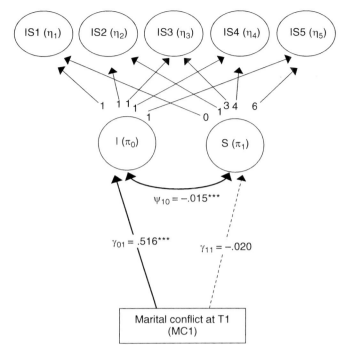

FIGURE 4.20 A Curve-of-Factors Model (CFM) with a Time-Invariant Observed Covariate.

Note: Unstandardized coefficients are shown. The measurement model of the LCFA is not shown in the figure for simplicity. Bold lines indicate statistically significant covariate paths. Broken lines indicate non-significant paths. I = Intercept. S = Slope. χ^2 (df) = 185.978(84), $p < .001$; CFI/TLI = .974/.963; RMSEA (90% CI) = .056 (.045, .067); SRMR = .063. *** $p < .001$.

predictors. That is, a CFA (representing the measurement model of a CFM) can be specified as:

$$x_{ji} = \tau_j + \lambda_j \times P_{ji} + \delta_{ji}, \qquad \delta_{ji} \sim NID\left(0, \sigma_j^2\right) \qquad (4.9)$$

where the subscript j refers to particular indicator j and the subscript i (= 1, 2, 3,…, n) refers to the individual. P represents an exogenous latent factor, and δ is a residual of the exogenous indicator (x), which is normally and independently distributed (NID) with a mean of zero and a variance of σ^2. The equations for the structural model of a CFM are as follows with an exogenous latent variable predictor (P):

$$\pi_{0i} = \mu_{00} + \gamma_{01} \times P + \zeta_{0i} \qquad (4.10)$$

$$\pi_{1i} = \mu_{10} + \gamma_{11} \times P + \zeta_{1i} \qquad (4.11)$$

where P is a latent time-invariant covariate that is uncorrelated with the residuals.

Illustrative Example 4.10: Adding a Multiple-Indicator Latent Factor as a CFM Predictor

Family economic problems (FEP) at T_0 and T_1 (FEP0 and FEP1) were used to create a latent variable assessing adolescents' perception of family economic problems. γ_{01} is the regression coefficient for the time-invariant latent covariate. ζ_{0i} and ζ_{1i} are residuals. This latent exogenous factor was then added to the CFM as a predictor.

In order to add an exogenous latent factor to the model as a TIC, the marker variable approach was used by fixing the loading of FEP0 (i.e., the marker variable) to 1. However, given the small number of indicators (two indicators) used to produce the latent variable, for model identification purposes the mean of the exogenous latent factor (FEP) was fixed to 0. The syntax below can be added to the MODEL command of the unconditional CFM.

FEP BY FEP0 FEP1;
I_IS S_IS ON FEP;

FEP0 and FEP1 represent family economic problems at Time 0 and Time 1. They are the manifest indicators of the latent exogenous factor, family economic hardship (FEP), which is specified as a time-invariant latent covariate of the intercept and slope of IS.

The results for this TIC are shown in Figure 4.21 and an interpretation of these results is included below.

A one unit increase in family economic hardship (FEP) was associated with an expected .590 ($p < .001$) unit change (an increase) in the intercept, π_0, and with an expected -.063 ($p < .05$) unit change (a decline) in the slope, π_1. Both coefficients were statistically significant. These findings indicate that higher values of FEP were, on average, associated with higher initial levels of IS and with a decrease in IS over time. One-fifth (21.5%) of the intercept factor variance was explained by FEP, while 11.5% of the variance in the slope factor was explained. This explained variance increased the R^2 statistic of both the intercept and slope factors.

Predicting Both Time-Specific Latent Factors and Second-Order Growth Parameters

Time-invariant latent covariates can be predictors of both growth parameters (intercept and slope, or π_0 and π_1, respectively) and time-specific latent variables (η). By doing so, this model specification estimates all regression coefficients (γs) of the time-invariant latent covariates after controlling for the effects of growth parameters (i.e., intercept and slope factors in the CFM). The growth curve model parameters are also estimated after correcting for the effects of the time-invariant latent covariate (Stoel, van der Wittenboer, & Hox, 2004). For this model, the

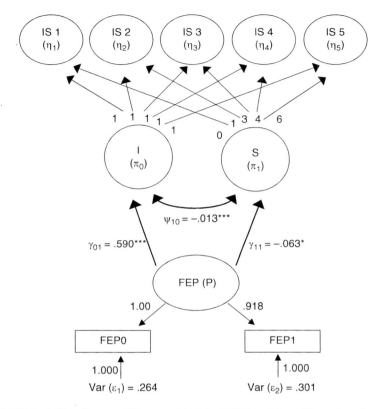

FIGURE 4.21 Curve-of-Factors Model (CFM) with a Time-Invariant Latent Covariate.

Note: Unstandardized coefficients are shown. The measurement model of the LCFA is not shown in the figure for simplicity. The marker variable approach for scale setting of the exogenous latent variable (FEP) was employed. I = Intercept. S = Slope. FEP0 = Family Economic Problems at Time 0. FEP1 = Family Economic Problems at Time 1. χ^2(df) = 192.933(198), p < .001; CFI/TLI = .978/.969; RMSEA (90% CI) = .047 (.037, .057); SRMR = .058. *** p < .001.

equations for individual i (= 1, 2, ..., n) at time point t (= 0, 1, 2, ..., t) are as follows:

$$\eta_{it} = \pi_{0i} + \lambda_t \times \pi_{1i} + \gamma_{01t} \times P + \gamma_{02t} \times P + \ldots + \gamma_{0mt} \times P + \zeta_{it},$$
$$\zeta_{it} \sim \mathrm{NID}\left(0, \Psi_t\right) \tag{4.12}$$

$$\pi_{0i} = \mu_{00} + \gamma_{01} \times P + \zeta_{0i}, \quad \zeta_{0i} \sim \mathrm{NID}\left(0, \Psi_{00}\right) \tag{4.13}$$

$$\pi_{1i} = \mu_{10} + \gamma_{11} \times P + \zeta_{1i}, \quad \zeta_{1i} \sim \mathrm{NID}\left(0, \Psi_{11}\right) \tag{4.14}$$

$$\Psi = \begin{bmatrix} \Psi_{00} & \\ \Psi_{10} & \Psi_{11} \end{bmatrix} \tag{4.15}$$

where $\gamma_{01t\sim0mt}$ are regression coefficients between the exogenous latent variable (P) and time-specific latent variables (η_{it}).

This conditional CFM can be understood as a mediation model, in which the growth factors (i.e., the intercept, π_0, and slope, π_1, latent factors) are hypothesized to completely mediate the effect of the time-invariant covariate on the first-order latent factors (Stoel et al., 2004). However, this assumption of complete mediation may not be reasonable. For example, the previous example model fixed all paths between time-specific latent variables (η_{it}) and time-invariant latent covariates. In other words, these paths were constrained to be zero. If fixing these paths is inappropriate, the model fit will be poor, and the estimated parameters will likely be biased (Preacher, Wichman, MacCallum, & Briggs, 2008). In order to estimate unbiased parameters, the time-invariant latent covariate can be specified as a predictor of both the growth parameters $(\pi_0$ and $\pi_1)$ and the time-specific latent variables (η_{it}). This full mediation model allows researcher(s) to investigate how time-invariant covariates influence residual variances of specific time periods (indicated by latent variables) after controlling for constant change over time (Preacher et al., 2008).

A disadvantage of modeling time-invariant covariates in this manner is that this alternative approach to mediation may lead to difficulties with model convergence and, consequently, the unavailability of standard errors. This convergence issue commonly happens in a SEM when too many unknown parameters are specified within a model.

Illustrative Example 4.11: Predicting Both Second-Order Growth Parameters and First-Order Latent Factors

In order to avoid convergence problems, we utilized a tentative approach. We specified one direct regression coefficient (γ_{0mt}) at a time between the time-invariant covariate (in our example, family economic problems, FEP) and time-specific latent variables $(\eta_{1\sim5})$ after controlling for the effects of FEP on the growth parameters. This approach allowed for the examination of the regression coefficient between FEP and the time-specific latent variables $(\eta_{1\sim5})$ for each time point individually. One statistically significant regression coefficient between FEP and IS was found with FEP predicting significant variation in IS at Time 5 (γ_{015}). The non-significant coefficients (effects of FEP on IS at Times 1–4) were then fixed to 0.

In order to investigate the influence of time-invariant covariates on both time-specific latent factors and second-order growth parameters, the syntax below can be added to the MODEL command of the unconditional CFM.

I_IS S_IS ON FEP;
IS5 ON FEP;

The results for our example model are shown in Figure 4.22. Interestingly, the association (γ_{11}) between FEP and the slope factor was not statistically significant after taking into account the direct effect of FEP on IS at Time 5. However, the results of this model indicated that FEP exerted a direct effect on IS at the fifth time point. This direct effect of -.340 is interpreted as a .340 decrease in the latent factor IS5 for each one unit increase in FEP after controlling for the influence of growth factors.

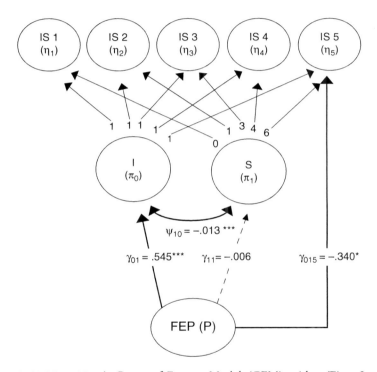

FIGURE 4.22 A Curve-of-Factors Model (CFM) with a Time-Invariant Latent Covariate Predicting both Time-Specific Latent Parameters and Second-Order Growth Parameters.

Note: Unstandardized coefficients are shown. I = Intercept. S = Slope. Regression coefficients (β) between FEP and the time-specific latent variables ($\eta_{1\sim4}$) were constrained to be 0. The measurement model of the LCFA is not shown in the figure for simplicity. The marker variable approach for scale setting of the exogenous latent variable (FEP) was employed. χ^2(df) = 187.082(97), $p < .001$; CFI/TLI = .979/.970; RMSEA (90% CI) = .046 (.036, .056); SRMR = .055. *$p < .05$. ***$p < .001$.

Predicting Distal Outcomes (D) of Second-Order Growth Factors

A CFM can also be used to predict distal outcomes, or consequences. As in the previous example model, latent distal outcomes (D) defined by multiple indicators (ys) or single indicator variables can be predicted by growth parameters (π_0 and π_1) as well as primary, time-specific, latent variables ($\eta_{1\sim5}$). We used a latent variable (D) defined by multiple indicators as the distal outcome for this illustration. Poor life quality (D) is defined by three manifest indicators: negative economic events, financial cutback, and poor work quality.

The outcome latent variable (D) can be specified as:

$$y_{ji} = \tau_j + \lambda_j \times D_i + \varepsilon_{ji}, \qquad \varepsilon_{ji} \sim \text{NID}\left(0, \sigma_j^2\right) \tag{4.16}$$

where the subscript j refers to particular indicator j and the subscript i ($= 1, 2, 3, \ldots, n$) refers to individual i. D is a latent factor with a variance, ψ_1, and ε is the residual indicator for y. ε is normally and independently distributed with a mean of zero and a variance of σ^2. The CFM with a single latent variable outcome, D, is then predicted as:

$$D_i = \mu_D + \beta_1 \pi_{0i} + \beta_2 \pi_{1i} + \beta_3 \eta_{it} + \zeta_{1i}, \qquad \zeta_{1i} \sim \text{NID}(0, \psi_1) \tag{4.17}$$

where $\beta_{1\sim2}$ are regression coefficients linking growth parameters (π_{0i} and π_{1i}, respectively) to the distal latent outcome (D). Likewise, β_3 is the regression coefficient linking the time-specific primary latent parameter (η_{it}) to the distal latent outcome, D. Similar to an OLS regression model, βs are interpreted as unique effects after controlling for the effects of other covariates.

In our example model, when both growth parameters and time-specific variables are used as predictors, there are convergence problems. That is, the model cannot be estimated when including regression paths (βs) linking both growth parameters (π_0 and π_1) and all time-specific latent variables ($\eta_{1\sim5}$) to the distal latent variable (D) simultaneously. Thus, as a tentative approach, we estimated one regression coefficient (β) at a time between the distal latent outcome (D) and the time-specific latent variables ($\eta_{1\sim5}$).

Illustrative Example 4.12: Predicting Distal Outcomes of Second-Order Growth Factors

In order to investigate how growth parameters in a CFM influence a distal outcome (poor life quality, or PLQ, in our example), the M*plus* syntax provided here can be added to the MODEL command for the unconditional CFM. Our tentative approach for estimating coefficients between the distal outcome and time-specific variables revealed one statistically significant effect. More specifically, IS4 was associated with PLQ. Consequently, this is the only time-specific effect estimated in this example.

PLQ BY NEE FC WQ;
PLQ ON I_IS S_IS IS4;

NEE (negative economic events), FC (financial cutback), and WQ (work quality) represent multiple indicators of PLQ. This particular syntax shows the estimation of coefficients between PLQ and both growth parameters (I_IS and S_IS) as well as a single time-specific latent factor (IS4).

The marker variable approach was used by fixing the loading of NEE (negative economic events) to 1 and the mean of the latent variable (PLQ) to 0 for model identification purposes. The results are shown in Figure 4.23. Interestingly, IS at

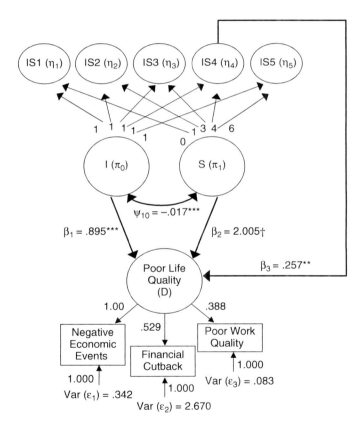

FIGURE 4.23 A Curve-of-Factors Model (CFM) with a Distal Latent Outcome. Note: Unstandardized coefficients are shown. I = Intercept. S = Slope. The measurement model of the LCFA is not shown in the figure for simplicity. The marker variable approach was used for scale setting of the endogenous latent variable (PLQ). Bold lines indicate statistically significant covariate paths. $\chi^2(\text{df}) = 214.327(113)$, $p < .001$; CFI/TLI = .977/.968; RMSEA (90% CI) = .045 (.036, .054); SRMR = .059. † $p < .010$. ** $p < .01$. *** $p < .001$.

T4 also had a unique positive effect on the distal outcome (PLQ) (not included in Equation 4.17), even after controlling for the effects of the IS growth factors on PLQ. This finding is an example of one of the strengths of a CFM, in that a CFM allows for the explanation of not only variance in growth factors of the global domain but also the explanation of unique variance of specific time-related components.

The regression coefficients can be interpreted as follows:

.895 = the change in PLQ for each one unit increase in the second-order *initial level* growth factor. That is, individuals with higher initial levels of IS generally reported greater PLQ than individuals with lower initial IS levels.

2.005 = the change in PLQ for each one unit increase in the second-order *slope* factor. In other words, across the sample, individuals with a more rapid rate of change in IS experienced poorer life quality than individuals with a slow or average rate of change in IS.

.257 = the change in PLQ for each one unit increase in latent factor IS4 after controlling for the effects of the second-order growth factors. Thus, there was a direct effect of IS at Time 4 on PLQ, with individuals who reported higher levels of IS at Time 4 experiencing poorer life quality than those with lower levels of IS at Time 4 regardless of their IS growth factors (initial level and slope).

If the distal outcome (D) is a manifest variable, the distal outcome should be directly specified in the unconditional CFM. For example, using NEE (negative economic events) as the distal outcome, the following syntax is added to the unconditional CFM:

NEE ON I_IS S_IS IS4;

Time-Varying Covariate (TVC) Model

So far, we have introduced how time-invariant covariates (TIC) can be incorporated into a CFM. This model framework is reasonable if researchers want to examine how early experiences/factors, as time-invariant covariates, influence trajectories of the outcome variable. However, the covariate may be time-varying (TVC) and have contemporaneous effects on the time-varying outcome at each time point. There are two approaches for incorporating a TVC into a CFM: (a) include the TVC as a direct predictor of the manifest outcome or (b) model the TVC as a parallel process to the CFM.

Time-Varying Covariate (TVC) as a Predictor of Manifest Outcomes

The first method for incorporating a predictor variable (Z) with repeated measures is to regress the repeated measures of the outcome variable (Y) on the predictor (Z) at each time point. Thus, level 1, level 2, and level 3 equations for a CFM can be written as:

Level 1 Equations (time-specific latent factors):

$$Y_{ikt} = \tau_{kt} + \lambda_{kt} \times \eta_{it} + \gamma_{kt} \times Z_{it} + \varepsilon_{ikt}^{(y)}, \qquad \varepsilon_{ikt}^{(y)} \sim \text{NID}\,(0, \sigma_{kt}^{(y)2}) \tag{4.18}$$

Level 2 Equations (first-order global factors):

$$\eta_{it} = \pi_{0i} + \lambda \times \pi_{1i} + \zeta_{it}, \qquad \zeta_{it} \sim \text{NID}\left(0, \Psi_{t}\right) \tag{4.19}$$

$$Z_{it} = \tau_{kt} + \varepsilon_{ikt}^{(z)}, \qquad \varepsilon_{ikt}^{(z)} \sim \text{NID}\,(0, \sigma_{kt}^{(z)2}) \tag{4.20}$$

$$\sigma_{k}^{(z)} = \begin{bmatrix} \sigma_{k11}^{(z)^2} & & & \\ \sigma_{k21}^{(z)} & \sigma_{k22}^{(z)^2} & & \\ \vdots & \vdots & \ddots & \\ \sigma_{kt1}^{(z)} & \cdots & \cdots & \sigma_{ktt}^{(z)^2} \end{bmatrix} \tag{4.21}$$

Level 3 Equations (second-order factors):

$$\pi_{0i} = \mu_{00} + \zeta_{0i}, \qquad \zeta_{1i} \sim \text{NID}\left(0, \ \Psi_{00}\right) \tag{4.22}$$

$$\pi_{1i} = \mu_{10} + \zeta_{1i}, \qquad \zeta_{1i} \sim \text{NID}\left(0, \ \Psi_{11}\right) \tag{4.23}$$

$$\Psi = \begin{bmatrix} \Psi_{00} & \\ \Psi_{10} & \Psi_{11} \end{bmatrix} \tag{4.24}$$

where subscript i refers to individual (= 1,2,3,…, n), t refers to time (= 1,2,3,…, t), and k refers to the specific subdomain of the outcome (i.e., marital dissatisfaction [MD] and marital unhappiness [MU]). σ refers to the variance-covariance structure among the exogenous TVC. The regression coefficient γ reflects the influence of the time-varying *exogenous* variable, Z, on the time-varying *endogenous* variable, Y, at each time point. The errors of the endogenous TVC, Y, are assumed to have a mean of zero and a single variance. It is also assumed that these errors are uncorrelated with the latent time-specific factors (η_t), growth factors (π_0 and π_1), and the exogenous TVCs (Bollen & Curran, 2006; Grimm, 2007).

Illustrative Example 4.13: Incorporating a time-varying covariate as a direct predictor of manifest indicators

In order to specify the TVC model, we used two time-varying covariates (i.e., marital unhappiness [MU] and marital dissatisfaction [MD]) from both husbands and wives to create latent variables of marital conflict (MC) at three time points (from Time 1 to Time 3). That is, as shown in Figure 4.24, the time-varying

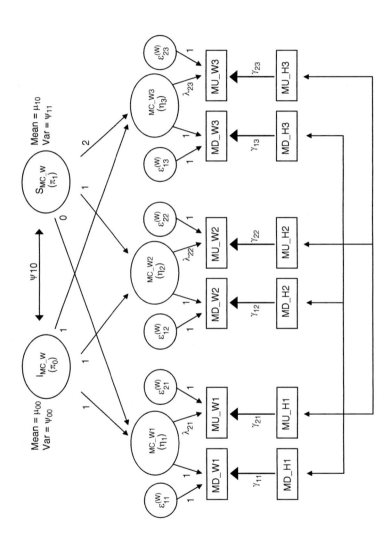

FIGURE 4.24 A Curve-of-Factors (CF) Model with Time-Varying Covariates (TVC).

Note: MD = Marital Dissatisfaction. MU = Marital Unhappiness. MC = Marital Conflict. H = Husband. W = Wife. I = Intercept. S = Slope. The marker variable scale setting approach was used.

predictor may also be from different subdomains. The path diagram of this unconditional TVC model is shown in Figure 4.24.

The corresponding M*plus* syntax is as follows:

```
MODEL:
MC_W1 BY MD_W1
          MU_W1 (L1);
MC_W2 BY MD_W2
          MU_W2 (L1);
MC_W3 BY MD_W3
          MU_W3 (L1);
[MD_W1-MD_W3@0];
[MU_W1-MU_W3](T1);

I_W S_W | M1@0 M2@1 M3@2;
[I_W S_W];
```

MD_W1 ON MD_H1 (P11);
MD_W2 ON MD_H2 (P12);
MD_W3 ON MD_H3 (P13);

MU_W1 ON MU_H1 (P21);
MU_W2 ON MU_H2 (P22);
MU_W3 ON MU_H3 (P23);

MD_H1 WITH MD_H2 MD_H3;
MD_H2 WITH MD_H3;
MU_H1 WITH MU_H2 MU_H3;
MU_H2 WITH MU_H3;

MODEL TEST:
P11=P12;
P12=P13;

The bold portions of the syntax can be added to estimate exogenous TVC effects
(husbands' marital dissatisfaction and marital unhappiness) on endogenous TVCs
(wives' marital dissatisfaction and marital unhappiness). In a SEM framework, we
can directly test whether these TVC effects are constant over time by assigning
labels, or names, in parentheses to all TVC parameters. We have used P11 to P23
to name these paths, but any names could be assigned. By specifying MODEL
TEST with P11=P12; P12=P13; we can test whether the effects of all TVCs (the
contemporaneous effect of husbands' marital dissatisfaction on wives' marital dissatisfaction at all three time points) are constant across time (i.e., P11=P12=P13).
A significant Wald test (i.e., $\Delta\chi^2$) indicates that the TVCs have different effects
across time, while a non-significant Wald test indicates non-varying effects
over time.

A Parallel Process Second-Order Model Using a Dyadic Model Framework as an Example

Another method for incorporating repeated measures of exogenous variables is to model a parallel process. For illustrative purposes, we use a time-varying covariate of a dyad member. This dyadic model is known as a "parallel process dyadic model" (Cheong, MacKinnon, & Khoo, 2003) and is a natural extension of a CFM (Wittaker, Beretvas, & Falbo, 2014). For example, a researcher may be interested in dyadic growth modeling using time-varying marital conflict of both husbands and wives. This kind of model commonly contains two sets of growth parameters and estimates covariances (or directional paths) among the intercept and slope factors. The equations for the measurement model are:

(y = wife's response; z = husband's response):

$$y_{jit} = \tau_{jty} + \lambda_{jt} \times \eta_{it} + \varepsilon_{jity}, \qquad \varepsilon_{jity} \sim \text{NID}\left(0, \sigma_{jty}^2\right) \tag{4.25}$$

$$z_{jit} = \tau_{jtz} + \lambda_{jt} \times \omega_{it} + \varepsilon_{jitz}, \qquad \varepsilon_{jitz} \sim \text{NID}\left(0, \sigma_{jtz}^2\right) \tag{4.26}$$

$$\sigma_{jt}^2 = \begin{bmatrix} \sigma_{jty}^2 & \\ \sigma_{jtzy} & \sigma_{jtz}^2 \end{bmatrix} \tag{4.27}$$

where the subscript j refers to particular indicator j, the subscript i (= 1, 2, 3, ..., n) refers to individual i, the subscript t (= 1, 2, 3, ..., t) refers to time, and the subscripts y and z refer to manifest indicators y and z, respectively. η and ω are the time-specific latent variables in the first and second set (corresponding to wives and husbands, respectively) of the CFM, respectively. σ_{jt}^2 represents the variance–covariance structure of residual (ε). Releasing covariance parameters (σ_{jtzy}^2) between dyadic indicators commonly leads to convergence problems. Thus, we assume there is no covariance between dyadic indicators by fixing these values to 0.

For the structural model, equations are as follows:

$$\eta_{it} = \pi_{0i} + \lambda_t \times \pi_{1i} + \zeta_{it}, \qquad \zeta_{it} \sim \text{NID}\left(0, \Psi_{\eta t}\right) \tag{4.28}$$

$$\omega_{it} = \pi_{2i} + \lambda_t \times \pi_{3i} + \zeta_{it} \qquad \zeta_{it} \sim \text{NID}\left(0, \Psi_{\omega t}\right) \tag{4.29}$$

$$\pi_{0i} = \mu_{00} + \zeta_{0i}, \qquad \zeta_{0i} \sim \text{NID}\left(0, \Psi_{00}\right) \tag{4.30}$$

$$\pi_{1i} = \mu_{10} + \zeta_{1i}, \qquad \zeta_{1i} \sim \text{NID}\left(0, \Psi_{11}\right) \tag{4.31}$$

$$\pi_{2i} = \mu_{20} + \zeta_{2i}, \qquad \zeta_{2i} \sim \text{NID}\left(0, \Psi_{22}\right) \tag{4.32}$$

$$\pi_{3i} = \mu_{30} + \zeta_{3i}, \qquad \zeta_{3i} \sim \text{NID}\left(0, \Psi_{33}\right) \tag{4.33}$$

$$\Psi = \begin{bmatrix} \Psi_{\eta t} & \\ \Psi_{\omega \eta t} & \Psi_{\omega t} \end{bmatrix} \tag{4.34}$$

$$\Psi_t = \begin{bmatrix} \Psi_{00} \\ \Psi_{10} & \Psi_{11} \\ \Psi_{20} & \Psi_{21} & \Psi_{22} \\ \Psi_{30} & \Psi_{31} & \Psi_{32} & \Psi_{33} \end{bmatrix} \tag{4.35}$$

where Ψ is the variance-covariance structure of the time-specific latent variables, and Ψ_t is the variance-covariance structure of growth parameters.

Illustrative Example 4.14: Incorporating a Time-Varying Covariate as a Parallel Process

An example incorporating a time-varying covariate as a parallel process is shown in Figure 4.25.

In order to build this parallel process model, also known as a Dyadic-CFM (D-CFM), we used two repeated indicators (i.e., marital dissatisfaction and marital unhappiness for husband and wives at $T_1 \sim T_3$) to create time-specific latent variables capturing marital conflict for husbands and wives separately (η_t and ω_t). Given the characteristics of our dyadic growth model (utilizing the same indicators for both husbands and wives over time), it is possible to investigate (a) whether the change in the construct of interest (marital conflict) over time is due to actual changes in the construct itself or changes in indicators and (b) whether the change in the construct of interest across measurement occasions is similar for each dyad member.

Dyadic measurement invariance can also be considered for longitudinal dyadic modeling by constraining the same parameters to be equal across dyad member groups (Wittaker et al., 2014) and conducting a nested model comparison ($\Delta\chi^2$). Following the suggestion of Wittaker et al. (2014), we applied this constraint in the current example model and confirmed strong invariance between husbands and wives, which indicates that factor loadings and intercepts of the same indicators between husbands and wives are equal across time. For consistency, the marker variable approach was used. The M*plus* syntax for a parallel CFM is shown here with marital dissatisfaction and marital unhappiness as two indicators of a latent construct of marital conflict (MD = marital dissatisfaction, MU = marital unhappiness, MC = marital conflict, H = husband, and W = wife).

MODEL:
MC_H1 BY MD_H1
     MU_H1 (L1);
MC_H2 BY MD_H2
     MU_H2 (L1);

```
MC_H3 BY MD_H3
        MU_H3 (L1);
MC_W1 BY MD_W1
        MU_W1 (L1);
MC_W2 BY MD_W2
        MU_W2 (L1);
MC_W3 BY MD_W3
        MU_W3 (L1);
[MD_H1-MD_ H3@0];
[MU_ H1-MU_ H3](T1);
[MD_W1-MD_W3@0];
[MU_W1-MU_W3](T1);
I_H S_H | MC_H1@0 MC_H2@1 MC_H3@2;
I_W S_W | MC_W1@0 MC_W2@1 MC_W3@2;
[I_H-S_W];
```

Note that in the current model, dyadic measurement invariance was assumed. Thus, factor loadings and mean parameters between husbands and wives were set to be equal across both husbands and wives and across time simultaneously. These equality constraints would typically be tested using a $\Delta\chi^2$ test statistic (i.e., for weak invariance, strong invariance, strict invariance; Bollen, 1989).

As shown in Figure 4.25, intercept covariance (ψ_{20}; cross-sectional covariance) and slope covariance (ψ_{31}) between husbands and wives were positive, which indicates parallel influences whereby marital conflict between husbands and wives moves "hand-in-hand." However, all other covariances (ψ_{10}, ψ_{12}, ψ_{30}, and ψ_{32}) between the intercept and slope factors were negative, reflecting that individuals who have lower initial level values of MC experience a greater increase in MC compared to individuals who have higher values of MC at the initial time point.

Below we provide the M*plus* syntax for utilizing the effect coding approach within a D-CFM for the purpose of comparing the two approaches.

```
MC_H1 BY MD_H1*(L1)
        MU_H1 (L2);
MC_H2 BY MD_H2*(L1)
        MU_H2 (L2);
MC_H3 BY MD_H3*(L1)
        MU_H3 (L2);
MC_W1 BY MD_W1*(L1)
        MU_W1 (L2);
MC_W2 BY MD_W2*(L1)
        MU_W2 (L2);
MC_W3 BY MD_W3*(L1)
        MU_W3 (L2);
```

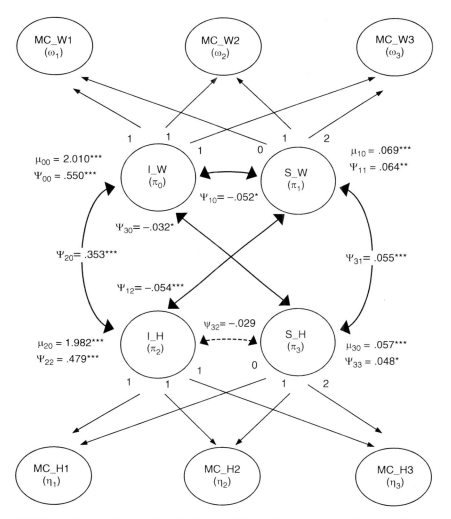

FIGURE 4.25 A Dyadic Parallel Process Model Extension of a Curve-of-Factors Model (D-CFM).

Note: Unstandardized coefficients are shown. I = Intercept. S = Slope. MC_W = Wives' Marital Conflict. MC_H = Husbands' Marital Conflict. The measurement model of the LCFA is not shown in the figure for simplicity. The marker variable scale setting approach was used. Bold lines indicate statistically significant paths. Broken lines indicate non-significant paths.

χ^2(df) = 138.544 (56), $p < .001$; CFI/TLI = .976/.971; RMSEA (90% CI) = .058 (.046, .071); SRMR = .054. * $p < .05$. * $p < .01$. *** $p < .001$.

[MD_H1-MD_H3](T1); [MU_H1-MU_H3](T2);
[MD_W1-MD_W3](T1); [MU_W1-MU_W3(T2);
I1 S1| MC_H1@0 MC_H2@1 MC_H3@2;
I2 S2| MC_W1@0 MC_W2@1 MC_W3@2;
[I1](M1); [I2](M2); [S1 S2];
MODEL CONSTRAINT: NEW (TAU1 TAU2 TAU3 TAU4);
L1 = 2 – L2; T1 = 0 – T2;
TAU1 = T1 + (L1*M1);
TAU2 = T2 + (L2*M1);
TAU3 = T1 + (L1*M2);
TAU4 = T2 + (L2*M2);

Chapter 4 Exercises

The exercise sample dataset contains depression, anxiety, and hostility symptoms of mothers at T1~T4. In the M*plus* syntax, indicate the variable order as follows:

> **DATA**: File is exercise_Ch.4.dat;
> **VARIABLE:** names are DEP1M DEP2M DEP3M DEP4M ANX1M ANX2M ANX3M ANX4M HOS1M HOS2M HOS3M HOS4M FC1_M NLE4M;

Note that DEP1M~DEP4M indicate mothers' depressive symptoms at T1~T4. ANX1M~ANX4M indicate mothers' anxiety symptoms at T1~T4. HOS1M~HOS4M indicate mothers' symptoms of hostility at T1~T4. FC1_M represents financial cutbacks reported by mothers at T1. NLE4M represents negative life events reported by mothers at T4.

1. Estimate a simple CFA model using three manifest indicators (i.e., DEP1M, ANX1M, and HOS1M) of a latent factor (internalizing symptoms, or IS). Use the marker variable approach to scale setting with ANX1M as the marker variable.
2. Using ANX1M~ANX4M, DEP1M~DEP4M, and HOS1M~HOS4M, estimate a longitudinal confirmatory analysis (LCFA) model. Indicate ANX1M~ANX4M as the marker variables. In order to estimate unbiased parameters, specify autocorrelated errors in the LCFA model.

   A. Test the various levels of longitudinal invariance, and complete Exercise Table 4.1 to illustrate the results.
   B. Which level of invariance can you confirm (i.e., weak or strong invariance)?

EXERCISE TABLE 4.1 Measurement Invariance in a Longitudinal CFA Model.

		$\Delta\chi^2 (\Delta df)$,					
$\chi^2 (df)$	Model Comparison	p -value	CFI	ΔCFI	RMSEA	SRMR	BIC
Configural LCFA model (M1)							
LCFA with weak invariance (M2)	M2 vs. M1						
LCFA with strong invariance (M3)	M3 vs. M2						

Note: M = Model. LCFA = Longitudinal Confirmatory Factor Analysis. All of these LCFA models included autocorrelated errors.

3. Estimate a curve-of-factors model (CFM) for internalizing symptoms based on the LCFA in Question 2. Specify factor loadings for linear change (t = 0, 1, 2, and 3), and interpret the global mean growth parameters (i.e., μ_{00} and μ_{10}).

4. Following the steps below, estimate a conditional CFM for internalizing symptoms using time-invariant observed and latent covariates.

   A. Specify FC1_M as a time-invariant observed covariate predicting global growth factors (π_0 and π_1) in the unconditional CFM for internalizing symptoms. Interpret all statistically significant path coefficients (unstand- · ardized coefficients).

   B. Using the conditional CFM specified in Question 4A, investigate whether FC1_M influences any of the time-specific latent variables (η). In order to avoid potential model convergence problems, test one or two direct paths at a time. Interpret the statistically significant path coefficients for both global growth factors and time-specific latent variables in your final model (unstandardized coefficients).

5. Incorporate mothers' negative life events (NLE4M) as a distal outcome in the unconditional CFM.

   A. Predict NLE4M as a distal latent outcome of global growth factors (π_0 and π_1) for internalizing symptoms. Interpret the statistically significant path coefficients (unstandardized coefficients).

   B. Using the conditional CFM specified in Question 5A, investigate whether any of the time-specific latent variables (η) predict the distal outcome NLE4M. In order to avoid model convergence problems, specify one or two direct regression coefficients at a time. For your final model, interpret the statistically significant path coefficients for both global growth factors and time-specific latent variables (unstandardized coefficients).

References

Bollen, K. A. (1989). *Structural equations with latent variables.* New York: John Wiley & Sons, Inc.

Bollen, K. A., & Curran, P. J. (2006). *Latent curve models: A structural equation approach.* Wiley Series on Probability and Mathematical Statistics. New Jersey: John Wiley & Sons.

Brown, T. A. (2006). *Confirmatory factor analysis for applied research.* New York: Guilford Press.

Chen, F. F., Sousa, K. H., & West, S. G. (2005). Testing measurement invariance of second-order factor models. *Structural Equation Modeling, 12*(3), 471–492.

Cheong, J., MacKinnon, D. P., & Khoo, S. T. (2003). Investigation of mediational processes using parallel process latent growth curve modeling. *Structural Equation Modeling, 10*(2), 238–262.

Cheung, G. W., & Rensvold, R. B. (2002). Evaluating goodness-of-fit indexes for testing measurement invariance. *Structural Equation Modeling, 9*, 233–255.

Ferrer, E., Balluerka, N., & Widaman, K. F. (2008). Factorial invariance and the specification of second-order latent growth models. *Methodology: European Journal of Research Methods for the Behavioral and Social Sciences, 4*(1), 22–36.

Geiser, C., Eid, M., Nussbeck, F. W., Courvoisier, D. S., & Cole, D. (2010). Analyzing true change in longitudinal multitrait-multimethod studies: Application of a multimethod change model on depression and anxiety in children. *Developmental Psychology*, 46(1), 29–45.

Geiser, C., & Lockhart, G. (2012). A comparison of four approaches to account for method effects in latent state-trait analyses. *Psychological methods*, 17(2), 255–283.

Grilli, L., & Varriale, R. (2014). Specifying measurement error correlations in latent growth curve models with multiple indicators. *Methodology: European Journal of Research Methods for the Behavioral and Social Sciences*, 10(4), 117–125.

Grimm, K. J. (2007). Multivariate longitudinal methods for studying developmental relationships between depression and academic achievement. *International Journal of Behavioral Development*, 31(4), 328–339.

Harring, J. R. (2009). A nonlinear mixed effects model for latent variables. *Journal of Educational and Behavioral Statistics*, 34(2), 293–318.

Kim, E. S., & Willson, V. L. (2014). Measurement invariance across groups in latent growth modeling. *Structural Equation Modeling*, 21(3), 408–424.

Little, T. D. (2013). *Longitudinal structural equation modeling*. New York, NY: Guilford.

Little, T. D., Siegers, D. W., & Card, N. A. (2006). A non-arbitrary method of identifying and scaling latent variables in SEM and MACS models. *Structural Equation Modeling*, 13(1), 59–72.

Matsunaga, M. (2010). How to factor-analyze your data right: Do's, don'ts, and how-to's. *International Journal of Psychological Research*, 3(1), 97–110.

Meade, A. W., Johnson, E. C., & Braddy, P. W. (2008). Power and sensitivity of alternative fit indices in tests of measurement invariance. *Journal of Applied Psychology*, 93(3), 568–592.

Meredith, W. (1964). Notes on factorial invariance. *Psychometrika*, 29(2), 177–186.

Meredith, W., & Horn, J. (2001). The role of factorial invariance in modeling growth and change. In A. Sayer & L. Collins (Eds.), *New methods for the analysis of change* (pp. 203–240). Washington, DC: American Psychological Association.

Muthén, L. K., & Muthén, B. O. (2007–2010). *Mplus user's guide* (6th ed.). Los Angeles: Authors.

Preacher, K. J., Wichman, A. L., MacCallum, R. C., & Briggs, N. E. (2008). *Latent growth curve modeling*. Thousand Oaks, CA: Sage Publications.

Stoel, R., van den Vittenboer, G., & Hox, J. (2004). Including time-invariant covariates in the latent growth curve model. *Structural Equation Modeling*, 11(2), 155–167.

Thompson, M. S., & Green, S. B. (2006). Evaluating between-group differences in latent means. In G. R. Hancock & R. O. Mueller (Eds.), *Structural equation modeling: A second course* (pp. 119–169). Greenwich: CT: Information Age.

Wittaker, T. A., Beretvas, S. N., & Falbo, T. (2014). Dyadic curve-of-factors model: An introduction and illustration of model for longitudinal nonexchangeable dyadic data. *Structural Equation Modeling*, 21(2), 303–317.

5

EXTENDING A PARALLEL PROCESS LATENT GROWTH CURVE MODEL (PPM) TO A FACTOR-OF-CURVES MODEL (FCM)

Introduction

This chapter discusses the association of several first-order growth curves with data at the same time points. This type of model is known as a parallel process model (PPM) or parallel process latent growth curve model. Because a PPM is considered a primer of a second-order growth curve, this chapter demonstrates how a PPM can be extended to a second-order growth curve in an incremental manner. First, this chapter provides an introduction to parallel process models (PPMs), including the estimation and interpretation of a PPM and when a PPM is appropriate. Second, the chapter introduces different specifications of associations among first-order growth factors within a PPM. Third, the chapter demonstrates how a PPM can be extended to a second-order growth curve model known as a factor-of-curves model (FCM). A FCM uses first-order growth factors as the indicators of second-order growth factors reflecting underlying global factors (i.e., intercept and slope factors of a global domain, such as internalizing symptoms).

Parallel Process Latent Growth Curve Model (PPM)

Researchers often investigate distinct but related subdomains (or dimensions) of global domains, such as physical health, mental health, youth development, socioeconomic and work quality, entrepreneurship, marital quality, close relationships, and healthy lifestyle behaviors. For example, antisocial behavior and substance use among adolescents are distinct, but related behavioral subdomains, and research has shown a covariance between the growth factors of these two time-varying attributes (Jessor & Jessor, 1977). Similarly, antisocial behavior and depressive symptoms are distinct, but associated, time-varying attributes (Angold, Costello, & Erkanli, 1999; Kubzansky & Kawachi, 2000; Loeber, Stouthamer-Loeber, & White, 1999). Furthermore, mastery, self-esteem, and behavioral control are related developmental subdomains, whereas immediate memory and distal memory are related cognitive subdomains. Together, school attachment and educational performance may assess a broader global domain of socioeconomic potential, whereas income and assets could be measures of a larger global domain of socioeconomic achievement. Health measures, such as BMI, cortisol, pulse rate, and blood pressure, may also represent subdomains of a broader health domain. Change in one subdomain is often associated with change in another subdomain over time (covariance), which produces associated parallel growth curves of subdomains (a PPM). In this chapter, we use different subdomains of internalizing symptoms (IS) for illustrations (e.g., depressive symptoms [DEP], anxiety symptoms [ANX], and hostility symptoms [HOS]).

The expectation in investigating subdomain-specific growth factors and their associations is that separate but associated growth curves capturing the individual subdomains (defined by first-order growth parameters with an intercept and slope(s) for each subdomain) capture not only the within-subdomain variation but also allow for the examination of between-subdomain associations over time. While the first-order growth parameters describe mean levels and inter-individual variation in intra-individual changes within a specific subdomain, the between-subdomain covariations reflect developmental associations among different subdomains (the co-development of subdomains). Thus, a parallel process model (PPM) contains a set of first-order growth curves for associated subdomains.

POINT TO REMEMBER...

First-order growth curves that are specific to individual subdomains are referred to as **primary growth curves** and their growth factors (initial levels and slopes) are referred to as **primary growth factors**.

As previously noted, a PPM allows researchers to examine the quality of primary growth factors that are specific to the individual subdomains. That is, researchers can examine whether primary growth factors of subdomains have significant means (average developmental trends) and variances describing inter-individual differences in intra-individual trends within each subdomain (e.g., anxiety symptoms). More importantly, researchers can examine covariance among primary growth factors to understand the co-development of different attributes, capturing communality in primary growth factors. For example, as shown in Figure 5.1, primary growth factors of depressive symptoms and anxiety symptoms can be correlated *within* each subdomain (e.g., a correlation between the initial level and slope of depressive symptoms). Also correlations can be estimated among primary growth factors *between* different subdomains (e.g., a correlation between the slope of depressive symptoms and the slope of anxiety symptoms). These non-directional associations depict the co-occurrence of different attributes (subdomains) and their co-development over time.

Estimating a Parallel Process Model (PPM)

Figure 5.1 illustrates a parallel process model (PPM) with two primary growth curves (depressive symptoms and anxiety symptoms). All parameters are shown in the figure along with the structural equations and variance-covariance matrix of errors among the latent growth factors.

Figure 5.2 provides the M*plus* syntax for the PPM shown in Figure 5.1. Two primary linear growth curves are specified using the following M*plus* syntax:

I_ANX S_ANX | ANX1@0 ANX2@1 ANX3@3 ANX4@4 ANX5@6;
I_DEP S_DEP | DEP1@0 DEP2@1 DEP3@3 DEP4@4 DEP5@6;

The correlations among growth factors are specified using the following M*plus* syntax:

I_ANX I_DEP PWITH S_ANX S_DEP;
I_ANX S_ANX PWITH I_DEP S_DEP;

In specifying each of the primary growth curves, as in the case of a LGCM, the variance of each measure is separated into two parts consisting of: (a) an item-specific variance component and (b) a time-specific variance component. The item-specific variance component is considered the measurement error of the item, whereas the time-specific variance component is transmitted to the latent growth factors. For example, the use of multiple repeated measures of depressive symptoms to define primary growth factors of depressive symptoms separates the variance of each measure of depressive symptoms (e.g., DEP1, DEP2, DEP3, DEP4, and DEP5) into a measurement error component and a time-specific component contributing to the primary growth factor. A similar decomposition occurs for the variances of repeated measures of anxiety symptoms in the primary growth curve assessing anxiety symptoms.

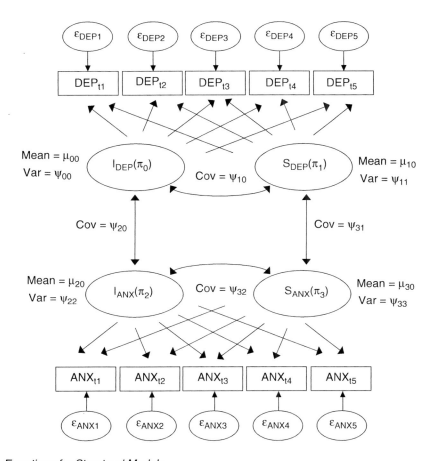

Equations for Structural Models

$$\pi_{0i} = \mu_{00} + \zeta_{0i}, \qquad \zeta_{0i} \sim NID(0, \psi_{00})$$
$$\pi_{1i} = \mu_{10} + \zeta_{1i}, \qquad \zeta_{1i} \sim NID(0, \psi_{11})$$
$$\pi_{2i} = \mu_{20} + \zeta_{2i}, \qquad \zeta_{2i} \sim NID(0, \psi_{22})$$
$$\pi_{3i} = \mu_{30} + \zeta_{3i}, \qquad \zeta_{3i} \sim NID(0, \psi_{33})$$

$$\psi = \begin{bmatrix} \psi_{00} & & & \\ \psi_{10} & \psi_{11} & & \\ \psi_{20} & 0 & \psi_{22} & \\ 0 & \psi_{31} & \psi_{32} & \psi_{33} \end{bmatrix}$$

FIGURE 5.1   Parallel Process Model (PPM).
Note: I = Initial Level. S = Slope. DEP = Depressive Symptoms. ANX = Anxiety Symptoms. i = individual (n = 1, 2,..., n). NID = Normally and Independently Distributed.

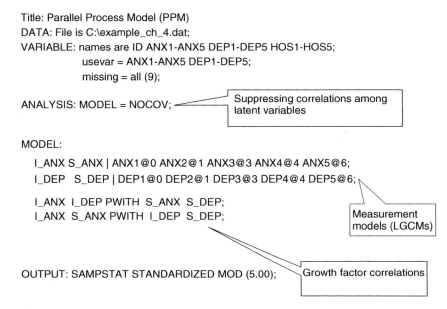

Title: Parallel Process Model (PPM)
DATA: File is C:\example_ch_4.dat;
VARIABLE: names are ID ANX1-ANX5 DEP1-DEP5 HOS1-HOS5;
          usevar = ANX1-ANX5 DEP1-DEP5;
          missing = all (9);

ANALYSIS: MODEL = NOCOV; ⟵ Suppressing correlations among latent variables

MODEL:

  I_ANX S_ANX | ANX1@0 ANX2@1 ANX3@3 ANX4@4 ANX5@6;
  I_DEP S_DEP | DEP1@0 DEP2@1 DEP3@3 DEP4@4 DEP5@6;

  I_ANX I_DEP PWITH S_ANX S_DEP;
  I_ANX S_ANX PWITH I_DEP S_DEP;   Measurement models (LGCMs)

OUTPUT: SAMPSTAT STANDARDIZED MOD (5.00);   Growth factor correlations

FIGURE 5.2 M*plus* Syntax for a Parallel Process Model (PPM).
Note: I = Initial Level. S = Slope. DEP = Depressive Symptoms. ANX = Anxiety Symptoms. LGCM = Latent Growth Curve Model.

Correlation of Measurement Errors in a PPM

A PPM allows for within-subdomain and between-subdomain correlations among measurement errors. For example, measurement errors of repeated measures of depressive symptoms or anxiety symptoms can be correlated *within* the subdomain (autocorrelations, e.g., depressive symptoms at t1 and t2) (see Figure 5.3). These correlations may capture unique variance components or methodological biases within a measure, such as a reporting bias among respondents.

Within-subdomain measurement errors can be correlated using the following M*plus* syntax, which is also shown in Figure 5.4:

ANX1 ANX2 ANX3 ANX4 PWITH ANX2 ANX3 ANX4 ANX5;
DEP1 DEP2 DEP 3 DEP4 PWITH DEP2 DEP3 DEP4 DEP5;

Also, as shown in Figure 5.5, measurement errors of repeated measures of depressive symptoms and anxiety symptoms can also be correlated *with each other*. Often these correlations can be attributed to time-specific patterns or time-specific reporting biases (e.g., reporting a lower level of symptoms at the first wave of data collection in a longitudinal study). Time-specific measurement errors across subdomains can be correlated using the following M*plus* syntax (see Figure 5.6 for the full syntax):

ANX1 ANX2 ANX3 ANX4 ANX5 PWITH DEP1 DEP2 DEP3 DEP4 DEP5;

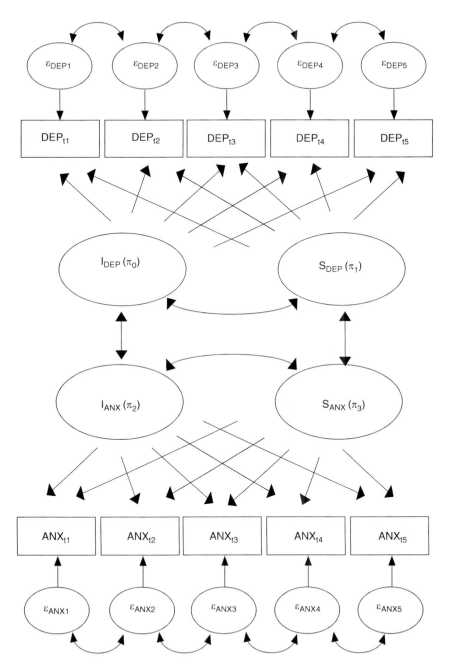

FIGURE 5.3 Parallel Process Model (PPM) with Within-Subdomain Correlations. Note: I = Initial Level. S = Slope. DEP = Depressive Symptoms. ANX = Anxiety Symptoms.

ANALYSIS: MODEL = NOCOV;

MODEL:

  I_ANX S_ANX | ANX1@0 ANX2@1 ANX3@3 ANX4@4 ANX5@6;
  I_DEP S_DEP | DEP1@0 DEP2@1 DEP3@3 DEP4@4 DEP5@6;

  I_ANX I_DEP PWITH S_ANX S_DEP;
  I_ANX S_ANX PWITH I_DEP S_DEP;

  ANX1 ANX2 ANX3 ANX4 PWITH ANX2 ANX3 ANX4 ANX5;
  DEP1 DEP2 DEP3 DEP4 PWITH DEP2 DEP3 DEP4 DEP5;

> Within-subdomain correlations

OUTPUT: SAMPSTAT STANDARDIZED MOD (5.00);

FIGURE 5.4 M*plus* Syntax for a Parallel Process Model (PPM).
Note: I = Initial Level. S = Slope. DEP = Depressive Symptoms. ANX = Anxiety Symptoms.

Influence of Growth Factors of One Subdomain on the Growth Factors of Other Subdomains

Furthermore, as shown in Figure 5.7, change in one subdomain may simultaneously predict changes in the other, which is evidence of co-development across the subdomains. That is, there can be directional influences between slope parameters of different subdomains. For example, the slope parameter of depressive symptoms may influence the slope parameter of anxiety symptoms. Such unidirectional, simultaneous influences provide insight into potential causal processes within the same time period. Using our example, the elevation of depressive symptoms may predict simultaneous growth in anxiety symptoms.

POINT TO REMEMBER...

For unidirectional model specifications, there should be strong theoretical and/or empirical support because the same magnitude of coefficients may exist regardless of the direction of effect specified. That is, specifying an effect in the reverse direction (the change in anxiety symptoms predicting change in depressive symptoms) may yield the same coefficient. Particularly in health studies, it is important to investigate such alternative explanations for symptom co-occurrence because each potential direction of effect has distinct etiological, prevention, and treatment implications.

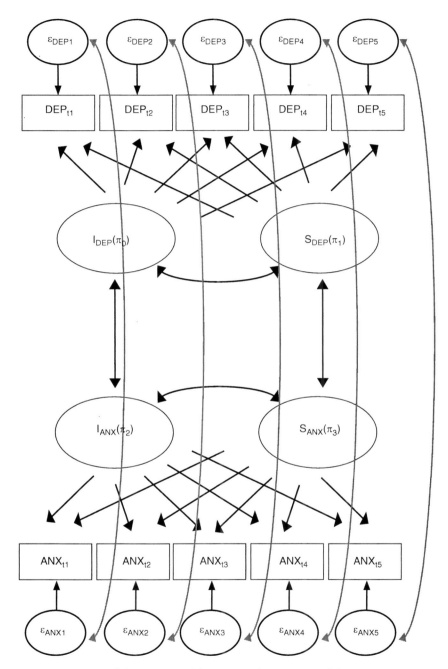

FIGURE 5.5 Parallel Process Model (PPM) with Between-Subdomain Measurement Error Correlations.

Note: I = Initial Level. S = Slope. DEP = Depressive Symptoms. ANX = Anxiety Symptoms.

ANALYSIS: MODEL = NOCOV;

MODEL:

   I_ANX S_ANX | ANX1@0 ANX2@1 ANX3@3 ANX4@4 ANX5@6;
   I_DEP S_DEP | DEP1@0 DEP2@1 DEP3@3 DEP4@4 DEP5@6;

   I_ANX I_DEP PWITH S_ANX S_DEP;
   I_ANX S_ANX PWITH I_DEP S_DEP;

   ANX1 ANX2 ANX3 ANX4 ANX5 PWITH DEP1 DEP2 DEP3 DEP4 DEP5;

> Between-subdomain correlations

OUTPUT: SAMPSTAT STANDARDIZED MOD (5.00)

FIGURE 5.6 M*plus* Syntax for a Parallel Process Model (PPM) with Between-Subdomain Measurement Error Correlations.
Note: I = Initial Level. S = Slope. DEP = Depressive Symptoms. ANX = Anxiety Symptoms.

Directional paths from anxiety (ANX) growth factors to depression (DEP) growth factors can be specified using the following M*plus* syntax, which is also provided in Figure 5.8:

I_DEP ON I_ANX;
S_DEP ON S_ANX;

As shown in Figure 5.9, directional longitudinal influences can be specified between growth parameters of different subdomains. That is, researchers can investigate if the *level* of one subdomain influences the *rate of change* in the other subdomain. For example, an increase in the early level of anxiety symptoms (the initial level) may influence the future growth rate of depressive symptoms (slope). Statistically, this is an interaction effect between time and the initial level of anxiety symptoms influencing depressive symptoms. Such unidirectional, longitudinal influences detect a cumulative effect over time; that is, researchers can discern whether the influence of depressive symptoms on anxiety symptoms gains strength over time. It is possible to assess both non-directional and directional associations among primary growth parameters in a PPM. Longitudinal directional paths from the initial level of anxiety symptoms (ANX) to the slope of depressive symptoms (DEP) can be specified using the following M*plus* syntax, which is shown in its entirety in Figure 5.10:

S_DEP ON I_ANX;

In a PPM, it is also possible that a more dynamic process exists between the two growth models. That is, there can be reciprocal influences among

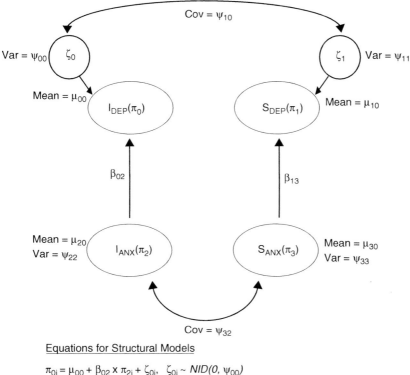

FIGURE 5.7   Parallel Process Model (PPM) with Uni-Directional Effects.
Note: I = Initial Level. S = Slope. DEP = Depressive Symptoms. ANX = Anxiety Symptoms. NID = Normally and Independently Distributed. i = individual (n = 1, 2,…, n). Measurement models (LGCMs) are not shown.

subdomains. For example, as shown in Figure 5.11, the initial levels of *both* subdomains (e.g., depressive symptoms and anxiety symptoms) may predict the slope of the other subdomain (i.e., longitudinal cross-lagged influences between the two subdomains) (Caron & Rutter, 1991). These reciprocal influences result in increasing trends in both depressive symptoms and anxiety symptoms, contributing to the co-development of subdomains over time. As shown in Figure 5.12, reciprocal, longitudinal, directional paths from the initial level of anxiety symptoms (ANX) to the slope of depressive symptoms

ANALYSIS: MODEL = NOCOV;

MODEL:

    I_ANX S_ANX | ANX1@0 ANX2@1 ANX3@3 ANX4@4 ANX5@6;
    I_DEP S_DEP | DEP1@0 DEP2@1 DEP3@3 DEP4@4 DEP5@6;

    I_ANX I_DEP PWITH S_ANX S_DEP;

    _I_DEP ON I_ANX;_   ⟵ | Uni-directional effects |
    _S_DEP ON S_ANX;_

OUTPUT: SAMPSTAT STANDARDIZED MOD (5.00);

FIGURE 5.8 M*plus* Syntax for a Parallel Process Model (PPM) with Directional Effects.
Note: I = Initial Level. S = Slope. DEP = Depressive Symptoms. ANX = Anxiety Symptoms.

(DEP) and from the initial level of DEP to the slope of ANX can be specified using the following M*plus* syntax:

S_DEP ON I_ANX;
S_ANX ON I_DEP;

Modeling Sequentially Contingent Processes over Time

There can also be longitudinal directional influences between change parameters of different subdomains (see Figure 5.13). That is, change in one subdomain over time e.g., from (t_1-t_5) can predict later changes in the other subdomain e.g., from (t_6-t_{10}). For example, a lagged slope parameter of anxiety can be modeled to influence a subsequent slope parameter of depressive symptoms. In our example, this could appear as an elevation in anxiety predicting a subsequent increase in the rate of change in depressive symptoms. Such unidirectional longitudinal influences help us to understand sequential causal processes over time, which act as a cascade effect. This type of model is no longer a parallel process model (PPM) reflecting co-development; instead it is a growth curve model reflecting a *sequentially contingent process*. These sequential models are typically most appropriate for a life-course study over an extended period of time. For example, primary growth factors of delinquency in adolescence influence primary growth factors of antisocial behavior in emerging adulthood, which, in turn, influence primary growth factors of relationship violence in young adulthood. A sequential longitudinal process involving early ANX $(t_1$ to $t_5)$ and later DEP $(t_6$ to $t_{10})$ can be specified using the following M*plus* syntax (see Figure 5.14 for the full syntax):

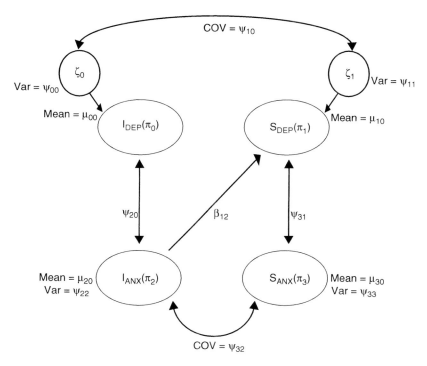

Equations for Structural Models

$$\pi_{0i} = \mu_{00} + \zeta_{0i}, \qquad\qquad \zeta_{0i} \sim NID(0, \psi_{00})$$
$$\pi_{1i} = \mu_{10} + \beta_{12} \times \pi_{2i} + \zeta_{1i}, \qquad \zeta_{1i} \sim NID(0, \psi_{11})$$
$$\pi_{2i} = \mu_{20} + \zeta_{2i}, \qquad\qquad \zeta_{2i} \sim NID(0, \psi_{22})$$
$$\pi_{3i} = \mu_{30} + \zeta_{3i}, \qquad\qquad \zeta_{3i} \sim NID(0, \psi_{33})$$

$$\psi = \begin{bmatrix} \psi_{00} & & & \\ \psi_{10} & \psi_{11} & & \\ \psi_{20} & 0 & \psi_{22} & \\ 0 & \psi_{31} & \psi_{32} & \psi_{33} \end{bmatrix}$$

FIGURE 5.9 Parallel Process Model (PPM) with a Cumulative Time Interaction Effect. Note: I = Initial Level. S = Slope. DEP = Depressive Symptoms. ANX = Anxiety Symptoms. NID = Normally and Independently Distributed. i = individual (n = 1, 2,…, n). Measurement models (LGCMs) are not shown.

```
I_ANX S_ANX | ANX1@0 ANX2@1 ANX3@3 ANX4@4 ANX5@6;
I_DEP S_DEP | DEP6@0 DEP7@2 DEP8@4 DEP9@6 DEP10@8;
I_ANX I_DEP PWITH S_ANX S_DEP;
S_DEP ON I_ANX;
S_ANX ON I_DEP;
```

ANALYSIS: MODEL = NOCOV;

MODEL:

    I_ANX S_ANX | ANX1@0 ANX2@1 ANX3@3 ANX4@4 ANX5@6;

    I_DEP S_DEP | DEP1@0 DEP2@1 DEP3@3 DEP4@4 DEP5@6;

    I_ANX I_DEP PWITH S_ANX S_DEP;

    I_ANX S_ANX PWITH I_DEP S_DEP;

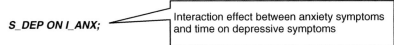

    S_DEP ON I_ANX; ⟵ Interaction effect between anxiety symptoms and time on depressive symptoms

OUTPUT: SAMPSTAT STANDARDIZED MOD (5.00);

FIGURE 5.10 M*plus* Syntax for a Parallel Process Model (PPM) with a Cumulative Time Interaction Effect.
Note: I = Initial Level. S = Slope. DEP = Depressive Symptoms. ANX = Anxiety Symptoms.

Extending a Parallel Process Latent Growth Curve Model (PPM) to a Factor-of-Curves Growth Curve Model (FCM)

Significant covariance among primary growth factors can be represented by underlying secondary growth factors, forming a second-order growth curve model known as a *factor-of-curves model* (FCM). For example, the developmental course of antisocial behavior across adolescence appears to be intertwined with substance use. A common global factor, such as the tendency to engage in unconventional behaviors, appears to underlie both antisocial behavior and substance use with both behaviors influenced by a common cause (Jessor & Jessor, 1977). Also, adolescent depressive symptoms and anxiety symptoms are related to a global factor of internalizing behaviors, which may stem from a common cause, such as a stressful early life context.

    In this extension of a PPM, the covariances among first-order growth factors are hypothesized to be explained by the second-order growth factors. For example, the covariance between the initial level of depressive symptoms and the initial level of anxiety symptoms can define a second-order initial level factor, while the covariance between the slope of depressive symptoms and the slope of anxiety symptoms can define a second-order slope factor. Together, these second-order growth factors form a second-order growth curve representing a global domain of internalizing symptoms (IS) comprised of two subdomains (symptoms of depression and anxiety).

    A first-order growth model for a specific subdomain k is defined as:

$$y_{ikt} = \pi_{0ik} + \lambda_t \times \pi_{1ik} + \varepsilon_{ikt}, \qquad \varepsilon_{ikt} \sim \mathrm{NID}(0, \sigma_{kt}^2) \qquad (5.1)$$

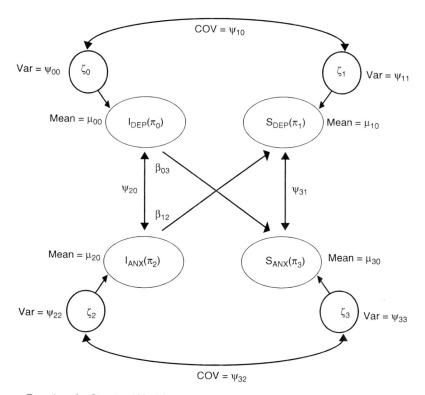

Equations for Structural Models

$\pi_{0i} = \mu_{00} + \zeta_{0i},$ $\zeta_{0i} \sim NID(0, \psi_{00})$
$\pi_{1i} = \mu_{10} + \beta_{12} \times \pi_{2i} + \zeta_{1i},$ $\zeta_{1i} \sim NID(0, \psi_{11})$
$\pi_{2i} = \mu_{20} + \zeta_{2i},$ $\zeta_{2i} \sim NID(0, \psi_{22})$
$\pi_{3i} = \mu_{30} + \beta_{03} \times \pi_{0i} + \zeta_{3i},$ $\zeta_{3i} \sim NID(0, \psi_{33})$

$$\psi = \begin{bmatrix} \psi_{00} & & & \\ \psi_{10} & \psi_{11} & & \\ \psi_{20} & 0 & \psi_{22} & \\ 0 & \psi_{31} & \psi_{32} & \psi_{33} \end{bmatrix}$$

FIGURE 5.11 Parallel Process Model (PPM) with Reciprocal Effects.
Note: I = Initial Level. S = Slope. DEP = Depressive Symptoms. ANX = Anxiety
Symptoms. NID = Normally and Independently Distributed. i = individual (n = 1,
2,…, n). Measurement models (LGCMs) are not shown.

where k = 1, 2, … is the subdomain, i is 1, 2, …, n, respondents, and, ε_{it} is the
error term for the ith individual at time t (t = 1, 2, 3, …). For a linear change with
four equally-spaced measurements t can be 0, 1, 2, and 3. Residuals at each point
in time are assumed to be normally and independently distributed (NID) with
means of zero. As in a LGCM, for each individual, equations link y measurements
of subdomain k at the repeated time points using an intercept (π_{0ik}) and a slope
(π_{1ik}). In our example PPM using anxiety and depressive symptoms trajectories

ANALYSIS: MODEL = NOCOV;

MODEL:

    I_ANX S_ANX | ANX1@0 ANX2@1 ANX3@3 ANX4@4 ANX5@6;
    I_DEP S_DEP | DEP1@0 DEP2@1 DEP3@3 DEP4@4 DEP5@6;

    I_ANX I_DEP PWITH S_ANX S_DEP;
    I_ANX S_ANX PWITH I_DEP S_DEP;

    S_DEP ON I_ANX;
    S_ANX ON I_DEP;

> Reciprocal effects between the level and slope of anxiety and depressive symptoms

OUTPUT: SAMPSTAT STANDARDIZED MOD (5.00);

FIGURE 5.12 *Mplus* Syntax for a Parallel Process Model (PPM) with Reciprocal Effects.
Note: I = Initial Level. S = Slope. DEP = Depressive Symptoms. ANX = Anxiety Symptoms.

as the subdomains, the equations for the two subdomain primary growth curves (LGCMs) can be written as:
For the primary growth curve of depressive symptoms $(k = 1)$,

$$y_{i1t} = \pi_{0i1} + \lambda_t \times \pi_{1i1} + \varepsilon_{i1t}, \quad \varepsilon_{i1t} \sim \text{NID}\ (0, \sigma_{1t}^2) \tag{5.2}$$

$$\pi_{0i1} = \mu_{001} + \zeta_{0i1}, \qquad \zeta_{0i1} \sim \text{NID}\ (0, \psi_{0i11}) \tag{5.3}$$

$$\pi_{1i1} = \mu_{101} + \zeta_{0i1}, \qquad \zeta_{0i1} \sim \text{NID}\ (0, \psi_{1i11}) \tag{5.4}$$

For the primary growth curve of anxiety symptoms $(k = 2)$,

$$y_{i2t} = \pi_{0i2} + \lambda_t \times \pi_{1i2} + \varepsilon_{i2t}, \qquad \varepsilon_{i2t} \sim \text{NID}\ (0, \sigma_{2t}^2) \tag{5.5}$$

$$\pi_{0i2} = \mu_{002} + \zeta_{0i2}, \qquad \zeta_{0i2} \sim \text{NID}\ (0, \psi_{0022}) \tag{5.6}$$

$$\pi_{1i2} = \mu_{102} + \zeta_{0i2}, \qquad \zeta_{0i2} \sim \text{NID}\ (0, \psi_{1122}) \tag{5.7}$$

where i is 1, 2, ..., n, respondents, and ε_{it} is the error term for the ith individual at time t.

Residuals at each point in time and in both levels are assumed to be normally and independently distributed (NID) with means of zero.

Second-Order Growth Factors

A factor-of-curves model (FCM) is illustrated in Figure 5.15. In this FCM, the primary growth curve factors of different subdomains (that is, their intercepts

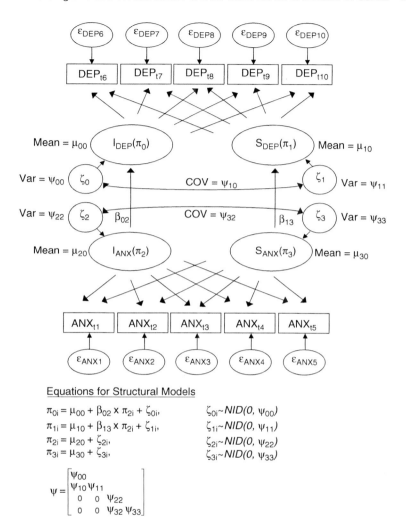

FIGURE 5.13 Parallel Process Model (PPM) with Longitudinal Directional Effects. Note: I = Initial Level. S = Slope. DEP = Depressive Symptoms. ANX = Anxiety Symptoms. NID = Normally and Independently Distributed. i = individual (n = 1, 2,..., n).

(π_{0ik}) and slopes (π_{1ik})) are used as indicators for second-order factors (η_{0i} and η_{1i}) to estimate the global domain of IS.

For primary latent growth factor structures for any subdomain k,

$$\pi_{0ik} = \mu_{00k} + \lambda_{0k0} \times \eta_{0i} + \zeta_{0ik}, \qquad \zeta_{0ik} \sim \text{NID } (0, \psi_{00kk}) \tag{5.8}$$

$$\pi_{1ik} = \mu_{10k} + \lambda_{1k1} \times \eta_{1i} + \zeta_{1ik}, \qquad \zeta_{1ik} \sim \text{NID } (0, \psi_{11kk}) \tag{5.9}$$

ANALYSIS: MODEL = NOCOV;

MODEL:

    I_ANX S_ANX | ANX1@0 ANX2@1 ANX3@3 ANX4@4 ANX5@6;
    I_DEP S_DEP | DEP6@0 DEP7@2 DEP8@4 DEP9@6 DEP10@8;

    I_ANX I_DEP PWITH S_ANX S_DEP;

    S_DEP ON I_ANX;
    S_ANX ON I_DEP; ————| Longitudinal directional effects |

OUTPUT: SAMPSTAT STANDARDIZED MOD (5.00);

FIGURE 5.14 *Mplus* Syntax for a Parallel Process Model (PPM) with Longitudinal Directional Effects.
Note: I = Initial Level. S = Slope. DEP = Depressive Symptoms. ANX = Anxiety Symptoms.

where subdomain k = 1, 2, … (e.g., symptoms of depression [DEP], anxiety [ANX], and hostility [HOS]).

For our example model, the equations are as follows:

For primary latent growth factor structures of depressive symptoms (k = 1 for DEP),

$$\pi_{0i1} = \mu_{001} + \lambda_{010} \times \eta_{0i} + \zeta_{0i1}, \quad \zeta_{0i1} \sim \text{NID} (0, \psi_{0011}) \tag{5.10}$$

$$\pi_{1i1} = \mu_{101} + \lambda_{111} \times \eta_{1i} + \zeta_{1i1}, \quad \zeta_{1i1} \sim \text{NID} (0, \psi_{1111}) \tag{5.11}$$

For primary latent growth factor structures of anxiety symptoms (k = 2 for ANX),

$$\pi_{0i2} = \mu_{002} + \lambda_{020} \times \eta_{0i} + \zeta_{0i2}, \quad \zeta_{0i2} \sim \text{NID} (0, \psi_{0022}) \tag{5.12}$$

$$\pi_{1i2} = \mu_{102} + \lambda_{121} \times \eta_{1i} + \zeta_{1i2}, \quad \zeta_{1i2} \sim \text{NID} (0, \psi_{1122}) \tag{5.13}$$

For estimating the second-order global factor structure (IS),

$$\eta_{0i} = \alpha_{00} + \zeta_{0i}, \quad \zeta_{0i} \sim \text{NID} (0, \psi_{00}) \tag{5.14}$$

$$\eta_{1i} = \alpha_{10} + \zeta_{1i}, \quad \zeta_{1i} \sim \text{NID} (0, \psi_{11})$$

$$\psi = \begin{bmatrix} \psi_{00} & \\ \psi_{10} & \psi_{11} \end{bmatrix} \tag{5.15}$$

Note that here the α denotes the mean of a second-order latent factor (i.e., a global factor). Now, researchers may be interested in investigating whether a second-order factor structure of a global domain (e.g., IS) describes the associations among primary growth factors of different subdomains. For illustration, in the subsequent example, we use three subdomains of symptoms (k = 3, with three

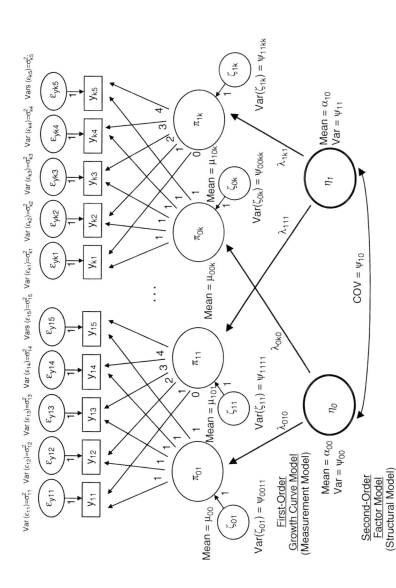

FIGURE 5.15 A Factor-of-Curves Model (FCM).
Note: Var = Variance.

primary growth curves representing symptoms of depression, anxiety, and hostility). The covariances among the primary growth factors are thought to be explained by the higher-order growth factors (η_{0i} and η_{1i}). That is, secondary growth factors of internalizing symptoms (IS) may explain the primary growth factors of symptoms of depression, anxiety, and hostility (i.e., the intercepts (π_{0ik}) and slopes (π_{1ik}) of these three subdomains). As previously noted, the primary growth factors of different subdomains are used as multiple indicators of higher-order factors. The structure of this part of the model is similar to the structure of a confirmatory · factor analysis (CFA), in which indicators define latent factors of a global domain. This type of second-order latent growth model is known as a *factor-of-curves model (FCM) because it first estimates growth curves and then defines factors of these curves* (Duncan & Duncan, 1996). In doing so, a FCM reflects the degree to which primary latent growth curve factors combine to form secondary latent factors (Duncan & Duncan, 1996).

As shown in Figure 5.15, in a FCM, the covariances among the primary growth factors are not specified (i.e., fixed to zero) to produce the second-order growth factors. In addition, factor loadings of similar primary growth factors to the second-order factor are constrained to be equal across different subdomains (i.e., $\lambda_{010} = \lambda_{111}$ and $\lambda_{0k0} = \lambda_{1k1}$, respectively). In our example, the loadings between the primary growth factors (i.e., intercept and slope) of depressive symptoms and second-order growth factors were constrained to be equal. In the same manner, the loadings between primary growth factors of anxiety symptoms and second-order growth factors were also constrained to be equal. Furthermore, for the same reasons that we discussed in Chapter 3 in regards to fitting a CFM, a FCM also has scaling issues that must be considered when producing parameters of global factors (α_0 and α_1). Because the marker variable method is the default scale setting approach in M*plus*, unless otherwise specified, we utilized the marker variable method across all example models by fixing the mean parameters (e.g., μ_{001} and μ_{101}) of one subdomain's growth factors (e.g., growth factors for depressive symptoms) to zero. By fixing these parameters to zero the global factors take on the scale of the depressive symptoms measure.

Unlike composite measures, which are computed assuming equal contributions (weights) of different subdomains (e.g., adding depressive symptoms and anxiety symptoms to create a composite measure of IS), higher-order modeling frameworks, such as a FCM, allow different subdomains to "behave" differently over time by forming primary growth factors that are specific to each subdomain. The primary growth factors may differentially contribute to the second-order factors of a FCM. For example, in a FCM, the primary growth factors of adolescent depressive symptoms and anxiety symptoms may differentially contribute to second-order growth factors of adolescents' internalizing symptoms. However, growth curve models utilizing composite measures estimate fewer parameters. Consequently, if the study sample size is relatively small, growth curve models with composite measures may be more appropriate because the parsimony reduces the likelihood of encountering model convergence problems.

Chapter 5 Exercises

You are provided with a panel dataset consisting of the following variables from mothers:

- DEP1M~DEP4M, which indicate mothers' depressive symptoms at T1~T4.
- ANX1M~ANX4M, which indicate mothers' anxiety symptoms at T1~T4.
- HOS1M~HOS4M, which indicate mothers' symptoms of hostility at T1~T4.

1. Draw separate growth curves for mothers' depressive, anxiety, and hostility symptoms using DEP1M~DEP4M, ANX1M~ANX4M, and HOS1M~HOS4M, respectively. Specify factor loadings for linear change (t = 0, 1, 2, and 3). Label all covariance parameters using conventional symbols.
2. Draw a parallel process model (PPM) using the three primary growth curves from Question 1. Label all of the parameters using conventional symbols.
3. Draw a factor-of-curves model (FCM) based on the PPM from Question 2. In order to estimate second-order global growth factors (η), use the intercept and slope growth factors of the anxiety LGCM as the marker variables and assign labels to constrain factor loading parameters. Label the constrained factor loadings for global parameters and the second-order global factor parameters.

References

Angold, A., Costello, E. J., & Erkanli, A. (1999). Comorbidity. *Journal of Child Psychology and Psychiatry, 40*(1), 57–87.

Caron, C., & Rutter, M. (1991). Comorbidity in child psychopathology: Concepts, issues and research strategies. *Journal of Child Psychology and Psychiatry, 32*(7), 1063–1080.

Duncan, S. C., & Duncan, T. E. (1996). A multivariate latent growth curve analysis of adolescent substance use. *Structural Equation Modeling, 3*(4), 323–347.

Jessor, R., & Jessor, S. L. (1977). *Problem Behavior and Psychosocial Development.* New York: Academic Press.

Kubzansky, L. D., & Kawachi, I. (2000). Going to the heart of the matter: Do negative emotions cause coronary heart disease? *Journal of Psychosomatic Research, 48*(4), 323–337.

Loeber, R., Stouthamer-Loeber, M., & White, H. R. (1999). Developmental aspects of delinquency and internalizing problems and their associations with persistent juvenile substance abuse between ages 7 and 18. *Journal of Clinical Child Psychology, 28*(3), 322–332.

6

ESTIMATING A FACTOR-OF-CURVES MODEL (FCM) AND ADDING COVARIATES

Introduction

The previous chapter introduced a factor-of-curves model (FCM), which is a second-order growth curve model building on a parallel process model (PPM). The present chapter serves as a practical guide to estimating a factor-of-curves model (FCM) in a step-by-step manner. The chapter also discusses advantages of a FCM over a conventional latent growth curve model (LGCM). Detailed instructions for adding covariates (predictors and outcomes) to a FCM are provided using example models. We also discuss the estimation of a multiple-group FCM and a multivariate FCM. Lastly, the chapter demonstrates how a higher-order FCM can be built from a curve-of-factors model (CFM). For each model, figures, M*plus* syntax, and outputs are presented.

Estimating a Factor-of-Curves Model (FCM)

There are three primary expectations when estimating a factor-of-curves model (FCM). First, we expect that primary, or first-order, growth factors (e.g., the initial level and slope) of a subdomain describe the developmental change in that specific subdomain over time. An example of a subdomain is "depressive symptoms," which, along with similar subdomains, such as "anxiety symptoms" and "hostility symptoms," make up a more general, or global, domain known as "internalizing symptoms." Second, we expect that the covariations between growth factors of different subdomains reflect a parallel process involving different subdomains over time. That is, the slopes of each subdomain should run parallel to each other; as one increases, so do the others. Third, we expect higher-order factors (also referred to as second-order or global factors) to accurately reflect the variances and covariances among the primary growth factors. The first- and second-order growth factors can be differentially associated with covariates; that is, empirical evidence suggests that there may be different antecedents and sequels (or consequences) for first- and second-order growth factors.

POINT TO REMEMBER...

Variances for first-order growth factors reflect the inter-individual differences in within-subdomain growth trajectories, while variances for second-order growth factors reflect differences in between-subdomain growth factors.

Similar to our incremental approach for building a curve-of-factors model (CFM) discussed in the Chapter 3, a FCM is also built in incremental steps:

Step One
> Investigate the correlation patterns among indicators within each subdomain over time to examine the possibility of forming primary growth curves for each subdomain.

Step Two
> Estimate a parallel process growth curve model (PPM) comprised of primary growth curves for each subdomain with correlations among primary growth factors and measurement errors.

Step Three
> Incorporate secondary growth factors, if appropriate, to form a FCM using the primary growth factors from the PPM as indicators.

Step Four
> Estimate a conditional FCM incorporating time-invariant and/or time-varying covariates.

Steps 1 and 2 refer to the primary latent growth models (i.e., measurement model), and Steps 3 and 4 address the estimation of unconditional and conditional second-order latent models (i.e., FCM).

Investigating the Longitudinal Correlation Patterns Among Repeated Measures of Each Subdomain (Step One)

In contrast to a CFM, a FCM has a measurement model consisting of multiple primary growth curves at the first level. As we discussed in Chapter 2, for the successful estimation of primary growth curves, the associations among repeated indicators at adjacent occasions should be consistently higher than associations at non-adjacent measurement occasions. (Lorenz, Wickrama, & Conger, 2004).

Illustrative Example 6.1: Investigating the Longitudinal Correlation Patterns Among Repeated Measures of Each Subdomain

As can be seen in Table 4.1 in Chapter 4, our sample correlation matrix indicated that correlation coefficients between two adjacent occasions (t and t+1) for each symptom subdomain were higher than correlations between non-adjacent occasions (anxiety rs ranged from .48 to .57; depression rs ranged from .40 to .61; hostility rs ranged from .33 to .57). Also, off-diagonal correlations of the same repeated measure over time (see the boxes in Table 4.1) were quite different from each other. Consistent with Equation 2.11 which utilized covariances, these correlations imply that significant slope variation may exist for each subdomain trajectory model.

Estimating a Parallel Process Growth Curve Model (PPM) (Step Two)

A parallel process model (PPM) allows researchers to examine the associations among primary growth parameters for multiple subdomain trajectories (e.g., depressive, hostility, and anxiety symptoms; Duncan, Duncan, & Strycker, 2006). The first step for developing this model is to confirm that each primary growth curve model has been successfully modeled independently (i.e., checking model fit). Then, it is essential to determine from each of the primary growth models whether there is sufficient inter-individual variation in the initial levels and slopes. Significant variations in the intercepts and slopes provide evidence that forming a PPM is a possibility. For example, univariate primary latent growth curve models (LGCMs) can be estimated using three types of symptoms (i.e., symptoms of anxiety, depression, and hostility). Then, growth parameters (i.e., intercept and slope variance) of each model can be correlated.

Illustrative Example 6.2: Estimating a Parallel Process Growth Curve Model (PPM)

M*plus* syntax for this PPM is as follows (loadings for the linear slope (S) are 0, 1, 3, 4, and 6 and correspond to the spacing of the measurement intervals):

```
USEVAR = DEP1-DEP5 ANX1-ANX5 HOS1-HOS5;
MODEL:
I_DEP S_DEP | DEP1@0 DEP2@1 DEP3@3 DEP4@4 DEP5@6;
I_ANX S_ANX | ANX1@0 ANX2@1 ANX3@3 ANX4@4 ANX5@6;
I_HOS S_ HOS | HOS1@0 HOS2@1 HOS3@3 HOS4@4 HOS5@6;
```

For reasons similar to those that we introduced in Chapter 4 regarding a CFM, the current PPM also incorporates autocorrelated measurement errors in order to reduce the likelihood of model misspecification, which leads to biased model parameter estimates. The main purpose of a PPM is to investigate longitudinal associations across the primary growth factors of multiple subdomains. In order to take into account associations between subdomains at the same time point, between-subdomain time-specific measurement errors can be specified in the PPM (see Figure 6.1).

The M*plus* syntax for specifying between-subdomain time-specific measurement error correlations in the MODEL statement is as follows:

DEP1 WITH ANX1 HOS1; ANX1 WITH HOS1;
DEP2 WITH ANX2 HOS2; ANX2 WITH HOS2;
DEP3 WITH ANX3 HOS3; ANX3 WITH HOS3;
DEP4 WITH ANX4 HOS4; ANX4 WITH HOS4;
DEP5 WITH ANX5 HOS5; ANX5 WITH HOS5;

The model results showed that all between-subdomain autocorrelated errors were statistically significant and were within the acceptable bounds, ranging from .493 (ANX1 with HOS1) to .717 (DEP4 with ANX4). Also, the model fit was acceptable (χ^2(df) = 223.956(78), p < .001; CFI/TLI = .965/.953; RMSEA = .066; SRMR = .059).

For the same reasons that we used modification indices when estimating a CFM in Chapter 4, we suggest using modification indices to identify potential

DATA: File is example_ch_6.dat;
VARIABLE: names are ANX1-ANX5 DEP1-DEP5 HOS1-HOS5;

    usevar are ANX1-ANX5 DEP1-DEP5 HOS1-HOS5
    missing = all (9);
MODEL:
  I_DEP S_DEP | DEP1@0 DEP2@1 DEP3@3 DEP4@4 DEP5@6; ⎤ Parallel
  I_ANX S_ANX | ANX1@0 ANX2@1 ANX3@3 ANX4@4 ANX5@6; ⎬ process
  I_HOS S_HOS | HOS1@0 HOS2@1 HOS3@3 HOS4@4 HOS5@6; ⎦ model (PPM)

  DEP1 WITH ANX1 HOS1; ANX1 WITH HOS1;
  DEP2 WITH ANX2 HOS2; ANX2 WITH HOS2;
  DEP3 WITH ANX3 HOS3; ANX3 WITH HOS3;   ← Autocorrelated errors
  DEP4 WITH ANX4 HOS4; ANX4 WITH HOS4;
  DEP5 WITH ANX5 HOS5; ANX5 WITH HOS5;
  DEP3 WITH DEP4; HOS3 WITH HOS4;

OUTPUT: SAMPSTAT STANDARDIZED MOD (5.00);

FIGURE 6.1 M*plus* Syntax for a Parallel Process Model (PPM).
Note: I = Initial Level. S = Slope. DEP = Depressive Symptoms. ANX = Anxiety Symptoms. HOS = Hostility Symptoms.

covariances among manifest indicators in this model. The inclusion of these covariances may improve model fit indices. M*plus* will compute modification indices if MOD (3.84) is specified in the OUTPUT command. The results revealed that specifying two additional error correlations would improve the model fit indices (DEP3 with DEP4 and HOS3 with HOS4). As mentioned in Chapter 4, comorbidity within the subdomains at different measurement occasions is common. Because these modifications are theoretically and empirically justifiable, a PPM was re-specified to include these two additional correlations. Results of the reanalysis yielded a statistically and substantially better fitting model (χ^2(df) = 190.991(76), $p < .001$; CFI/TLI = .973/.962; RMSEA = .059; SRMR = .057). The PPM results provide evidence for the existence of both between- and within-subdomain autocorrelated error structures. The results of this final PPM are shown in Table 6.1.

All growth parameters (i.e., means and variances for both the intercept and slope) across the three primary growth curves (curves for symptoms of anxiety, depression, and hostility) were statistically significant. The covariances among the primary growth factors were also all statistically significant, which is evidence for the existence of a parallel process of growth across the subdomains. More importantly, correlations (that is, standardized covariances) *among* the intercepts of the three subdomains were higher than the correlations *between* the intercepts and slopes (see the bold and italic numbers in Table 6.1). Similarly, correlations *among* the slopes of the three subdomains were higher than the correlations *between* the intercepts and slopes. These relatively strong correlations among the same type of growth parameters across the three subdomains imply the existence of significant global (second-order) growth factors (η_0 and η_1) in a FCM.

POINT TO REMEMBER...

It should be noted that, for all of the models we introduce, when examining output files, all of the meaningful comparisons within a model depend on the common metric and are best compared in terms of correlations (standardized solution).

Estimating a Factor-of-Curves Model (FCM) (Step Three)

To obtain empirical evidence of the successful estimation of second-order growth factors (η_0 and η_1) for a FCM, the covariances (or correlations) among primary growth factors, which are the indicators of the second-order growth curve, should be examined first. Importantly, even if the second-order model explains all of the covariations among the primary growth factors, the model fit indices may

TABLE 6.1 Results of the Parallel Process Model (PPM) Using Three Symptoms of Internalizing Symptoms.

Symptoms	Intercept growth factors (Unstandardized)		Slope growth factors (Unstandardized)	
	Mean	Variance	Mean	Variance
Depressive symptoms	1.557***	.221***	-.014**	.004***
Anxiety	1.405***	.150***	-.032***	.004***
Hostility	1.498***	.211***	-.029***	.005***

Correlations among growth factors (Standardized)

	INT_{DEP}	INT_{ANX}	INT_{HOS}	SLP_{DEP}	SLP_{ANX}	SLP_{HOS}
INT_{DEP}	—					
INT_{ANX}	.855***	—				
INT_{HOS}	.710***	.786***	—			
SLP_{DEP}	-.598***	-.553***	-.472***	—		
SLP_{ANX}	-.523***	-.634***	-.506***	.910***	—	
SLP_{HOS}	-.390***	-.523***	-.764***	.629***	.697***	—

Notes. Autocorrelations among subdomains at the same time are not shown. INT = Intercept. SLP = Slope. DEP = Depressive Symptoms. ANX = Anxiety Symptoms. HOS = Hostility Symptoms. χ^2(df) = 190.991(76), $p < .001$; CFI/TLI = .973/.962; RMSEA (90% CI) = .059 (.049, .069); SRMR = .057.
**$p < .01$. ***$p < .001$.

not improve over those of the corresponding primary growth curve models. However, if the model fit indices for the second-order growth model are close to those of the corresponding primary growth models, Duncan and Duncan (1996) argue that a FCM can still be considered an acceptable model because it is more parsimonious.

Illustrative Example 6.3: Estimating a Factor-of-Curves Model (FCM)

As shown in Table 6.1, in the estimated PPM, the covariances (expressed in their standardized form as correlations) among the intercept growth factors across the subdomains were moderately high (ranged from .710 to .855, $p < .001$). Additionally, the correlations among the slope factors of the subdomains were also moderately high (ranged from .629 to .910, $p < .001$). These results imply the existence of significant second-order growth factors. However, it is interesting that the results also showed moderately high correlations between intercept and slope growth factors within the subdomains. For example, the correlation between the intercept and slope factors for hostility symptoms was -.764, $p < .001$. These high correlations may imply the need to incorporate correlations between the intercept and slope factors of each of the primary growth models in a FCM. Similar to the CFM in Chapter 4, in this FCM, we fixed the loadings and means of the depressive symptoms trajectories to 1 and 0, respectively, in order to estimate means of the second-order intercept and slope factors using the scale of depressive symptoms (i.e., marker variable approach). The M*plus* syntax of this FCM is shown in Figure 6.2.

We have described most parts of this FCM M*plus* syntax when introducing previous models except for the syntax specifying factorial loading invariance. As can be seen in the M*plus* syntax, the second-order global factors (I and S of internalizing symptoms [IS]) are loaded by the respective intercept and slope latent factors from the three primary growth models. We assume that the primary growth factors contribute equally to the second-order growth factors of the global domain (IS) (Howe, Hornberger, Weihs, Moreno, & Neiderhiser, 2012). For this reason, we constrained these loading to be equal; for example, the intercept loading of depressive symptoms on the internalizing symptoms (IS) intercept was constrained to equal the loading of the depressive symptoms slope factor on the IS slope. As a result, each subdomain contributes the same proportion of variance to both the global intercept and the global slope factor.

This FCM syntax fixes all means of observed indicators to zero for model identification purposes (Duncan et al., 2006). The PWITH statement was used to estimate pair correlations between intercept and slope factors for each primary factor model (the growth curve for each subdomain).

The M*plus* output showing model fit indices and estimated parameters is provided in Tables 6.2 and 6.3. As can be seen in Table 6.2, the model fit indices were acceptable (χ^2(df) = 134.632(83), CFI/TLI = .978/.972, RMSEA = .038,

MODEL:

I_DEP S_DEP | DEP1@0 DEP2@1 DEP3@3 DEP4@4 DEP5@6;
I_ANX S_ANX | ANX1@0 ANX2@1 ANX3@3 ANX4@4 ANX5@6;
I_HOS S_HOS | HOS1@0 HOS2@1 HOS3@3 HOS4@4 HOS5@6;

DEP1 WITH ANX1 HOS1; ANX1 WITH HOS1;
DEP2 WITH ANX2 HOS2; ANX2 WITH HOS2;
DEP3 WITH ANX3 HOS3; ANX3 WITH HOS3;
DEP4 WITH ANX4 HOS4; ANX4 WITH HOS4;
DEP5 WITH ANX5 HOS5; ANX5 WITH HOS5;
DEP3 WITH DEP4; HOS3 WITH HOS4;

I BY I_DEP@1
   I_ANX (L1)
   I_HOS (L2);

By placing names in parentheses after the parameter (e.g., L1 and L2), *factor loadings* of the same subdomain are invariant (i.e., held constant)

S BY S_DEP@1
   S_ANX (L1)
   S_HOS (L2);

By specifying @1 to I_DEP and S_DEP, intercept and slope factors of depressive symptoms are used as *marker variables* to produce variance-covariance structures of the global factors (I and S)

[I_DEP-S_DEP@0];

By specifying @0 to I_DEP and S_DEP in square brackets, intercept and slope factors of depressive symptoms are used as *marker variables* to produce mean structures of the global factors (I and S)

[I_ANX I_HOS];
[S_ANX S_HOS];
[I S];

Estimating mean parameters of the intercept and slope factors for *primary and global factors*

I_DEP I_ANX I_HOS PWITH I_DEP I_ANX I_HOS;

Estimating *correlations between intercept and slope factors* of each primary growth model

OUTPUT: SAMPSTAT STANDARDIZED MOD (5.00);

FIGURE 6.2 *Mplus* Syntax for a Factor-of-Curves Model (FCM).
Note: I = Initial Level. S = Slope. DEP = Depressive Symptoms. ANX = Anxiety Symptoms. HOS = Hostility Symptoms. The marker variable scale setting approach was used.

SRMR = .059), which indicates that the FCM fits the observed data well. Also, Table 6.3 shows that the factor loadings for all of the primary growth factors on the global factors were high and statistically significant, which indicates that each of the primary growth factors contributed significantly to defining the global factors. The mean and variance of the global intercept were 1.555 and .188, respectively, while the mean and variance of the global slope factor were -.014 and .004, respectively. All of these global parameters were statistically significant at the $p < .05$ level. R-square statistics for each growth factor indicated that the . global intercept factor accounted for approximately 81.6%, 90.7%, and 62.0% of the variation in the primary intercept factors for symptoms of depression, anxiety, and hostility, respectively. Furthermore, approximately 90.6%, 86.5%, and 53.1% of the variation in the primary slope factors for symptoms of depression, anxiety, and hostility, respectively, was accounted for by the global IS slope factor. The negative IS slope factor indicates an overall decline in global IS over time. The model results also showed that the growth parameters (i.e., means and variances) of the global intercept and slope factors (I and S) were statistically significant, suggesting that significant inter-individual variation existed for the second-order intercept and slope factor means. This statistically significant variation across

TABLE 6.2 M*plus* Output: Model Fit Indices of the FCM.

MODEL FIT INFORMATION	
Chi-Square Test of Model Fit	
Value	134.632*
Degrees of Freedom	83
P-Value	0.0003
Scaling Correction Factor for MLR	1.5154
RMSEA (Root Mean Square Error Of Approximation)	
Estimate	0.038
90 Percent C.I.	0.026 0.049
Probability RMSEA <= .05	0.962
CFI/TLI	
CFI	0.978
TLI	0.972
Chi-Square Test of Model Fit for the Baseline Model	
Value	2455.827
Degrees of Freedom	105
P-Value	0.0000
SRMR (Standardized Root Mean Square Residual)	
Value	0.059

TABLE 6.3 M*plus* Output: Unstandardized Parameter Estimates for the FCM.

		Estimate	S.E.	Est./S.E.	Two-Tailed P-Value
I	BY				
I_DEP		1.000	0.000	999.000	999.000
I_ANX		0.840	0.061	13.727	0.000
I_HOS		0.803	0.080	10.027	0.000
S	BY				
S_DEP		1.000	0.000	999.000	999.000
S_ANX		0.840	0.061	13.727	0.000
S_HOS		0.803	0.080	10.027	0.000
I_DEP	WITH				
S_DEP		−0.003	0.002	−1.539	0.124
I_ANX	WITH				
S_ANX		−0.001	0.002	−0.867	0.386
I_HOS	WITH				
S_HOS		−0.011	0.003	−4.040	0.000
S	WITH				
I		−0.018	0.004	−4.219	0.000
Means					
I		1.555	0.028	55.580	0.000
S		−0.014	0.006	−2.323	0.020
Intercepts					
I_DEP		0.000	0.000	999.000	999.000
S_DEP		0.000	0.000	999.000	999.000
I_ANX		0.098	0.091	1.082	0.279
S_ANX		−0.020	0.004	−5.500	0.000
I_HOS		0.249	0.122	2.045	0.041
S_HOS		−0.017	0.005	−3.496	0.000
Variances					
I		0.188	0.028	6.817	0.000
S		0.004	0.001	4.413	0.000
Residual Variances					
I_DEP		0.043	0.011	3.755	0.000
S_DEP		0.000	0.001	0.766	0.444
I_ANX		0.014	0.007	1.926	0.054
S_ANX		0.000	0.000	1.446	0.148
I_HOS		0.074	0.015	4.798	0.000
S_HOS		0.002	0.001	3.716	0.000

TABLE 6.3 (*cont.*)

R-SQUARE				
I_DEP	0.816	0.042	19.481	0.000
S_DEP	0.906	0.110	8.239	0.000
I_ANX	0.907	0.051	17.819	0.000
S_ANX	0.865	0.091	9.509	0.000
I_HOS	0.620	0.077	8.018	0.000
S_HOS	0.531	0.102	5.221	0.000

individual trajectories may be systematically associated with theoretically important covariates. Thus, this significant variation in the second-order growth factors is important because it justifies the examination of a conditional FCM with covariates.

Interestingly, our example models showed that the variances of most primary growth factors were statistically significant, especially for the hostility symptoms growth factors. Also, significant covariance between the intercept and slope of the hostility symptoms growth model was observed even after controlling for internalizing symptoms (IS) growth factors ($-.011, p < .001$). These results may indicate the existence of unique variance in hostility after controlling for secondary IS factors.

Often, researchers are unsure whether a curve-of-factors model or a factor-of-curves model is best for their investigation. When selecting between a CFM or a FCM, researchers should consider (a) substantive meaning and interpretability of parameters/results, (b) model fit, and (c) model parsimony. Model fit statistics and global factor parameters between these two competing models (CFM and FCM) can be compared. This comparison will provide empirical support for the better fitting model. Of course, consideration should also be given to the interpretability of results and theoretical support available for each model. The following example illustrates the comparison of a CFM (estimated in Chapter 4) and a FCM (estimated in this chapter). Such a comparison is possible because both models used the same repeated measures.

Illustrative Example 6.4: Comparing Two Competing Models Empirically

A comparison of model fit and global factor parameters between a curve-of-factors model (CFM) and a competing factor-of-curves model (FCM) is provided in Table 6.4. As can be seen, most parameters were similar between the two models, and the model fit indices also revealed similarities. Thus, comparing the parameters and model fit indices does not provide much support for one model over the other, but the factor-of-curves model with 83 degrees of freedom (df) is

TABLE 6.4 Comparison of Global Factor Parameters Between the Curve-of-Factors and Factor-of-Curves Approaches.

	Curve-of-Factors Model			Factor-of-Curves Model		
	Coeff.	SE	t-value	Coeff.	SE	t-value
Means						
Intercept	1.589	.026	60.133	1.555	.028	55.580
Slope	−.026	.005	−5.656	−.014	.006	-2.323
Variances						
Intercept	.178	.019	9.408	.188	.028	6.817
Slope	.004	.001	5.334	.004	.001	4.413
Covariance	−.017	.003	−5.758	−.018	.004	-4.219
Model fit						
χ^2 (df)	166.598 (71)			134.632 (83)		
AIC / BIC	5257.156 / 5518.125			5270.572 / 5482.609		
CFI / TLI	.977 / .966			.978 / .972		
RMSEA (90% CI)	.056 (.045, .067)			.038 (.026, .049)		
SRMR	.062			.059		

Note: Unstandardized coefficients are shown. Coeff. = Coefficient. Both models used the marker variable scale setting approach.

more parsimonious than the curve-of-factors model (df = 71). However, consideration should also be given to the interpretability of the results and the theoretical support available for each model.

Estimating a Conditional FCM (Step Four)

It is important to examine covariates that have systematic associations with the second-order growth factors. Similar to the process of estimating a conditional CFM, both time-invariant covariates (TICs) and time-variant covariates (TVCs) can easily be incorporated in a FCM. Following the same modeling procedures from our presentation of a CFM in Chapter 4, we first introduce how to use TICs in a FCM before turning to the utilization of TVCs.

Adding Time-Invariant Covariates (TICs) to a FCM

Covariates may exist for both second-order global growth factors (ηs) and primary subdomain growth factors (πs). Also, some predictors may only be associated with certain growth factors and not others. Like a CFM, several conditional FCMs can be investigated.

Predicting both first-order and second-order growth factors: Like a CFM, a FCM can also estimate the direct and indirect effects of a latent predictor, P, on primary growth factors (π) by specifying paths (γ) to both subdomain-specific latent factors and second-order growth factors. The equation with a latent predictor variable (P) consisting of multiple indicators (x's) can be written as:

$$x_{ji} = \tau_j + \lambda_j \times P_i + \delta_{ji}, \quad \delta_{ji} \sim NID\left(0, \sigma_j^2\right) \tag{6.1}$$

where the subscript j refers to particular indicator j, and the subscript i (= 1, 2, 3,..., n) refers to the individual. P represents an exogenous latent factor, and δ refers to residuals of the exogenous indicators (x). Residuals are assumed to be normally and independently distributed (NID) with a mean of zero and a variance of σ^2.

Next, the equations for the primary growth factors (i.e., the direct effect of the latent predictor on the primary growth factors, πs) can be written as:

$$\pi_{0ik} = \mu_{00k} + \lambda_{0k0} \times \eta_{0i} + \gamma_{0k} \times P_i + \zeta_{0ik}, \quad \zeta_{0ik} \sim NID\ (0,\ \psi_{00kk}) \tag{6.2}$$

$$\pi_{1ik} = \mu_{10k} + \lambda_{1k0} \times \eta_{1i} + \gamma_{1k} \times P_i + \zeta_{1ik}, \quad \zeta_{1ik} \sim NID\ (0,\ \psi_{11kk}) \tag{6.3}$$

where subscript k refers to a specific subdomain (k = 1, 2, ..., k).

Also, the equations for the second-order global factor structures are now written as:

$$\eta_{0i} = \alpha_{00} + \gamma_0 \times P_i + \zeta_{0i}, \quad \zeta_{0i} \sim NID\ (0, \psi_{00}) \tag{6.4}$$

$$\eta_{1i} = \alpha_{10} + \gamma_1 \times P_i + \zeta_{1i}, \quad \zeta_{1i} \sim NID\ (0, \psi_{11}) \tag{6.5}$$

where the variances and covariance of the global factors are represented as:

$$\psi = \begin{bmatrix} \Psi_{00} & \\ \Psi_{10} & \Psi_{11} \end{bmatrix} \tag{6.6}$$

When estimating many paths (γ) at the same time, model convergence problems often arise, which can be reflected in negative variances or correlations greater in absolute value than 1. These convergence problems often lead to the unavailability of standard error estimates. In this FCM, if the effect(s) of a predictor on the growth factors of each primary subdomain (π_{0k} or π_{1k}) are investigated individually (e.g., investigate the growth factors for depressive symptoms, *then* anxiety symptoms, and *then* hostility symptoms) before estimating a FCM, convergence problems can often be avoided. The following example illustrates the incorporation of a time-invariant covariate into a FCM. In order to compare the results with those from previous models, for consistency, we use the same exogenous latent variable (family economic problems, or FEP) that we have used in previous examples.

Illustrative Example 6.5: Adding Time-Invariant Covariates (TIC) to a FCM

The M*plus* syntax for a FCM with FEP as an exogenous latent variable is shown in Panels A and B of Figure 6.3. Panel A corresponds to predicting global growth

MODEL:

   ...

DEP5 WITH ANX5 HOS5; ANX5 WITH HOS5;
DEP3 WITH DEP4; HOS3 WITH HOS5;

Creating a latent predictor variable (FEP) using a BY statement

FEP BY FEP0 FEP1;

Estimating the FEP effect on only global growth factors (I and S)

I S ON FEP;

OUTPUT: SAMPSTAT STANDARDIZED MOD (5.00);

Panel A. Predicting Global Factors (η_0 and η_1)

ANALYSIS: MODEL = NOCOV;

Suppressing correlations among latent variables

MODEL:

   ...

DEP5 WITH ANX5 HOS5; ANX5 WITH HOS5;
DEP3 WITH DEP4; HOS3 WITH HOS4;
FEP BY FEP0 FEP1;

Estimating the FEP effect on both global growth factors (I and S) and primary growth factors (i.e., DEP)

I S I_DEP S_DEP ON FEP;

OUTPUT: SAMPSTAT STANDARDIZED MOD (5.00);

Panel B. Predicting Both Global Growth Factors (η_0 and η_1) and Primary Growth Factors (π)

FIGURE 6.3 M*plus* Syntax for a Factor-of-Curves Model (FCM) with a Time-Invariant Predictor.
Note: I = Intercept. S = Slope. DEP = Depressive Symptoms. ANX = Anxiety Symptoms. HOS = Hostility
Symptoms. The marker variable scale setting approach was used.

factors, while Panel B corresponds to predicting both primary and global growth factors. The bold and italic portions of the syntax add the covariate FEP to the existing unconditional FCM. We constrained the covariance between the global growth factors (η_0 and η_1) to be zero for model identification purposes in Panel B.

The results are shown in Panels A and B of Figure 6.4. As can be seen in Panel A of Figure 6.4, all estimated parameters were similar to those of the CFM we estimated previously. Recall that we used the same repeated measures for both models (see Figure 4.21). For example, the path between FEP and the global intercept growth factor was of a similar magnitude in the CFM ($\gamma_0 = .597$). Interestingly, the results showed the unique influence of FEP on the slope of one primary growth factor (i.e., depressive symptoms) in the FCM. The results showed that FEP was associated with an increase in depressive symptoms over time after controlling for the effects of the global domain (IS) ($\gamma_{11} = .064$).

Primary and secondary growth factors of the FCM as predictors of latent distal outcomes (D): A latent distal outcome (D) can also be included in the FCM. The equation for incorporating a latent distal outcome is written as:

$$y_{ji} = \tau_j + \lambda_j \times D_i + \varepsilon_{ji}, \qquad \varepsilon_{ji} \sim NID\left(0, \sigma_j^2\right) \tag{6.7}$$

where the subscript j refers to particular indicator j, and the subscript i ($= 1, 2, 3, \ldots, n$) refers to individual i. ε represents the residual indicator for y and is normally and independently distributed with a mean of zero and a variance of σ^2.

The equation for a FCM predicting a latent variable outcome, D, is then written as:

$$D_i = \mu_D + \beta_1 \eta_{0i} + \beta_2 \eta_{1i} + \beta_{0k} \pi_{0ik} + \beta_{1k} \pi_{1ik} + \zeta_{1i} \tag{6.8}$$

where subscript k refers to a specific subdomain ($k = 1, 2, \ldots, k$), and $\beta_{1\sim2}$ represent regression coefficients linking primary and secondary growth parameters (η_{0i} and η_{1i}, and πs) to the latent distal outcome (D).

Illustrative Example 6.6: Incorporating a Latent Distal Outcome into a FCM

In this illustration, we used poor life quality (PLQ) as a latent distal outcome (D). D can also be an observed variable (i.e., a single indicator) rather than a latent outcome, but when D is an observed variable Equation 6.7 should be ignored. The M*plus* syntax for specifying a FCM with an endogenous latent distal variable is as follows:

PLQ BY NEE01 FC01 WQ01;
PLQ ON I S I_DEP;

This syntax can be added to the existing unconditional FCM (see Figure 6.2). Also, we included one additional covariance between the slope of depressive symptoms

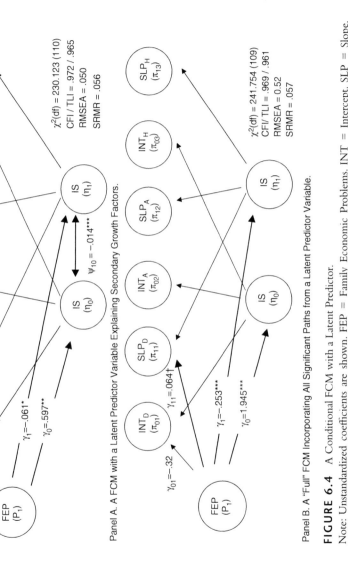

Panel A. A FCM with a Latent Predictor Variable Explaining Secondary Growth Factors.

Panel B. A "Full" FCM Incorporating All Significant Paths from a Latent Predictor Variable.

FIGURE 6.4 A Conditional FCM with a Latent Predictor.

Note: Unstandardized coefficients are shown. FEP = Family Economic Problems. INT = Intercept. SLP = Slope. D = Depressive Symptoms. A = Anxiety Symptoms. H = Hostility Symptoms. Measurement models are not shown. Covariances between intercept and slope variance of primary growth factors are not shown. The marker variable scale setting approach was used. † p <.10. *p < .05. **p < .01. ***p <.001.

and the slope of anxiety symptoms (SLP_D and SLP_A) by specifying S_DEP with S_ANX in the MODEL statement based on the modification indices. The results of this FCM with a latent distal outcome are shown in Figure 6.5. Overall, the results are similar to those of the conditional CFM discussed in Chapter 4 (see Figure 4.23). However, path coefficients are of greater magnitude in the FCM than the CFM. In our example FCM (see Figure 6.5), the effects of the global growth factors (intercept and slope of IS) on the distal outcome (PLQ) were estimated after controlling for the effects of the primary growth factors. The secondary intercept factor (η_0) not only directly influenced the latent distal outcome (PLQ), but also indirectly influenced PLQ through an intercept factor from a primary growth model (depressive symptoms).

POINT TO REMEMBER…

While a CFM allows researchers to control for the effects of time-specific global effects (η_t) on distal outcomes, a FCM allows researchers to control for the effects of domain-specific growth factors (π) on distal outcomes when estimating the effects of secondary growth factors on distal outcomes.

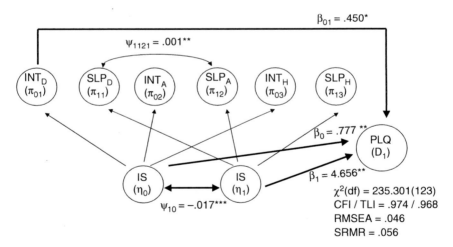

FIGURE 6.5 A FCM with a Latent Distal Outcome (D).
Note: Unstandardized coefficients are shown. PLQ = Poor Life Quality. INT = Intercept. SLP = Slope. D = Depressive Symptoms. A = Anxiety Symptoms. H = Hostility Symptoms. Measurement models are not shown. The marker variable scale setting approach was used. Covariances between intercept and slope variance of primary growth factors are not shown.
*p < .05. **p < .01. ***p <.001.

We have presented results from two models in Figures 6.4 and 6.5; Figure 6.4 presents a model with the latent predictor, and Figure 6.5 presents the model with the latent outcome. Although it is possible to incorporate a latent predictor and latent outcome in a FCM simultaneously, we encountered convergence problems when estimating these effects simultaneously (the results are not shown). This is likely due to the model's complexity.

Adding Time-Varying Covariates (TVC) to a FCM

In the previous section, we described how time-invariant covariates (TICs) can be incorporated into a factor-of-curves model (FCM) and discussed how to interpret the resulting parameter coefficients. Now, we introduce how time-varying covariates (TVCs) can be incorporated into a FCM. Similar to the incorporation of a TVC into a CFM, two common approaches can be used to incorporate a TVC into a FCM. A TVC can be incorporated as (1) a predictor of indicators of the primary growth curves at the first level or (2) a growth curve of a TVC (primary or secondary) correlated with the factor-of-curves model.

A time-varying covariate (TVC) as a direct predictor of indicators: A TVC can be specified as a predictor in the first-level equation. For the purpose of easily describing this approach, here we incorporate a TVC into a univariate LGCM. However, this simple model can be easily expanded to specify a TVC in a multivariate LGCM, such as a PPM and a FCM.

The regression coefficient γ reflects the influence of a *time-varying exogenous variable* (Z) on a *time-varying endogenous variable* (Y) at each time point. Thus, the equations for estimating the LGCM with an unconditional TVC can be written as follows:

Level 1 Equation:

$$Y_{it} = \pi_{0i} + \lambda_t \times \pi_{1i} + \gamma_t \times Z_{it} + \varepsilon_{yit}, \qquad \varepsilon_{yit} \sim \text{NID} \ (0, \ \sigma_{yt}^2) \tag{6.9}$$

$$Z_{it} = \tau_t + \varepsilon_{zit}, \qquad \varepsilon_{zit} \sim \text{NID} \ (0, \sigma_{zt}^2) \tag{6.10}$$

$$\sigma_{zt}^2 = \begin{bmatrix} \sigma_{11}^2 & & & \\ \sigma_{21} & \sigma_{22}^2 & & \\ \vdots & \vdots & \ddots & \\ \sigma_{t1} & \sigma_{t2} & \cdots & \sigma_{tt}^2 \end{bmatrix} \tag{6.11}$$

Level 2 Equation:

$$\pi_{0i} = \mu_{00} + \zeta_{0i}, \qquad \zeta_{0i} \sim \text{NID} \ (0, \psi_{00}) \tag{6.12}$$

$$\pi_{1i} = \mu_{10} + \zeta_{1i}, \qquad \zeta_{1i} \sim \text{NID} \ (0, \psi_{11}) \tag{6.13}$$

$$\psi = \begin{bmatrix} \psi_{00} & \\ \psi_{10} & \psi_{11} \end{bmatrix} \tag{6.14}$$

where subscript i refers to individual (= 1, 2, 3,..., n), and t refers to time (= 1, 2, 3,..., t). σ^2_{zt} represents the variance-covariance structure of the *exogenous* TVC, Z. In the Level 1 equation, the errors of the *endogenous* TVC, Y, are assumed to be uncorrelated with both the growth factors and the *exogenous* TVC, Z.

Illustrative Example 6.7: Incorporating a Time-Varying Covariate (TVC) as a Direct Predictor

In this illustration, we used a composite measure of FEP as a single-indicator, time-varying covariate from Time 1 to Time 5 and specified associations with the indicators in the FCM. (Note that in this illustration, we only show how a TVC is contemporaneously associated with the indicators of one primary growth curve, for example depressive symptoms.) Thus, the research questions that can be answered by this TVC model are:

(1) How does a time-varying covariate (FEP) influence indicators of a primary growth curve for a symptom subdomain (depressive symptoms) contemporaneously while controlling for the time trends?
(2) How do the subdomain symptoms (depressive symptoms) change over time when controlling for changes in FEP?

The path diagram of this unconditional TVC model is shown in Figure 6.6.

The M*plus* syntax for this model with the direct predictor is as follows:

MODEL:
I_DEP S_DEP | DEP1@0 DEP2@1 DEP3@3 DEP4@4 DEP5@6;
DEP1 ON FEP1 (P1); DEP2 ON FEP2 (P2);
DEP3 ON FEP3 (P3); DEP4 ON FEP4 (P4);
DEP5 ON FEP5 (P5);
[FEP1-FEP5];
FEP1 WITH FEP2-FEP5;
FEP2 WITH FEP3-FEP5;
FEP3 WITH FEP4-FEP5;
FEP4 WITH FEP5;
MODEL TEST:
P1=P2; P2=P3;
P3=P4; P4=P5;

Note that this syntax can be extended to incorporate a TVC as a direct predictor of indicators for multiple subdomains simultaneously (e.g., symptoms of depression, anxiety, and hostility) in a FCM.

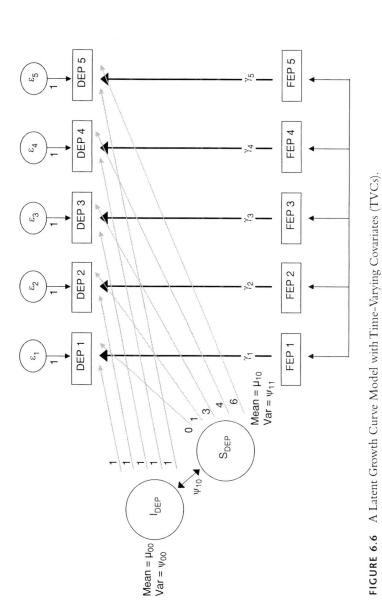

FIGURE 6.6 A Latent Growth Curve Model with Time-Varying Covariates (TVCs).

Note: DEP = Depressive Symptoms. FEP = Family Economic Problems. I = Intercept. S = Slope.

The bolded portions of the syntax can be added to the traditional LGCM to estimate the effects of a time-varying variable (family economic hardship [FEP]) on another time-varying variable (depressive symptoms [DEP]). In a SEM framework, we can directly test whether these TVC effects are constant over time. In order to test this, we used parentheses in the syntax to assign names, or labels, to all TVC parameters. Our names ranged from P1 to P5. By specifying MODEL TEST in the syntax with "P1=P2; P2=P3; P3=P4; P4=P5;" we can test whether all of the TVCs have constant effects across time (i.e., P1=P2=P3=P4=P5). Because these are nested models, we can examine the statistical significance of the Wald chi-square test (i.e., $\Delta\chi^2$) to determine which TVC effects are significantly different from each other. A statistically significant Wald chi-square test indicates path coefficients vary in magnitude, whereas a non-significant test statistic indicates TVCs with constant effects across time.

Incorporating a TVC as a secondary growth curve: A second-order parallel process dyadic model. As we illustrated with a CFM in Chapter 4, a time-varying covariate can also be incorporated into a FCM using a dyadic modeling framework. For illustrative purposes, we introduce this concept using a dyadic FCM (also known as a parallel process dyadic-FCM; D-FCM) using the same couple variables from Chapter 4. The equations for the measurement model can be written as follows where y represents the wives' responses and z represents the husbands' responses:

For wives:

$$y_{ikt} = \pi_{0ik}^{(y)} + \lambda_{kt}^{(y)} \times \pi_{1ik}^{(y)} + \varepsilon_{ikt}^{(y)}, \qquad \varepsilon_{ikt}^{(y)} \sim \mathrm{NID}\left(0, \sigma_{kt}^{(y)^2}\right) \tag{6.15}$$

$$\sigma_{kt}^{(y)^2} = \begin{bmatrix} \sigma_{1111}^{(y)^2} & & & \\ \sigma_{1121}^{(y)} & \sigma_{1122}^{(y)^2} & & \\ \vdots & \vdots & \ddots & \\ \sigma_{k1t1}^{(y)} & \sigma_{k1t2}^{(y)} & \cdots & \sigma_{kktt}^{(y)^2} \end{bmatrix} \tag{6.16}$$

For husbands:

$$z_{ikt} = \pi_{0ik}^{(z)} + \lambda_{kt}^{(z)} \times \pi_{1ik}^{(z)} + \varepsilon_{ikt}^{(z)}, \qquad \varepsilon_{ikt}^{(z)} \sim \mathrm{NID}\left(0, \sigma_{kt}^{(z)^2}\right) \tag{6.17}$$

$$\sigma_{kt}^{(z)^2} = \begin{bmatrix} \sigma_{1111}^{(z)^2} & & & \\ \sigma_{1121}^{(z)} & \sigma_{1122}^{(z)^2} & & \\ \vdots & \vdots & \ddots & \\ \sigma_{k1t1}^{(z)} & \sigma_{k1t2}^{(z)} & \cdots & \sigma_{kktt}^{(z)^2} \end{bmatrix} \tag{6.18}$$

Variance-covariance for husbands and wives:

$$\sigma^2_{kt} = \begin{bmatrix} \sigma^{(y)^2}_{kt} & \\ \sigma^{(zy)}_{kt} & \sigma^{(z)^2}_{kt} \end{bmatrix} \tag{6.19}$$

where the subscript k refers to a specific subdomain indicator, the subscript i (= 1, 2, 3, …, n) refers to individual, the subscript t (= 1, 2, 3, …, t) refers to time, and the subscripts y and z refer to manifest indicators y and z, respectively. $\sigma^{(y)^2}_{kt}$ and $\sigma^{(z)^2}_{kt}$ represent the variance-covariance structure of residuals for manifest indicators y and z, respectively (autocorrelated errors). σ^2_{kt} represents the variance-covariance structure of the residuals involving dyadic indicators. Releasing covariance parameters ($\sigma^{(zy)}_{kt}$) between dyadic indicators often leads to convergence problems. Thus, we assumed no covariance between dyadic indicators by fixing these values to 0.

For primary latent growth factor structures for manifest indicators y and z (husband and wife, respectively), equations are:

$$\pi^{(y)}_{0ik} = \mu^{(y)}_{00k} + \lambda^{(y)}_{0k0} \times \eta^{(y)}_{0i} + \zeta^{(y)}_{0ik}, \qquad \zeta^{(y)}_{0ik} \sim \text{NID}\,(0, \psi^{(y)}_{00kk}) \tag{6.20}$$

$$\pi^{(y)}_{1ik} = \mu^{(y)}_{10k} + \lambda^{(y)}_{1k1} \times \eta^{(y)}_{1i} + \zeta^{(y)}_{1ik}, \qquad \zeta^{(y)}_{1ik} \sim \text{NID}\,(0, \psi^{(y)}_{11kk}) \tag{6.21}$$

$$\pi^{(z)}_{0ik} = \mu^{(z)}_{00k} + \lambda^{(z)}_{0k1} \times \eta^{(z)}_{0i} + \zeta^{(z)}_{0ik}, \qquad \zeta^{(z)}_{0ik} \sim \text{NID}\,(0, \psi^{(z)}_{00kk}) \tag{6.22}$$

$$\pi^{(z)}_{1ik} = \mu^{(z)}_{10k} + \lambda^{(z)}_{1k1} \times \eta^{(z)}_{1i} + \zeta^{(z)}_{1ik}, \qquad \zeta^{(z)}_{1ik} \sim \text{NID}\,(0, \psi^{(z)}_{11kk}) \tag{6.23}$$

$$\Psi_{kk} = \begin{bmatrix} \psi^{(y)}_{0011} & & & & \\ \psi^{(y)}_{1011} & \psi^{(y)}_{1111} & & & \\ \psi^{(y)}_{0021} & \psi^{(y)}_{0121} & \psi^{(y)}_{0022} & & \\ \vdots & \vdots & \vdots & \ddots & \\ \psi^{(zy)}_{10k1} & \psi^{(zy)}_{11k1} & \cdots & \cdots & \psi^{(z)}_{11kk} \end{bmatrix} \tag{6.24}$$

where π represents the primary growth factors for y and z indicators, respectively. Ψ_{kk} is the variance-covariance structure of the primary growth factor variables (π). Equations for the second-order global factor structures are:

$$\eta^{(y)}_{0i} = \alpha^{(y)}_{00} + \zeta^{(y)}_{0i}, \qquad \zeta^{(y)}_{0i} \sim \text{NID}\,(0, \psi^{(y)}_{00})$$

$$\eta^{(y)}_{1i} = \alpha^{(y)}_{10} + \zeta^{(y)}_{1i}, \qquad \zeta^{(y)}_{1i} \sim \text{NID}\,(0, \psi^{(y)}_{11})$$

$$\eta^{(z)}_{0i} = \alpha^{(z)}_{00} + \zeta^{(z)}_{0i}, \qquad \zeta^{(z)}_{0i} \sim \text{NID}\,(0, \psi^{(z)}_{00})$$

$$\eta^{(z)}_{1i} = \alpha^{(z)}_{10} + \zeta^{(z)}_{1i}, \qquad \zeta^{(z)}_{1i} \sim \text{NID}\,(0, \psi^{(z)}_{11}) \tag{6.25}$$

$$\psi = \begin{bmatrix} \psi^{(y)}_{00} & & \\ \vdots & \ddots & \\ \psi^{(zy)}_{10} & \cdots & \psi^{(z)}_{11} \end{bmatrix} \tag{6.26}$$

where η represents global growth factors for primary growth factor variables (π). ψ is the variance-covariance structure of the global growth parameters.

In order to build a dyadic-FCM (D-FCM), we used two repeated indicators (i.e., marital dissatisfaction and marital unhappiness for husbands and wives at T1 – T3) to estimate primary growth curve factors ($\pi^{(h)}_{0-1}$ and $\pi^{(w)}_{0-1}$) capturing marital conflict for husbands and wives separately. Like dyadic invariance in a D-CFM, dyadic factorial invariance also needs to be confirmed in a D-FCM by constraining the same factor loading parameters on global factors ($\eta^{(h)}_{0-1}$ and $\eta^{(w)}_{0-1}$) to be equal across dyads (Wittaker, Beretvas, & Falbo, 2014). We applied this constraint for the current example model and were able to confirm factorial · loading invariance (i.e., tau-equivalence across groups) between husbands and wives. Tau-equivalence indicates that factor loadings of the same primary growth factor between husbands and wives are identical. This process can be confirmed using a nested model comparison of model fit indices (e.g., $\Delta\chi^2$). (Note that we will return to the process of testing group invariance later in this chapter after illustrating the incorporation of a time-varying variable into a FCM through a second-order parallel process model [D-FCM].)

Illustrative Example 6.8: Incorporating a Time-Varying Predictor as a Parallel Process

The M*plus* syntax for a D-FCM is shown below with marital dissatisfaction (MD) and marital unhappiness (MU) as two indicators of a primary growth factor of marital conflict (MC) (H = husband and W = wife).

```
ANALYSIS: MODEL=NOCOV;
MODEL:
I1 S1 | MD_H1@0 MD_H2@1 MD_H3@2;        ! primary GC of MD of husbands
I2 S2 | MU_H1@0 MU_ H2@1 MU_H3@2;       ! primary GC of MU of husbands
I3 S3 | MD_W1@0 MD_W2@1 MD_W3@2;        ! primary GC of MD of wives
I4 S4 | MU_W1@0 MU_W2@1 MU_W3@2;        ! primary GC of MU of wives
I_F BY I1@1 I2(L1); S_F BY S1@1 S2(L1);   ! for MC of husbands' FCM
I_M BY I3@1 I4(L1); S_M BY S3@1 S4(L1);   ! for MC of wives' FCM
[MU_F90-MD_M92@0];
[I1-S1@0]; [I3-S3@0]; [I2 S2]; [I4 S4];
[I_F S_F I_M S_M];
I_F WITH S_F; I_M WITH S_M;
I_F WITH I_M; S_F WITH S_M;
I_F WITH S_M; I_M WITH S_F;
MU_H1 WITH MD_ H1; MU_H2 WITH MD_H2;
MU_H3 WITH MD_H3;
MU_W1 WITH MD_W1; MU_W2 WITH MD_W2;
MU_W3 WITH MD_W3;
MU_H1 WITH MU_W1; MU_H2 WITH MU_W2;
```

MU_H3 WITH MU_W3;
MD_H1 WITH MD_W1; MD_H2 WITH MD_W2;
MD_H3 WITH MD_W3;

Note, including an exclamation point (!) before text in M*plus* syntax tells the program to disregard this text. This is a helpful tool for leaving "notes" in your syntax files.

The results are shown in Figure 6.7. Most estimated parameters, especially the mean and variance of the global factors, in this D-FCM are close in value to those of the D-CFM we estimated in Chapter 4 (see Figure 4.25). Also, most model fit indices are very similar to each other. The most notable difference is found for the degrees of freedom (df). As mentioned earlier, a FCM can be viewed as a more parsimonious model compared to a CFM. This example D-FCM estimated 17 fewer parameters than the D-CFM.

A Multiple-Group FCM (Multi-Group Longitudinal Modeling)

When investigating longitudinal growth, it is not only of interest to determine whether true change exists over time, but it is also often of interest to determine whether the change in the construct of interest differs across groups, such as men and women.

Group invariance can be imposed in a second-order growth model by con-straining model parameters to be equal across group members (e.g., men and women). Procedures look similar to the standard equality constraint tests we have utilized previously, in that, model constraints can be used to constrain target parameters to be equal and then an equality test can be conducted to determine if these constraints significantly worsen the model fit. A poorer model fit suggests that the model fit varies depending on group membership. Thus, strong invariance (equal loadings and equal intercepts) are essential for all multi-group longitudinal models. In some sense, if the group invariance in a longitudinal model is a key theoretical focus, at least strong invariance should be achieved for equality of the latent construct across groups and across time (i.e., multi-group longitudinal modeling; Little, 2013). However, this does not mean that the assumption of strong factorial invariance between groups has to be met for all multi-group longitudinal models. If the assumption of strong group invariance is achieved across time, but not across groups, this suggests that the overall model has the same parameters across *time*, but the parameters vary across *groups*.

Chen, Sousa, and West (2005) applied an approach similar to the method used for assessing group invariance in a CFA model (Brown, 2006). First, they used step-wise procedures to test for equality among all first-order factor parame-ters (measurement invariance). Then, they tested for similarity in the parameters

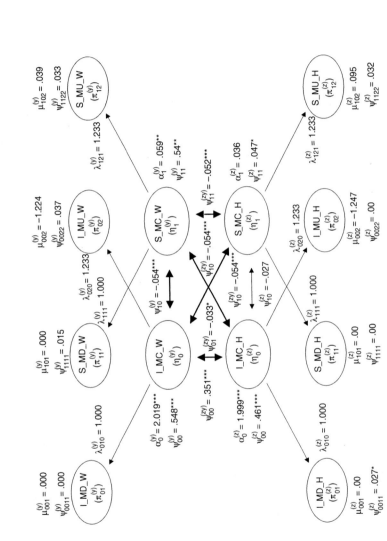

FIGURE 6.7 A Dyadic Parallel Process Model Extension of a Factor-of-Curves Model (D-FCM).

Note: Unstandardized coefficients are shown. I = Intercept. S = Slope. MC = Marital Conflict. MD = Marital Dissatisfaction. MU = Marital Unhappiness. The primary growth curve model is not shown. The marker variable scale setting approach was used. Covariances between intercept and slope variance of primary growth factors are not shown. χ^2 (df) = 93.525 (39), $p < .001$; CFI/TLI = .984/.973; RMSEA (90% CI) = .057 (.042, .072); SRMR = .051.* $p < .05$. ** $p < .01$.*** $p < .001$.

comprising the second-order factors (population heterogeneity). The following example illustrates the estimation of a multi-group FCM imposing group invariance.

Illustrative Example 6.9: Estimating a FCM for Multiple Groups

The Mplus syntax for the unconstrained model (i.e., baseline model, M1) assessing gender differences is shown in Figure 6.8. Table 6.5 shows our suggested testing procedure sequence along with the results of our tests for group invariance.

The GROUPING command was used to identify the variable in the data set that denotes the groupings that will be used in the multi-group FCM. The first model command is the same as that of an unconditional FCM (i.e., single group). Thus, only specified parameters will be freely estimated in both models. However, Mplus imposes an equality constraint for some parameters by default, specifically, the factor loadings and intercepts of primary growth factors. In the FCM, the time factor loadings have already been constrained to be equal across the primary growth models $(0, 1, 2, \ldots,$ and t) and all intercepts have been fixed to 0 (the assumption of strong measurement invariance) for model identification purposes. Thus, in the measurement model of this FCM, the assumption of strong invariance is met. In addition, the factor loadings on global factors (λs) and the mean parameters (α) of global factors impose an equality constraint between men and women by default if those parameters are not specified in the female model. In order to allow different loadings and mean parameters of global factors (i.e., a completely unconstrained model) to be estimated between the male and female models, these parameters should be freely estimated. This can be accomplished by specifying loadings and mean parameters separately in the FEMALE MODEL: command. However, note that the loadings should be constrained within the female and male model. Thus, we assigned different labels (L3 and L4), to allow constraints for each factor loading to be equal in the female model only. Given that some indicators were positively skewed in both groups, the maximum likelihood robust (MLR) estimation in Mplus was used across all group invariance tests by specifying ESTIMATION=MLR under the ANALYSIS command.

First, we tested whether the assumption of strict invariance in the measurement model (equal residual variances and covariances, M1-1) is met. Thus, we assigned labels to each residual variance and covariance. Mplus syntax for the strict measurement model is as follows:

DEP1–DEP5 (V111–V115); ANX1–ANX5 (V121–V125);
HOS1–HOS5 (V131–V135);
DEP1 WITH ANX1 (COV111); DEP1 WITH HOS1 (COV112);
ANX1 WITH HOS1 (COV113); DEP2 WITH ANX2 (COV114);

VARIABLE: names are G ANX1 ANX2 ANX3 ANX4 ANX5;

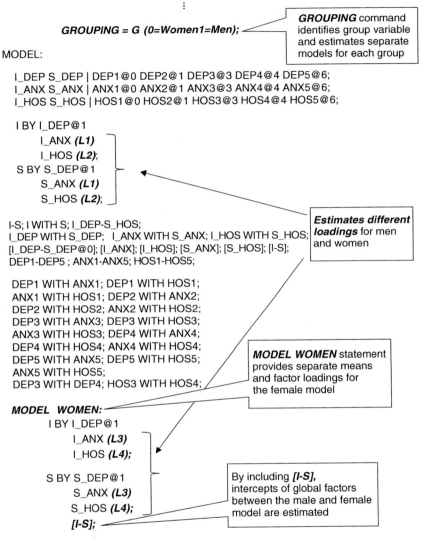

FIGURE 6.8 A Measurement Invariance FCM (Baseline Model) between Men and Women.

Note: I = Intercept. S = Slope. DEP = Depressive Symptoms. ANX = Anxiety Symptoms. HOS = Hostility Symptoms. The marker variable scale setting approach was used.

TABLE 6.5 Tests of Group Invariance of Factor Loadings (λ) on Global Intercept and Slope Factors (η_0 and η_1) Between Men and Women.

	χ^2	df	Model Comparison	$\Delta\chi^2$ (df)	RMSEA (90% CI)	CFI / TLI	ΔCFI	SRMR	AIC	BIC
Group Invariance										
Base line model with free estimation of residual errors and covariance (**M1**)	225.531	166			.041 (.026, .053)	.978 / .972		.068	4961.501	5385.576
Female (n = 233)	118.315									
Male (n = 203)	107.216									
Base line model with **strict invariance measurement** model (**M1-1**)	326.432	198	M1-1 vs. M1	100.901 (32)	.055 (.044, .065)	.952 / .949	.026	.095	5090.852	5384.442
Equal Loadings (M2)	*232.986*	*168*	*M2 vs. M1*	*7.455 (2)*	*.042 (.028, .055)*	*.976 / .970*	*.002*	*.073*	*4969.488*	*5385.407*
+ Primary Factor Intercepts (**M3**)	286.525	172	M3 vs. M2	53.539 (4)	.055 (.044, .066)	.957 / .948	.019	.080	5031.237	5430.846
+ Primary Factor Variances (**M4**)	297.552	178	M4 vs. M3	11.027 (6)	.056 (.044, .066)	.956 / .948	.001	.082	5033.122	5408.265
+ Global Factor Variances (**M5**)	303.256	180	M5 vs. M4	5.704 (2)	.056 (.045, .067)	.954 / .947	.002	.101	5040.310	5407.298
+ Global Factor Means (**M6**)	317.333	182	M6 vs. M5	14.077 (2)	.058 (.048, .069)	.950 / .942	.004	.094	5055.887	5414.720

Note: Primary intercept and slope growth parameters of depressive symptoms were used as marker variables. Baseline model assumed equal residual variance and co-variance of manifest indicators.

DEP2 WITH HOS2 (COV115); ANX2 WITH HOS2 (COV116);
DEP3 WITH ANX3 (COV117); DEP3 WITH HOS3 (COV118);
ANX3 WITH HOS3 (COV119); DEP4 WITH ANX4 (COV1110);
DEP4 WITH HOS4 (COV1111); ANX4 WITH HOS4 (COV1112);
DEP5 WITH ANX5 (COV1113); DEP5 WITH HOS5 (COV1114);
ANX5 WITH HOS5 (COV1115);
DEP3 WITH DEP4 (COV1116); HOS3 WITH HOS4 (COV1117);

where DEP1-DEP5 indicate residual variances of depressive symptoms from Time 1 through Time 5, and the WITH options indicate autocorrelated errors. Note that we arbitrarily assigned the above labels. M*plus* allows any label name to be used in naming (and constraining) the parameters.

The equality constraint test is usually performed through a traditional nested chi-square statistic (Bollen, 1989). However, as we discussed in Chapter 4, chi-square statistics are sensitive to sample size. Thus, instead of relying solely on this statistic ($\Delta\chi^2$), we used several overall model fit indices to evaluate model fit. The results of this model are shown in Table 6.5. The overall model fit of the baseline model was acceptable (χ^2(df) = 225.531(166); CFI/TLI = .978/.972; SRMR = .068; and AIC/BIC = 4961.501/5385.576).

In general, M*plus* does not provide model fit statistics separately for each group (men and women in our example). Instead, when estimating the model fit indices, M*plus* only provides the combined model fit indices. For example, in Table 6.5, the χ^2 statistic for the overall model is 225.531 (= 118.315 [female] + 107.216 [male]). If the model fit indices, such as CFI/TLI, are desired for each group, those would need to be calculated by fitting an appropriate model for each group. As expected, the strict invariance model (M1-1) had a generally poor model fit. Also, compared to the model with unconstrained residual error variances and covariances (M1), the model fit indices showed that M1-1 was a significantly poorer fit to the data. Thus, we failed to meet the requirements necessary to assume strict invariance between men and women.

The M*plus* syntax for the next model tested (the equal loadings model testing the assumption of weak invariance, M2) is shown in Figure 6.9. By assigning all loading parameters the same label (L1), the intercept factor loadings for anxiety in the male model are now not only equal to the loading of the slope factors, but the intercept factor loadings for anxiety are also equal to the intercept and slope factor loadings for women. The results of this weak invariance model are shown in Table 6.5. Although the increased chi-square value indicates that the model has a poorer fit than the unconstrained model, other model fit indices, including the Δ CFI (which is less than .01; Cheung & Rensvold, 2002) and the lower AIC, indicate that the weak invariance model with equal factor loadings is preferred over the unconstrained model. Next, we compared the weak invariance model

VARIABLE: names are ID G ANX1 ANX2 ANX3 ANX4 ANX5

⋮

GROUPING = G (0=WOMEN 1=MEN);

MODEL:

I_DEP S_DEP | DEP1@0 DEP2@1 DEP3@3 DEP4@4 DEP5@6;
I_ANX S_ANX | ANX1@0 ANX2@1 ANX3@3 ANX4@4 ANX5@6;
I_HOS S_HOS | HOS1@0 HOS2@1 HOS3@3 HOS4@4 HOS5@6;

I BY I_DEP@1
I_ANX (L1)
I_HOS (L2);
S BY S_DEP@1
S_ANX (L1)
S_HOS (L2);

I-S; I WITH S; I_DEP-S_HOS;
I_DEP WITH S_DEP;   I_ANX WITH S_ANX; I_HOS WITH S_HOS;
[I_DEP-S_DEP@0]; [I_ANX]; [I_HOS]; [S_ANX]; [S_HOS]; [I-S];
DEP1-DEP5; ANX1-ANX5; HOS1-HOS5;

DEP1 WITH ANX1; DEP1 WITH HOS1;
ANX1 WITH HOS1; DEP2 WITH ANX2;
DEP2 WITH HOS2; ANX2 WITH HOS2;
DEP3 WITH ANX3; DEP3 WITH HOS3;
ANX3 WITH HOS3; DEP4 WITH ANX4;
DEP4 WITH HOS4; ANX4 WITH HOS4;
DEP5 WITH ANX5; DEP5 WITH HOS5;
ANX5 WITH HOS5;
DEP3 WITH DEP4; HOS3 WITH HOS4;

MODEL FEMALE:
I BY I_DEP@1
I_ANX (L1)
I_HOS (L2);

S BY S_DEP@1
S_ANX (L1)
S_HOS (L2);
[I-S];

Assigning the same loadings for men and women

FIGURE 6.9 A FCM with Equal Loadings Specified for Men and Women (Weak Invariance).
Note: I = Intercept. S = Slope. DEP = Depressive Symptoms. ANX = Anxiety Symptoms. HOS = Hostility Symptoms. The marker variable scale setting approach was used.

(M2) with the strong invariance model (equal factor loadings and intercepts of primary growth factors, M3).

In the same manner that we tested the weak invariance model (see Figure 6.10), M*plus* can be utilized to test the assumption of strong invariance by simply assigning labels to all intercepts. M*plus* syntax for the strong invariance model is as follows:

[I_DEP-S_DEP@0];
[I_ANX] (T1);
[I_HOS] (T2);
[S_ANX] (T3);
[S_HOS] (T4);
[I-S];

where brackets in the syntax [] are used to estimate variable means. By assigning labels such as T1 though T4, all intercepts of the subdomain growth factors were constrained to be identical between men and women. The comparison between the models testing the assumptions of weak invariance (M2) and strong invariance (M3) can be seen in Table 6.5. Model fit indices ($\Delta \chi^2$ (df) = 53.539(4) and Δ CFI = .019) showed that M3 was a significantly worse fit to the data. Thus, we selected the weak invariance model (with equal loadings, M2) as the optimal model. Figure 6.10 shows the model parameter estimates from this model. Statistically significant means were observed for both global factors (η_0 and η_1) in the female model. However, in the model for men, non-significant means were observed for both global factors (η_0 and η_1). The opposite pattern was observed for slope means (α_{11}, α_{12}, and α_{13}). Statistically significant means were observed for all specific subdomain growth factors (π_{01} though π_{13}) for men but not women. Regarding a comparison of the variances between the models for men and women, in both models, there was statistically significant variation in the global factors (both the intercept and slope of IS). Furthermore, unique variances of growth factors (π_0 and π_1) for hostility symptoms were observed for both men and women. However, unique variances of intercept factors for depressive symptoms were observed for women only. Based on our results, it seems that the three subdomain symptoms equally contribute to the global domain of internalizing symptoms (IS) for men and women, but it appears that the developmental changes vary for men and women.

Multivariate FCM

Previous studies suggest that more than one set of second-order growth factors is often needed to explain the covariation among primary growth factors (Krueger, 1999; Vollebergh et al., 2001). For example, Krueger, Caspi, Moffit, and Silva (1998) showed that ten different psychiatric disorders could be modeled with two higher-order latent constructs of internalizing and externalizing disorders.

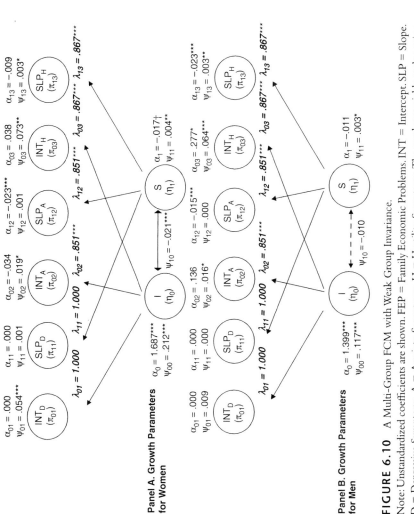

FIGURE 6.10 A Multi-Group FCM with Weak Group Invariance.

Note: Unstandardized coefficients are shown. FEP = Family Economic Problems. INT = Intercept. SLP = Slope. D = Depressive Symptoms. A = Anxiety Symptoms. H = Hostility Symptoms. The marker variable scale setting approach was used. Covariances between intercept and slope variance of primary growth factors are not shown. † $p < .10.$ *$p < .05.$ **$p < .01.$ ***$p < .001.$

Thus, several second-order growth factors can exist that account for variances and covariances among primary growth factors of different domains.

The identification of second-order factors involves locating the source of obliqueness among first-order primary factors (Hancock, Kuo, & Lawrence, 2001). For example, Measelle, Stice, and Hogansen (2006) demonstrated that more than one set of second-order growth factors is necessary to capture patterns of co-development among depressive symptoms, disordered eating, antisocial behavior, and substance use. They found that two sets of second-order growth factors · explained the covariations among this set of four primary growth factors. One set of secondary growth factors accounted for the co-development of depressive symptoms and disordered eating. The other set of secondary growth factors accounted for the co-development of antisocial behavior and substance abuse. Theories of affect disturbance and mood regulation support the second-order associations between depressive symptoms and eating behaviors (Stice, Burton, & Shaw, 2004), whereas theories of behavioral disinhibition and poor impulse control support the associations between antisocial behavior and substance abuse problems (Pennington, 2002). The following example illustrates the estimation of a multivariate FCM with our sample data.

Illustrative Example 6.10: Estimating a Multivariate FCM

In order to replicate the findings from Measelle et al. (2006), we used two externalizing symptoms as manifest indicators: substance use and antisocial behavior from Time 1 through Time 5. Five items were used to create the indicator of substance use, including items assessing tobacco and illegal drug use. The manifest indicator of antisocial behavior was created from five items, including items assessing theft and vandalism. These dichotomized items (yes/no) were summed. M*plus* syntax for this multivariate FCM is shown in Figure 6.11.

Note that ANT represents antisocial behavior, and SUB represents substance use. As this syntax illustrates, a multivariate FCM is specified by expanding on the syntax of an unconditional FCM to incorporate an additional FCM. The results of the final multivariate FCM are shown in Figure 6.12.

The overall model fit indices were acceptable (χ^2(df) = 486.755(255), $p < .001$; CFI/TLI = .946/.936; RMSEA (90% CI) = .046 (.039, .052); SRMR = .062). All parameters (means and variances) of the global factors were statistically significant. The results indicated that internalizing symptoms (depressive symptoms and disordered eating) decreased across time whereas externalizing symptoms (substance abuse and antisocial behavior) increased. Statistically significant variances for all global factors indicate the existence of inter-individual differences in the internalizing and externalizing developmental patterns. Furthermore, most of the covariances among global factors were statistically significant. Importantly, the covariances between the same global factors in different global domains were statistically significant. For example, the covariance between the intercept factors

DATA: File is *example_ch_6_4.dat*;
VARIABLE:
     names are ANX1-ANX5 DEP1-DEP5 HOS1-HOS5 ***ANT1-ANT5 SUB1-SUB5***;
     usevar are ANX1-ANX5 DEP1-DEP5 HOS1-HOS5 ***ANT1-ANT5 SUB1-SUB5***;
MODEL:
  I_DEP S_DEP | DEP1@0 DEP2@1 DEP3@3 DEP4@4 DEP5@6;
  I_ANX S_ANX | ANX1@0 ANX2@1 ANX3@3 ANX4@4 ANX5@6;
  I_HOS S_HOS | HOS1@0 HOS2@1 HOS3@3 HOS4@4 HOS5@6;

  I_ANT S_ANT | ANT1@0 ANT2@1 ANT3@3 ANT4@4 ANT5@6;
  I_SUB S_SUB | SUB1@0 SUB2@1 SUB3@3 SUB4@4 SUB5@6;

  I_INT BY I_DEP@1
        I_ANX (L1)
        I_HOS (L2);
  S_INT BY S_DEP@1
        S_ANX (L1)
        S_HOS (L2);

> Specification of a FCM using **two externalizing symptoms with weak invariance**

  I_EXT BY I_ANT@1
       I_SUB (L3);
  S_EXT BY S_ANT@1
       S_SUB (L3);

> Specifying **equal loadings (weak invariance)** in the externalizing FCM (constrained by label L3)

  I_DEP I_ANX I_HOS PWITH S_DEP S_ANX S_HOS;
  I_ANT I_SUB PWITH S_ANT S_SUB;

> Specification of **covariance between intercept and slope primary factors** in the externalizing domain

  I_DEP I_HOS WITH I_SUB I_ANT;
  S_DEP S_HOS WITH S_SUB S_ANT;

  [I_DEP-S_DEP@0];
  [I_ANX S_ANX]; [I_HOS S_HOS];
  [I_ANT-S_ANT@0];
  [I_SUB S_SUB]; [I_INT S_INT];
  [I_EXT S_EXT];

> Antisocial behaviors were used as marker variables by fixing the loadings at 1 and intercepts at 0

  DEP1 WITH ANX1 HOS1; ANX1 WITH HOS1;
  DEP2 WITH ANX2 HOS2; ANX2 WITH HOS2;
  DEP3 WITH ANX3 HOS3; ANX3 WITH HOS3;
  DEP4 WITH ANX4 HOS4; ANX4 WITH HOS4;
  DEP5 WITH ANX5 HOS5; ANX5 WITH HOS5;
  DEP3 WITH DEP4; HOS3 WITH HOS4;
  ANT1 WITH SUB1; ANT2 WITH SUB2;
  ANT3 WITH SUB3; ANT4 WITH SUB4;

> Autocorrelated errors **between different domains at the same measurement occasion**

  ANT5 WITH SUB5;
  ANT3 WITH ANT4 SUB4;

> Additional autocorrelations specified as suggested by the modification indices

OUTPUT: SAMPSTAT STANDARDIZED MOD (5.00);

FIGURE 6.11 A Multivariate FCM using Internalizing and Externalizing Symptoms.
Note: I = Intercept. S = Slope. DEP = Depressive Symptoms. ANX = Anxiety Symptoms. HOS = Hostility Symptoms. ANT = Antisocial Behavior. SUB = Substance Use. The marker variable scale setting approach was used.

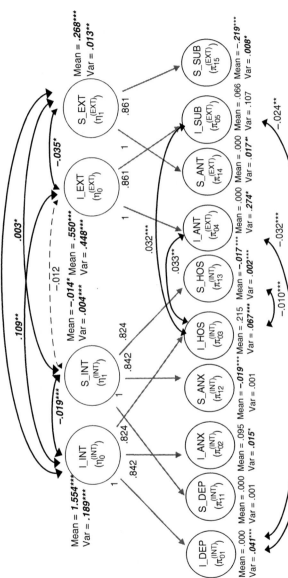

FIGURE 6.12 A Multivariate FCM.

Note: Unstandardized coefficients are shown. I = Intercept. S = Slope. INT = Internalizing Symptoms. EXT = Externalizing Symptoms. DEP = Depressive Symptoms. ANX = Anxiety Symptoms. HOS = Hostility Symptoms. ANT = Antisocial Behavior. SUB = Substance Use. The marker variable scale setting approach was used. Measurement models (primary growth models) are not shown. Only statistically significant covariances among primary growth factors are shown. The dotted line indicates a non-significant path. Bold and italic numbers indicate statistically significant paths. χ^2 (df) = 486.755(255), $p < .001$; CFI/TLI = .946/.936; RMSEA (90% CI) = .046 (.039, .052); SRMR = .062.

$*p < .05. **p < .01. ***p < .001.$

for the internalizing and externalizing domains was statistically significant (.109, $p < .01$). A similar pattern of significant covariance was observed between the internalizing and externalizing symptoms slope factors (.003, $p < .05$), which indicates the existence of a parallel process (co-development) between internalizing and externalizing symptoms at the global factor level. These results support the existence of general psychopathology modeled as a *factor-of-factor-of-curves model* (FFCM).

An important advantage of this multivariate FCM approach is the ability to investigate associations among growth factors in both primary growth factors and global growth factors. Although we demonstrated this multivariate modeling approach using two FCMs (internalizing and externalizing symptoms), the model can be extended to incorporate more global factors. Thus, this multivariate modeling approach can directly test multiple theoretical perspectives simultaneously, which is not possible when utilizing a single primary growth curve or a second-order growth curve modeling approach.

Model Selection: Factor-of-Curves vs. Curve-of-Factors

Although we briefly discussed the topic of model selection previously, now that both a CFM and a FCM have been introduced in detail (see Chapters 3 and 4 for a CFM), it is important to have a clear understanding of which approach is appropriate for the specific research question at hand. The difference in these two models can be most broadly identified from the different base models used to build the second-order models. While the basic form of a FCM is an extension of a parallel process model examining growth curves of different subdomains simultaneously, the basic form of a CFM is an extension of a longitudinal confirmatory factor analysis examining the global domain (IS) as a latent factor at multiple time points.

POINT TO REMEMBER…

With the FCM approach, primary growth factors of specific subdomains are explicitly estimated, so the primary growth factors as well as the covariations among these primary growth factors can be examined. Thus, for research questions aimed at investigating both primary growth factors and their covariations in an effort to understand the co-development of different attributes, the FCM approach is appropriate.

In particular, we suggest that the FCM approach may be most appropriate when investigating the co-development of time-varying attributes, such as depressive symptoms, anxiety symptoms, and hostility symptoms, during adolescence because it allows for the examination of both trajectories of specific subdomains and trajectories of the global, second-order domain during stages or phases of rapid development. That is, risks that are specific to a given subdomain and shared risks across subdomains can be examined simultaneously, which makes this approach more applicable to rapid developmental periods when attributes (e.g., socioeconomic, behavioral, psychological, and/or physiological character-istics) are thought to primarily change independently. However, this change may be associated enough to allow for the formation of second-order growth factors. ·

Because we estimated primary growth factors specific to each subdomain with a FCM approach, the notion of desired longitudinal covariance patterns of the repeated measures, as we discussed in the Chapter 2, can be applicable to each of the primary growth curves. Furthermore, in a FCM, these primary growth factors are indicators of second-order global latent factors. Thus, the covariances among these primary growth factors of different subdomains determine the "loadings" of primary growth factors as indicators of second-order global growth factors. The explained variance of each indicator is given by the square of the "loadings," which reflects the reliability of the primary growth factors as indicators of the second-order growth factors. If this empirical evidence (e.g., adequate loadings) is found, a FCM is empirically appropriate for examining co-development among different subdomains.

Alternatively, in the CFM approach, a single second-order growth curve is expected to capture the co-development among different subdomains (i.e., con-temporaneous covariances). The time-varying ("repeated") indicators of this sin-gle second-order growth curve are latent factors defined by subdomains. Thus, in the CFM approach, the primary growth factors of each individual subdomain are not explicitly estimated, and when utilizing this approach, we are unable to assess differential growth in the various subdomains over time or the associations among growth factors from different subdomains.

In the CFM approach, repeated confirmatory latent factors comprised of mul-tiple subdomains are modeled as indicators of second-order growth factors. If the correlations among the measures of different subdomains at each time point are high, these measures are most likely reliable and valid indicators of confirma-tory latent factors (Bollen, 1989). Furthermore, the correlations among confirma-tory latent factors provide evidence of the desired longitudinal patterns for a potential "good" second-order growth curve. If this evidence is found, the CFM approach is empirically appropriate for examining the co-development of differ-ent subdomains.

Thus, a CFM and a FCM use different approaches to explain higher-order changes in attributes over time. The results of empirical examples show that both approaches provide similar results, albeit not exactly identical, in relation to the global growth parameters and for the investigation of predictors and distal outcomes.

The FCM approach is conceptually different from the CFM approach. The latter assumes that the global domain comprised of multiple subdomain factors is established first and that these stable factors progress over time (factor invariance). But this assumption is not always accurate. For instance, a multi-subdomain phenomenon (e.g., internalizing symptoms) may not be established early in the measurement period (e.g., an early stage of life in our example). If these global factors are established at the early stage, this ensures time invariance of the latent factor and suggests that the CFM approach is more appropriate. Empirically, the success of these confirmatory factors depends on correlations high in magnitude among the different indicators (e.g., DEP1, ANX1, and HOS1) at each measuring occasion.

In summary, for successful implementation of the FCM approach, evidence of the desired longitudinal pattern of correlations among within-subdomain repeated measures and significant correlations among primary growth factors for the formation of second-order growth factors is needed. For successfully implementing the CFM approach, high correlations among time-specific measures of study subdomains and the necessary longitudinal pattern of correlations among confirmatory latent factors are needed. Both of these models (CFM or FCM) result in the creation of secondary growth parameters, often with similar results in secondary parameters. (There may be differences in the magnitudes and variances of growth parameters depending on the correlation patterns of indicators.) Fit indices can be used to compare the fit of the models to the data. Furthermore, a FCM is a more parsimonious model compared to a CFM because it estimates fewer parameters. Thus, choosing the appropriate model involves taking into consideration model fit statistics, parsimony, interpretability, and substantive or theoretical considerations.

Illustrative Example 6.11: Empirically Comparing CFM and FCM Approaches

Our example correlations among the three internalizing symptoms, which are shown in Table 6.6, are helpful for evaluating the data structure to determine the preferred modeling approach. Regarding the CFM approach, contemporaneous correlations (correlations among different indicators at the same occasion) should be relatively strong in magnitude and stable across time. In our example, all coefficients are high and stable across time. For example, the correlations among symptoms of depression, anxiety, and hostility at Time 1 were .75, .61, and .66, respectively (see Table 6.6). At Time 2, those correlations were .77, .72, and .70. At Time 3, those correlations were .76, .65, and .67. These moderately large and stable contemporaneous correlations support the estimation of a CFM. Moreover, in Table 6.6, longitudinal covariances among latent variables (IS) consisting of these three symptoms are presented. In general, adjacent covariances between two time points (t, t+1) were higher than other covariances. For example, the

TABLE 6.6 Standardized Correlations and Covariances Among Manifest Indicators and Among Subdomain Growth Factors.

Correlations among different indicators at the same occasion (contemporaneous correlations)

		DEP	ANX	HOS
Time 1	DEP	—		
	ANX	.75	—	
	HOS	.61	.67	—
Time 2	DEP	—		
	ANX	.77	—	
	HOS	.71	.70	—
Time 3	DEP	—		
	ANX	.77	—	
	HOS	.65	.67	—
Time 4	DEP	—		
	ANX	.78	—	
	HOS	.70	.74	—
Time 5	DEP	—		
	ANX	.71	—	
	HOS	.66	.60	—

Covariances among the same indicators at different occasions

	DEP1	DEP2	DEP3	DEP4	DEP5
DEP1	.31				
DEP2	**.20**	.36			
DEP3	.18	**.19**	.42		
DEP4	.15	.15	**.21**	.35	
DEP5	.09	.11	.15	**.13**	.28

	ANX1	ANX2	ANX3	ANX4	ANX5
ANX1	.21				
ANX2	**.13**	.26			
ANX3	.12	**.14**	.27		
ANX4	.07	.08	**.11**	.19	
ANX5	.05	.08	.09	**.08**	.14

	HOS1	HOS2	HOS3	HOS4	HOS5
HOS1	.31				
HOS2	**.19**	.38			
HOS3	.13	**.17**	.36		
HOS4	.10	.11	**.16**	.26	
HOS5	.06	.06	.08	**.07**	.19

Covariances among first-order latent factors

(baseline model is LCFA with partial strong invariance)

	IS1	IS2	IS3	IS4	IS5
IS1	.22				
IS2	.16	.28			
IS3	.14	.16	.30		
IS4	.09	.11	.15	.23	
IS5	.06	.08	.10	.10	.16

Correlations among primary growth factors

(baseline model is a PPM)

	INT_{DEP}	INT_{ANX}	INT_{HOS}	SLP_{DEP}	SLP_{ANX}	SLP_{HOS}
INT_{DEP}	—					
INT_{ANX}	.86	—				
INT_{HOS}	.71	.79	—			
SLP_{DEP}	-.60	-.55	-.47	—		
SLP_{ANX}	-.52	-.63	-.51	.91	—	
SLP_{HOS}	-.39	-.52	-.76	.63	.70	—

Note: DEP = Depressive Symptoms. ANX = Anxiety Symptoms. HOS = Hostility Symptoms. IS = Internalizing Symptoms. INT = Intercept. SLP = Slope. LCFA = Longitudinal Confirmatory Factor Analysis. PPM = Parallel Process Growth Curve Model.

correlations of global factors (i.e., IS) between two adjacent time points ranged from .10 to .16. These covariances were generally stronger than other longitudinal covariances (ranged from .07 to .14). This evidence supports the use of a CFM for investigating the longitudinal change of a second-order global domain of IS.

Conversely, in order to establish that the given dataset is appropriate for estimating a FCM, each primary growth model should first be confirmed by investigating adjacent correlations among the same indicators at different occasions. See the three boxes in Table 4.1, which indicate longitudinal correlations for each primary growth model. As shown in the table, adjacent correlations were generally greater than other correlations for the same indicator. For example, for depressive symptoms, correlations between two adjacent time points were .61, .49, .53, and .40, and these correlations were greater than other correlations among depressive symptom measures. Furthermore, as can be seen in Table 6.1, relatively large correlations among the same primary growth factors of different subdomains support the use of our dataset for a FCM as well. For example, correlations among intercept factors of different subdomains were .86, .71, and .79, and correlations among slope factors of different subdomains were .91, .63, and .70.

Thus, our data structure may fit well for both the CFM and FCM approaches according to statistical comparisons. Consequently, the appropriate model should be selected by taking into account both statistical and substantial considerations, as statistical considerations often do not provide a clear-cut indication of the preferred model. The correlations above provide one example of when statistical considerations do not suggest one approach over the other. Another example can be drawn from the model fit indices. It is possible that a CFM may have a lower AIC value than a FCM, but a FCM may be the "better" fitting model based on other fit statistics (Duncan et al., 2006). In general, it is useful to fit both alternative models, a CFM and a FCM, and compare common model parameters that describe change in the study phenomenon. Furthermore, a FCM is more parsimonious with fewer parameters to be estimated. More importantly, there may be interpretability and substantive reasons to select one model over the other.

Combining a CFM and a FCM: A Factor-of-Curves-of-Factors (FCF) Model

It is also possible to model several CFMs within a FCM. For example, one set of subdomains (e.g., depression, anxiety, and hostility symptoms) may form a CFM of internalizing symptoms and another set of subdomains (e.g., drug use, alcohol use, and smoking) may form a CFM of substance use. The second-order growth factors of these CFMs can be indicators of third-order growth factors (a FCM) reflecting the co-development of internalizing symptoms and substance use. This type of model is known as a factor-of-curves-of-factors model (FCFM).

Illustrative Example 6.12: Estimating a Factor-of-Curves-of-Factors (FCF) Model

Using dyadic couple data (file name: example_ch_6_2.dat.), the M*plus* syntax for a factor-of-curves-of-factors model (FCFM) is shown in Figure 6.13. For this type of model, two CFMs were fit: marital conflict (MC) for husbands and wives. For both partners, MC, as a latent variable, was defined by MD (marital dissatisfaction) and MU (marital unhappiness) repeatedly over three time points.

In the initial results, the female model estimated a negative variance for the slope of marital conflict (SLP_{MC}). Thus, we fixed the variance of this path to zero. The constrained model was not statistically different from the initial estimated model ($\Delta\chi^2$ (df) = .017(1), p = .896). Thus, we proceeded with the constrained model. The model results are shown in Figure 6.14. Overall, the model was a good fit to the data (χ^2(df) = 135.809(57), p < .001; CFI/TLI = .977/.973; RMSEA (90% CI) = .056 (.044, .069); SRMR = .052). As can be seen, all of the parameters

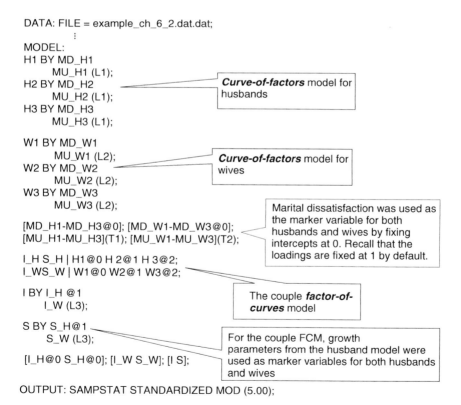

FIGURE 6.13 M*plus* syntax for a Factor-of-Curves-of-Factors Model (FCFM).
Note: I = Intercept. S = Slope. H = Husband. W = Wife. D = Depressive Symptoms.

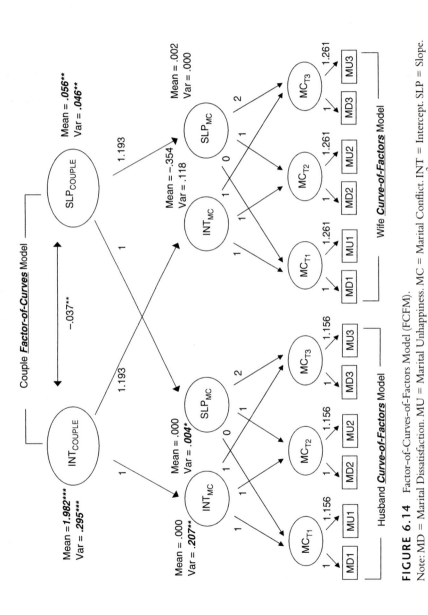

FIGURE 6.14 Factor-of-Curves-of-Factors Model (FCFM).

Note: MD = Marital Dissatisfaction. MU = Marital Unhappiness. MC = Marital Conflict. INT = Intercept. SLP = Slope. Unstandardized coefficients are shown. The marker variable scale setting approach was used. χ^2 (df) = 135.809(57), $p < .001$; CFI/TLI = .977/.973; RMSEA (90% CI) = .056 (.044, .069); SRMR = .052. *$p < .05$. **$p < .01$. ***$p < .001$.

for the global factors at the couple level were statistically significant. Interestingly, after controlling for global factors at the couple level, none of the covariances between primary growth factors were statistically significant, which implies that the global factors fully explain both husbands' and wives' experiences of marital conflict. Also, the model fit is acceptable even though we did not specify any autocorrelated errors among the manifest indicators. This suggests that multiple indicators (i.e., marital dissatisfaction and unhappiness) of marital conflict not only co-occur at the same measurement occasion (a curve-of-factors perspective) but also co-develop across time (a factor-of-curves perspective).

Chapter 6 Exercises

This is a hypothetical dataset consisting of composite mean scores for three internalizing indicators (depression, anxiety, and hostility symptoms) of target adolescents at T1~T4. The M*plus* syntax below indicates the variable order.

DATA: File is exercise_Ch.6.dat;
VARIABLE: names are DEP1 DEP2 DEP3 DEP4 ANX1 ANX2 ANX3 ANX4 HOS1 HOS2 HOS3 HOS4 FC1_M GHLT4_M;

Note that DEP1~DEP4 indicate composite mean scores for depressive symptoms of adolescents at T1~T4. ANX1~ANX4 indicate composite mean scores for anxiety symptoms of adolescents at T1~T4. HOS1~HOS4 indicate composite mean scores for hostility symptoms of adolescents at T1~T4. FC1_M represents financial cutback as reported by the mother at Time 1. NLE4M represents negative life events reported by the mother at Time 4.

1. Examine adjacent covariances (t, t+1) among the same repeated indicators to determine if they are higher (which supports a FCM) than non-adjacent covariances.
2. Next, specify a parallel process model (PPM). That is, using the same variables (DEP1-DEP4, ANX1-ANX4, and HOS1-HOS4), estimate primary growth curve models and release all possible covariances among the primary growth curve factors. Specify factor loadings for linear changes (t = 0, 1, 2, and 3).
   A. Investigate the model fit indices. Does this model fit the data well?
   B. If the model has a poor fit, you may think about including autocorrelated errors. In this chapter, we introduced two different types of autocorrelated error structures (correlated errors within subdomains and correlated errors between subdomains). Using model fit indices, identify the optimal (best fitting) error structure.

3. Next, build a FCM. Based on the PPM you identified as the optimal model in Question 2, specify global intercept and slope factors (η_0 and η_1). Use the intercept and slope growth factors for anxiety symptoms as the marker variables.

   A. Test loading invariance (equality) by assigning labels to constrain factor loadings (i.e., a weak invariance model). Complete Exercise Table 6.1. Do the primary growth factors equally contribute to the variances of the global factors (loading equality)?

EXERCISE TABLE 6.1 Loading Invariance (Equality) Test for the Factor-of-Curves Model (FCM).

	χ^2 (df)	Model comparison	$\Delta\chi^2$ (df), p-value
Uncontrained loadings (M1)			
Constrained loadings (M2)		M2 vs. M1	

   B. Interpret the means of the global factor growth parameters (η_0 and η_1).

4. Test the effect of mothers' reports of financial cutback as a predictor in your FCM.

   A. Specify the regression path between the predictor (FC1_M) and the global growth factors (η_0 and η_1) in the FCM to test a conditional FCM. Interpret all statistically significant unstandardized coefficients (γs).

   B. Based on the conditional FCM from Question 4A, investigate whether FC1_M influences any of the primary growth factors (πs). In order to avoid model convergence problems, you may decide to specify one or two direct regression coefficients at a time. Interpret all statistically significant unstandardized coefficients (γs).

5. Test the effect of adolescents' internalizing symptoms on mothers' subsequent poor physical health in your FCM.

   A. Specify regression paths between the global growth factors (η_0 and η_1) and the distal outcome (GHLT4_M). Interpret all statistically significant unstandardized coefficients (βs).

   B. Based on the conditional FCM from Question 5A, investigate whether any of primary growth factors (πs) influence the outcome (GHLT4_M). In order to avoid model convergence problem, you may decide to specify one or two direct regression coefficients at a time. Interpret all statistically significant unstandardized coefficients (βs).

References

Bollen, K. A. (1989). *Structural Equations with Latent Variables*. New York: John Wiley & Sons, Inc.

Brown, T. A. (2006). *Confirmatory factor analysis for applied research*. Spring, NY: Guilford.

Chen, F. F., Sousa, K. H., & West, S. G. (2005). Testing measurement invariance of second-order factor models. *Structure Equation Modeling*, 12(3), 471–492.

Cheung, G. W., & Rensvold, R. B. (2002). Evaluating goodness-of-fit indexes for testing measurement invariance. *Structural Equation Modeling*, 9(2), 233–255.

Duncan, S. C., & Duncan, T. E. (1996). A multivariate latent growth curve analysis of adolescent substance use. *Structural Equation Modeling*, 3(4), 323–347.

Duncan, T. E., Duncan, S. C., & Strycker, L. A. (2006). *An introduction to latent variable growth curve modeling. Concepts, issues and applications* (2nd ed.). Mahwah, NJ: Erlbaum.

Hancock, G. R., Kuo, W. L., & Lawrence, F. R. (2001). An illustration of second-order latent growth models. *Structural Equation Modeling*, 8(3), 470–489.

Howe, G. W., Hornberger, A. P., Weihs, K., Moreno, F., & Neiderhiser, J. M. (2012). Higher-order structure in the trajectories of depression and anxiety following sudden involuntary unemployment. *Journal of Abnormal Psychology*, 121(2), 325–338.

Krueger, R. F. (1999). The structure of common mental disorders. *Archives of General Psychiatry*, 56(10), 921–926.

Krueger, R. F., Caspi, A., Moffitt, T. E., & Silva, P. A. (1998). The structure and stability of common mental disorders (DSM–III–R): A longitudinal-epidemiological study. *Journal of Abnormal Psychology*, 107(2), 216–227.

Little, T. D. (2013). *Longitudinal Structural Equation Modeling*. New York, NY: Guilford.

Lorenz, F. O., Wickrama, K. A. S., & Conger, R. D. (2004). Modeling continuity and change in family relations with panel data. In R. D. Conger, F. O. Lorenz, & K. A. S. Wickrama (Eds.), *Continuity and change in family relations: Theory, methods, and empirical findings* (pp. 15–62). Mahwah, NJ: Erlbaum.

Measelle, J. R., Stice, E., & Hogansen, J. M. (2006). Developmental trajectories of co-occurring depressive, eating, antisocial, and substance abuse problems in female adolescents. *Journal of Abnormal Psychology*, 115(3), 524–538.

Pennington, B. F. (2002). *The development of psychopathology: Nature and nurture.* New York: Guilford Press.

Stice, E., Burton, E. M., & Shaw, H. (2004). Prospective relations between bulimic pathology, depression, and substance abuse: Unpacking comorbidity in adolescent girls. *Journal of Consulting and Clinical Psychology*, 72(1), 61–72.

Vollebergh, W. A. M., Iedema, J., Bijl, R. V., de Graaf, R., Smit, F., & Ormel, J. (2001). The structure and stability of common mental disorders. *Archives of General Psychiatry*, 58(8), 597–603.

Wittaker, T. A., Beretvas, S. N., & Falbo, T. (2014). Dyadic curve-of-factors model: An introduction and illustration of model for longitudinal nonexchangeable dyadic data. *Structural Equation Modeling*, 21(2), 303–317.

PART 2

Growth Mixture Modeling

7

AN INTRODUCTION TO GROWTH MIXTURE MODELS (GMMs)

Introduction

The preceding four chapters discussed a curve-of-factors model (Chapters 3 and 4) and a factor-of-curves model (Chapters 5 and 6) as higher-order extensions of conventional latent growth curve models, LGCMs. This chapter discusses another extension of latent growth curve modeling: growth mixture modeling (GMM). Mixture modeling is a latent class extension of latent growth curve modeling and is necessary for instances when trajectory heterogeneity exists (Muthén & Shedden, 1999). A LGCM is based on standard normal distributional assumptions, which are often violated when there are multiple trajectories within a sample due to the presence of multiple groups, or sub-populations, because trajectory heterogeneity results in categorical data. A GMM, which is based on a class of statistical models known as a generalized linear model (GLM), makes appropriate distributional assumptions to incorporate categorical data into the model. Because a GMM assumes a "mixture" of normal distributions (Feldman, Masyn, & Conger, 2009), it can accommodate the existence of sub-populations captured by latent classes of trajectories. That is, if sub-populations exist with different means, slopes, and associated variances, then the aggregate distribution will appear non-normal and particularly heterogeneous. Once the variability between sub-populations, or groups, is taken into account, the error variance that remains is more likely to be "normal."

First, this chapter discusses the possible existence of heterogeneity that may be characterized into distinct sub-populations of individual trajectories. For our discussion, we focus mainly on social-psychological and health attributes, but such clustering of attributes could extend to other study areas as well. Second, we explain why fitting a conventional latent growth curve model (LGCM) is problematic when trajectory heterogeneity exists. Third, this chapter discusses how the heterogeneity of individual trajectories is taken into account in a GMM by incorporating not only continuous latent variables that reflect growth parameters but also categorical latent variables that capture the heterogeneity of developmental trajectories. Fourth, we summarize the necessary steps for successfully estimating a GMM. For each model, figures and M*plus* syntax are presented. Finally, this chapter elaborates on several common issues that arise when estimating a GMM, including model convergence problems and choosing the · optimal number of classes. To conclude the chapter, we have incorporated an overview of our proposed model building strategy as a summary of the content presented in this chapter. Please see the following chapter (Chapter 8) for a discussion of how covariates (predictors and distal outcomes) can be incorporated into a GMM.

A Conventional Latent Growth Curve Model (LGCM)

As we discussed in Chapter 2, when fitting a conventional growth curve model (LGCM) it is assumed that all individuals come from a single population and a common growth function is applicable for all members of that population. That is, a LGCM assumes that the same trajectory form or shape applies to all individuals in the sample (e.g., a linear trajectory defined by an intercept and linear slope growth parameters). Although a LGCM allows the magnitudes of growth parameters (intercept and slope) to vary across individuals, these individual growth parameters are conceived of as random variables, which vary uniformly around the population mean (normal theory distributional assumptions). In a SEM-based growth curve model, these growth parameters are defined as continuous, and random, latent variables. Thus, conventional latent growth curve modeling is well-suited for studying change in attributes when the growth function for all members of the population is of the same form but where magnitudes of the growth parameters have uniform variations. It is also important to note that conventional growth curve modeling assumes that covariates affecting the growth parameter(s) influence each individual in the same way.

Potential Heterogeneity in Individual Trajectories

In some instances latent growth curve modeling is not well-suited for studying change because of the inherent heterogeneity that is found in trajectories of some individual attributes. For example, socioeconomic and health attributes of individuals are typically not uniformly distributed throughout the population and may not have a common growth function. Instead, qualitatively different trajectories may exist within the population. As a result, individual trajectories of various socioeconomic, developmental, and health attributes may be heterogeneous, or clustered. For instance, classes, or groups of individuals, may exist based on variations in social mobility, including classes of consistently advantaged and consistently disadvantaged individuals as well as classes with varying socio-economic conditions over time (e.g., climbing or falling mobility). Furthermore, there may be considerable differences between these classes.

In the longitudinal context, heterogeneity, or clustering, of individual trajectories may be characterized into distinct sub-populations of trajectories based on their patterns of growth for two reasons. First, the heterogeneity, or clustering, of individual trajectories can be based merely on the rates of change (slopes) across individuals even though all individuals share a common growth function (e.g., all respondents share a linear growth pattern, but there could be three [or more] clusters based on rate of change). For example, sub-populations with varying rates of change could include linear and increasing, linear and decreasing, and linear but constant. The heterogeneous rate of change across individuals produces different patterns resulting in unobserved variation. For an empirical example, consider individual linear health trajectories. Individuals may be categorized into sub-populations of those who are chronically ill, those with deteriorating health, and those who are generally healthy or are improving their health. Although all of these sub-populations share a common linear growth function, they have mean-ingfully different average rates of change (and/or different initial levels). These differences result in different patterns of growth, or trajectory heterogeneity. Second, sub-populations of individual trajectories for a given attribute (e.g., marital quality or alcohol use) may have different functional forms of growth (e.g., linear and curvilinear). These different functional growth forms produce unique growth patterns resulting in trajectory heterogeneity.

POINT TO REMEMBER…

Unlike a conventional growth curve model, where it is assumed that covariates (predictors and outcomes) are associated with the growth parameters of *all* individuals in the *same* way, the mixture modeling approach allows covariates to be differentially associated with the growth parameters of individuals depending on their trajectory class membership.

In the area of adolescent psychopathology, our data demonstrate that three distinct sub-populations for adolescent depressive symptoms trajectories exist. These sub-population trajectories represent normative (chronically low) levels of depressive symptoms, escalating depressive symptoms, and recovering (decreasing) depressive symptoms. Panel A of Figure 7.1 illustrates the observed individual trajectories of depressive symptoms for all individuals in the sample. This figure does not show a clear overall trend in depressive symptoms over time. Panel B of Figure 7.1 shows some randomly selected trajectories of depressive symptoms

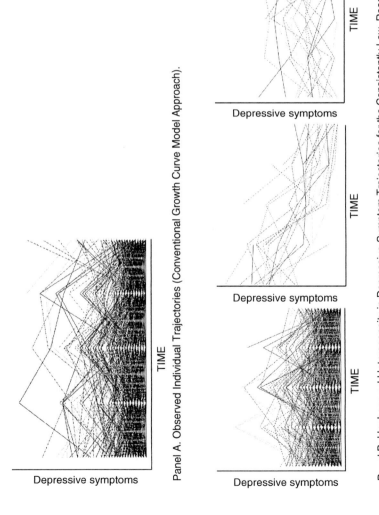

Depressive symptoms

TIME

Panel A. Observed Individual Trajectories (Conventional Growth Curve Model Approach).

Depressive symptoms

TIME

Depressive symptoms

TIME

Depressive symptoms

TIME

Panel B. Unobserved Heterogeneity in Depressive Symptom Trajectories for the Consistently Low, Recovering, and Escalating Classes.

FIGURE 7.1 Individual Depressive Symptom Trajectories for Members of Each Trajectory Class.

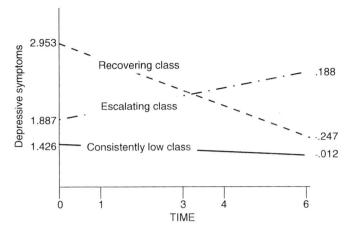

FIGURE 7.2   Estimated Mean Parameters of Latent Trajectory Classes Assessing Youth Depressive Symptoms.
Note: Unstandardardized coefficients are shown.

belonging to the three identified heterogeneous patterns, or trajectory classes. The individuals in these classes exhibited different patterns of change with some within-class variations. Figure 7.2 shows the estimated average trajectories of these identified trajectory groups or classes. These findings suggest that assuming the existence of a single population (rather than several sub-populations) and/or a common growth function is oversimplifying the heterogeneous growth of unique sub-populations.

Growth Mixture Modeling (GMM)

Recall that a conventional LGCM is based on standard normal distributional assumptions, which may be violated by combining sub-populations with different trajectories into a single LGCM. A GMM assumes that the population consists of distinct sub-groups of trajectories (that is, a "mixture" of distributions), thereby accommodating for heterogeneous trajectories or latent classes of trajectories. Thus, a GMM is more appropriate than a conventional latent growth curve model for the investigation of individual trajectories when there is heterogeneity in the trajectories of interest (Muthén & Muthén, 2000).

As shown in Figure 7.3, a GMM accommodates for this heterogeneity by extending a conventional growth curve model to incorporate categorical latent classes (C) in addition to continuous latent growth factors (level and slope) within the same modeling framework. (Note that C represents latent classes of individual trajectories.) Thus, a GMM fully captures inter-individual variation (i.e., within-class variation) in individual growth trajectories like a conventional

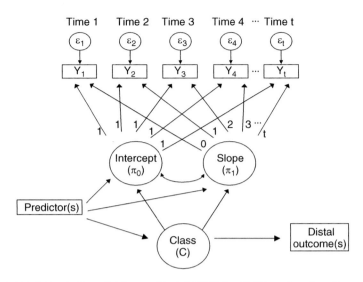

FIGURE 7.3 A Growth Mixture Model (GMM) with Time-Invariant Covariates (TICs).

growth model, while also taking into account between-class variation (i.e., unobserved heterogeneity; Jung & Wickrama, 2008). In other words, a GMM assumes that the mixture of within-class normal distributions corresponds to several classes that have between-class variation (Bauer & Curran, 2003; Feldman et al., 2009). Thus, as shown in Figure 7.3, the main advantage of the GMM approach is the ability to specify multiple covariates (i.e., predictors or distal outcomes) in order to explain both within- and between-class variability simultaneously. Because a GMM assumes a mixture of normal distributions, this allows for the existence of within-class normality (Bauer & Curran, 2003; Feldman et al., 2009).

However, under some circumstances, there may be insufficient variability to estimate the within-class variances for one or more classes. Additionally, estimating a GMM requires the estimation of many model parameters (e.g., means and variances of growth parameters across all classes) simultaneously. The large number of model parameters estimated often leads to estimation and/or convergence problems. Under these conditions, a restricted version of a GMM, known as a latent class growth analysis (LCGA) may be appropriate, and in some instances, a partially restricted version of a GMM may be appropriate. There are also theoretical reasons to prefer a LCGA over a GMM in some instances (Nagin, 1999).

Latent Class Growth Analysis (LCGA): A Simplified GMM

A latent class growth analysis (LCGA) can be recognized as a reduced nested model of a GMM as depicted by Figure 7.4. A LCGA eliminates all within-class variations ·(i.e., variances and covariances) by constraining those variations to zero (Nagin,

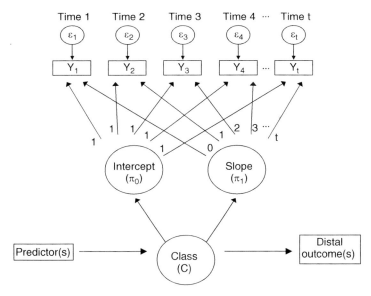

FIGURE 7.4   Latent Class Growth Analysis (LCGA) with a Time-Invariant Covariate (TIC).

1999). By setting the within-class variations to zero, a LCGA assumes within-class homogeneity; that is, all individuals within the same class share the same trajectory (Feldman et al., 2009). Thus, as shown in Figure 7.4, a LCGA only allows covariates (i.e., predictors or distal outcomes) to be associated with between-class variability because it assumes within-class homogeneity. This assumption of within-class homogeneity may be theoretically unrealistic (Muthén, 2002). Moreover, compared to a GMM, a LCGA often requires more classes to describe the underlying complexity of trajectories (i.e., spurious classes). This increased number of classes makes the generalizability of the sample classes to the full population questionable (Bauer & Curran, 2004; Shiyko, Ram, & Grimm, 2012).

In general, a GMM and a LCGA are similar; in that, the purpose of both is to describe unobserved heterogeneity; however, as discussed previously, the two approaches differ slightly in how the heterogeneity is defined and estimated. When selecting one approach over the other, it is important to carefully determine which approach is a better fit to the data and provides more substantively meaningful classes. We will provide more guidance on how to select the optimal class model later in this chapter.

Specifying a Growth Mixture Model (GMM)

As we previously discussed, a LGCM assumes a single population where (a) the trajectory shape or form is the same across individuals for a given attribute, and (b) predictors exert a similar effect on all individuals. In the following formulation, as we discussed in Chapter 2, we use a composite measure of internalizing symptoms (IS) as a personal attribute that changes over time. Recall the following equations for a LGCM for the total population:

$$\text{Internalizing Symptoms (IS)}_{it} = \pi_{0i} + \pi_{1i}\, t + \varepsilon_{it}, \quad \varepsilon_{it} \sim \text{NID}\,(0,\, \sigma_t^2) \quad (7.1)$$

where there are i = 1, 2, ..., n respondents, and ε_{it} = the error term for the ith individual at time t (assuming that error, ε, is normally and independently distributed with mean of 0 and a variance of σ^2). Each person's intercept (π_0) and · slope (π_1) can be combined with all other persons' intercepts and slopes (i.e., an inter-individual process) so that we have as many intercepts and slopes as we have study participants (n). The univariate (i.e., unconditional) version of this model can be written as:

$$\pi_{0i} = \mu_{00} + \zeta_{0i}, \quad \text{where } \zeta_{0i} \sim \text{NID}(0, \psi_{00}) \quad (7.2)$$

$$\pi_{1i} = \mu_{10} + \zeta_{1i}, \quad \text{where } \zeta_{1i} \sim \text{NID}\,(0, \psi_{11}) \quad (7.3)$$

$$\psi = \begin{bmatrix} \psi_{00} & \\ \psi_{10} & \psi_{11} \end{bmatrix} \quad (7.4)$$

where there are i = 1, 2, ..., n respondents in the study, μ_{00} and μ_{10} represent the average intercept and slope, and ζ_{0i} and ζ_{1i} indicate the errors of the estimated intercepts and slopes around their means (assuming that the errors are normally and independently distributed). ψ_{00} and ψ_{11} represent the error variance of the intercept and slope, respectively. ψ_{10} denotes the covariance between π_{0i} and π_{1i}. When heterogeneity (i.e., classes of individual trajectories) exists, the above equations have to be class-specific; that is, the equations are indexed by the class number.

Like a latent class analysis (LCA), a GMM assumes a categorical latent variable, C, that specifies each individual's membership in a certain class. In a GMM, individuals are assigned to one of the categories (or latent classes) of distinct growth patterns (intercept and slopes) according to their own growth pattern. This assignment is based on "posterior probabilities," which are similar to factor scores. That is, a GMM extends a LGCM by using a categorical latent variable "to represent a mixture of classes of sub-populations" (Li, Duncan, Duncan, & Acock, 2001, p.494). Posterior probabilities for each person for all of the classes are inferred from the data. In other words, a posterior probability is calculated for each class for each person, and the person is classified into the class for which he/she has the highest posterior probability. The equation for calculating posterior probabilities of latent class membership is shown in the *Point to Remember* box about posterior probabilities later in this chapter.

Specifying Trajectory Classes: Class-Specific Equations

The second-level equations (i.e., structural model for individual i in class k), are indexed by class k, and are written as:

$$\pi_{k0i} = \mu_{k00} + \zeta_{k0i}, \quad \text{where } \zeta_{k0i} \sim \text{NID}(0, \psi_{k00}) \tag{7.5}$$

$$\pi_{k1i} = \mu_{k10} + \zeta_{k1i}, \quad \text{where } \zeta_{k1i} \sim \text{NID}(0, \psi_{k11}) \tag{7.6}$$

$$\Psi_k = \begin{bmatrix} \psi_{k00} & \\ \psi_{k10} & \psi_{k11} \end{bmatrix} \tag{7.7}$$

where μ_{k00} and μ_{k10} represent the average intercept and slope in latent trajectory class k; ζ_{k0i} and ζ_{k1i} are the disturbances (i.e., errors) reflecting the variability of the estimated intercepts and slopes across individuals within the latent trajectory class k. These disturbances have a variance-covariance structure represented by ψ_{k10}. ψ_{k00} and ψ_{k11} are the error variances of the estimated intercept and slope factors, and ψ_{k10} is the covariance between the intercept and slope factors within the latent trajectory class k. The k subscript indicates that most parameters are allowed to vary between the estimated latent trajectory classes. Thus, each latent trajectory class could be defined by its own LGCM based on class-specific parameters consisting of (1) variance and covariance structures (ψ_k) of growth factors, (2) means (i.e., μ_{k00} and μ_{k10}) of growth factors, and (3) residual variances of manifest indicators (i.e., σ_{kt}^2). This GMM specification is shown in Figure 7.5.

Specifying a Latent Class Growth Analysis (LCGA)

As previously indicated, a latent class growth analysis (LCGA) is a special type of GMM, referred to as semiparametric group-based modeling (Nagin, 1999). This type of GMM only estimates means and fixes within-class variances to zero. The first-level equation is the same as that of a GMM, but the second-level equation (i.e., the structural model) is modified by removing the within-class variances:

$$\pi_{k0i} = \mu_{k00} \tag{7.8}$$

$$\pi_{k1i} = \mu_{k10} \tag{7.9}$$

Fixing these variance-covariance estimates to zero within each class assumes that all individual trajectories within a class are homogeneous. More importantly, the zero constraints on the variance estimates for all classes allow for faster model convergence with less computational burden (Kreuter & Muthén, 2007).

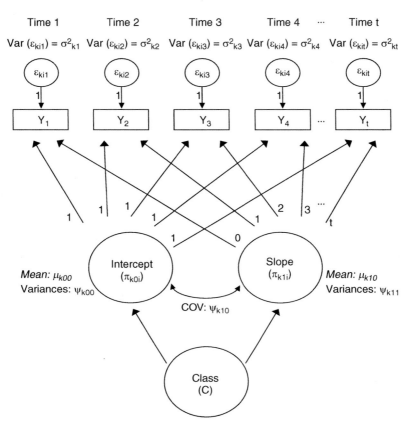

FIGURE 7.5 Specifying an Unconditional Growth Mixture Model (GMM).

POINT TO REMEMBER...

Posterior probabilities

The posterior probability, which is the probability of an individual belonging to any given class, is calculated by maximum likelihood estimation using observed variables. The posterior probability of an individual belonging to class k is calculated by the following equation (Vermunt & Magidson, 2002).

$$P(c = k \mid Y = y) = \frac{P(c = k)P(Y = y \mid c = k)}{P(Y = y)} \tag{7.10}$$

where y represents the growth pattern, and c represents a latent class variable with k levels (k = 1,2, ..., K). The left side of the equation P(c = k | Y = y) denotes the posterior probability of an individual with a growth pattern of y belonging to class k. This posterior probability can be obtained by the Bayes rule as shown in Equation 7.11.

- $P(c = k) \, P(Y = y \mid c = k)$ is the joint probability of belonging to class $k, P(c = k)$.

Individuals are assigned to the class for which they have highest posterior probability (Vermunt & Magidson, 2002, p. 95). Thus, equations for a traditional LGCM can be extended to incorporate latent classes of sub-populations by incorporating a multiplying factor $\sum_{k=1}^{k} P(c = k)$ to the equation.

Using our example model with internalizing symptoms (IS), the first-level LCGA equation (i.e., measurement model) for individual i at time t in class k can be written as:

$$(IS)_{kti} = \sum_{k=1}^{k} P(c = k) \, [\, \pi_{k0i} + \pi_{k1i} \, t + \varepsilon_{kti} \,], \quad \varepsilon_{kti} \sim NID \, (0, \, \sigma_{kt}^{2}) \quad (7.11)$$

The probability that individual i belongs to category $k, P(c = k)$ ranges between 0 and 1 with the constraint that their sum equals one. The equation in the square bracket represents a LGCM for class k with an intercept and slope of π_{k0i} and π_{k1i}, respectively. σ_{kt}^{2} represents the error variance at each point in latent trajectory class k. These error variances are assumed to be normally distributed with means of zero and are assumed to be uncorrelated with other residuals (i.e., a mixture of normal distributions; Bauer & Curran, 2003).

Building A Growth Mixture Model (GMM) Using M*plus*

In order to estimate unbiased model parameters, we recommend following the five steps below to build a GMM using M*plus*.

Step One
    Specify a traditional growth curve model (LGCM)
Step Two
    Perform a latent class growth analysis (LCGA)
Step Three
    Specify a growth mixture model (GMM)
Step Four
    Address estimation problems
Step Five
    Select the optimal unconditional model

Specify a Traditional Growth Curve Model (LGCM) (Step One)

First, before fitting a GMM to the data, we recommend estimating a conventional LGCM in order to determine if the LGCM assumptions of uniform variation with a common growth function are valid. That is, fitting a LGCM to the data allows the researcher to examine the overall fit of the data to a trajectory with a single shape or form before taking into account trajectory heterogeneity. We recommend examining two important model results from this stage: model fit indices and variances of the growth factors. Poor fit indices and statistically significant variances of growth factors suggest the existence of heterogeneous trajectories, or sub-populations, with distinct trajectories.

As discussed in Chapter 2, M*plus* syntax for a linear latent growth curve model · (LGCM) is written as:

DATA: FILE=example.ch_7.dat;
VARIABLE: NAMES = ID IS1-IS5;
MISSING ARE ALL (9.00);
MODEL:
I S | IS1@0 IS2@1 IS3@3 IS4@4 IS5@6;
OUTPUT: SAMPSTAT MOD (5.00);

Note that IS1–IS5 indicate internalizing symptoms at Times 1, 2, 3, 4, and 5. Factor loadings are specified for a linear change proportionate to the unequal time intervals of the current example data (t = 0, 1, 3, 4, and 6). For descriptive purposes, we specified the linear growth function model (without autocorrelated errors) as the optimal function of internalizing symptoms across time. However, as we discussed in Chapter 2, the growth function could be non-linear, such as a quadratic, cubic, or piecewise growth function. Thus, using a nested model comparison test ($\Delta\chi^2$), the optimal growth function can be identified. Regarding the model results, the model fit indices suggested that the LGCM fit the data poorly (χ^2(df) = 70.741(10), p < .001; RMSEA (90%) = .118 (.093, .145); CFI/ TLI = .892/.892; SRMR = .067). (Recall that recommended cut-off values for these indices are given in Chapter 2.) Furthermore, the variances for the growth factors were statistically significant (.165, p < .001 and .004, p <.001, for the intercept and slope factors, respectively), indicating significant variation in the growth function between individuals. Together the poor model fit and the significant variances provide evidence for the existence of heterogeneity in longitudinal changes of internalizing symptoms.

Estimating a Latent Class Growth Analysis (LCGA) (Step Two)

As previously indicated, a latent class growth analysis (LCGA) is a restricted or reduced version of a GMM, which estimates means, but fixes within-class variances to zero.

Title: Latent Class Growth Analysis (LCGA)
DATA: File is C:\example.ch_7.dat;
   LISTWISE=ON;
VARIABLE: names are ID IS1-IS5;
   usevar = IS1-IS5;
   CLASSES=C(2);
   IDVARIABLE=ID;
   missing = all (9);

SAVEDATA: FILE=C:\LCGA.txt;
   SAVE=CPROB;

ANALYSIS: **TYPE=MIXTURE;**
   STARTS = 500 10;
   STITERATIONS = 10 ;
   LRTBOOTSTRAP=50;

MODEL:**%OVERALL%**

I S | IS1@0 IS2@1 IS3@3 IS4@4 IS5@6;
I-S@0;

OUTPUT: SAMPSTAT STANDARDIZED MOD (5.00) **TECH7 TECH8 TECH11 TECH13 TECH14;**
PLOT: SERIES = IS1-IS5(S); TYPE=PLOT3;

LISTWISE = ON is available only when multivariate normality is tested (**TECH13**).

The **CLASSES** option defines "C" and specifies the number of classes in the mixture model. The **IDVARIABLE** option pertains to the **SAVE DATA** command.

The **SAVE DATA** command specifies the .txt file name to be created containing classification information (individual posterior probabilities and each subject (i)'s latent class number).

TYPE=MIXTURE is required for mixture models.

STARTS gives Mplus instructions about how many randomly selected starting values to use.

STITERATIONS is used to specify the maximum number of iterations allowed in the initial stage.

LRTBOOTSTRAP is used in conjunction with the TECH14 option to specify the number of bootstrap draws to be completed when estimating the *p*-value of the true 2 time log-likelihood distribution.

%OVERALL% specifies an overall mixture model.

An LCGA is specified by fixing variances of intercept and slope factor to zero.

TECH7 requests the sample statistics for each class.
TECH8 requests the optimization history (i.e., iteration process).
TECH11 requests the Lo-Mendell-Rubin likelihood ratio test that compares k-1 class model to the k class model.
TECH13 requests the multivariate skewness and kurtosis test.
TECH14 requests a parametric bootstrapped likelihood ratio test that compares the k-1 class model to k class model.

FIGURE 7.6 M*plus* Syntax for a Latent Class Growth Analysis (LCGA) Model.
Note: I = Initial Level. S = Slope. IS = Internalizing Symptoms.

Illustrative Example 7.1: Mplus Syntax for a Latent Class Growth Analysis (LCGA)

In M*plus*, an unconditional LCGA model can be specified using the syntax shown in Figure 7.6. The bold portions of the syntax are the parts that add the LCGA to the existing univariate LGCM.

The TYPE=MIXTURE option specifies the mixture model algorithm. By specifying CLASSES = C(2), a two-class model is examined. If the model converges successfully and the results are interpretable, the number of classes can be increased by specifying CLASSES = C(3) followed by CLASSES = C(4), CLASSES = C(5), and so on until the optimal class model is selected.

The STARTS and STITERATIONS syntax commands are not required for estimating a LCGA because M*plus* automatically sets these parameters at default values. Unless other values are specified, M*plus* assumes 20 random sets of starting values and 4 optimizations in the final stage. However, adjusting these values may aid in successful model convergence. These convergence issues will be discussed in detail in a later section (Step 4: Addressing estimation problems).

In the OUTPUT command, TECH7 can be specified to request sample statistics for each class. TECH8 provides the optimization history (i.e., iteration process), which is used in conjunction with STARTS and STITERATIONS. Specifying TECH11 requests the Lo-Mendell-Rubin likelihood ratio test (LMR-LRT) of model fit (Lo, Mendell, & Rubin, 2001). If TECH13 is specified, univariate, bivariate, and multivariate skewness and kurtosis values are provided in the output (i.e., Mardia's measure of multivariate normality; Muthén, 2003). TECH14 provides the bootstrapped likelihood ratio test (BLRT). We will return to these tests when we introduce how to select the optimal class model (Step 5: Selecting the optimal class model).

Finally, the PLOT command can be used to request graphical displays of the model results. The observed and estimated means and trajectories for each class can be viewed in the graphs provided in the M*plus* output. Using the dropdown menus in M*plus* allows the user to view the graphs specified by the PLOT syntax (Plot → View Plots → Estimated Means and Observed Individual Values).

Specifying a Growth Mixture Model (GMM) (Step Three)

In M*plus*, using our example model (i.e., composite mean scores of internalizing symptoms at multiple time points), a growth mixture model (GMM) can be specified by deleting the syntax constraining the intercept and slope variances (I-S@0) from the LCGA syntax (see Figure 7.6). Now, all growth parameters (i.e., means and growth factor variances and covariances) within the classes are freed and estimated.

However, M*plus* automatically constrains all variances and covariances of the growth factors to be equal across classes by default (i.e., variance-covariance equality). This variance-covariance equality of growth parameters may lead to problems. Simulation studies have shown that similar restrictions (i.e., equal variances-covariances) could result in the over-extraction of latent classes and biased parameter estimates (Bauer & Curran, 2004; Enders & Tofighi, 2008; Magidson & Vermunt, 2004). Thus, it may be necessary to free some (or all) of the within-class variances.

Illustrative Example 7.2: Mplus Syntax for a Growth Mixture Model (GMM)

A GMM can be estimated by freeing the variances and covariances (deleting "I-S@0;" from the LCGA syntax) and adding the syntax below to the model command of the LCGA syntax (see Figure 7.6):

MODEL: %OVERALL%
I S | IS1@0 IS2@1 IS3@3 IS4@4 IS5@6;
%C#1%
[I–S](M1–M2);
I-S (V1–V2);
I WITH S (COV1);
%C#2%
[I–S](M3–M4);
I-S (V3–V4);
I WITH S (COV2);
MODEL TEST:
M1=M3;

The bold commands allow the unique variances and covariances of the intercepts and slopes to be freely estimated for each class. Also, these syntax options make it possible to test hypotheses about all growth parameters (e.g., mean and variance-covariance parameter equality across classes or invariance tests of growth parameters) using a Wald chi-square test (i.e., likelihood ratio test, LRT; Morin et al., 2011). This test is one of the statistical tools often utilized to select the optimal modeling approach when choosing between a GMM and a LCGA or between a GMM with equal variances-covariances in all classes and one with freely estimated variances-covariances. In our example, we assigned names in parentheses to all growth parameters (M1, M2, V1, etc.). The MODEL TEST command indicates our intent to test a series of parameter constraints. The lines of text following the MODEL TEST command provide the specific constraints to be tested. For descriptive purposes, in the current example, we specify the null hypothesis that M1 = M3. This corresponds to the null hypothesis that the mean trajectories of IS are equal across two classes.

Also, similar to the LCGA approach, the variance and covariance of specific growth parameters can also be fixed to zero (a partially constrained GMM) by adding the appropriate syntax under each class specification. For example, to fix the growth parameters of Class 2 to zero the following syntax is added:

%C#2%
I-S@0; I WITH S@0;

By assigning labels to the variance and covariance parameters, these constrained variances-covariances can be tested using a Wald chi-square test (i.e., specifying MODEL TEST in the model command).

Regardless of the GMM specification approach employed, it is important to remember that specifying a model from an exploratory perspective does not always lead to theoretically consistent outcomes. Consequentially, when employing an exploratory approach to model specification, results should be treated circumspectly as decisions about class identification should be based on theory, interpretability, prior research, and practical considerations (e.g., model convergence and computational time speed).

Addressing Estimation Problems (Step Four)

Mixture models often experience certain estimation difficulties, which lead to a non-optimal or non-estimated class solution. Previous literature has reported several main reasons for these estimation problems with mixture modeling. First, the probability distribution of the indicators used for constructing the classes may not be normal (Bauer & Curran, 2003). Second, there can be a convergence problem related to a "local solution" (Hipp & Bauer, 2006). Third, the model may not converge (non-convergence) due to the inappropriateness of data and non-identification of the model (Collins & Lanza, 2010; Wurpts & Geiser, 2014).

Estimation Problems Related to a Non-normal Probability Distribution

When estimating a GMM, the number of latent classes identified may be incorrect due to the nature of the distribution of the observed data (e.g., multivariate skewness or non-normality; Bauer & Curran, 2003). In conventional growth modeling, it is assumed that the repeated measures are normally distributed (i.e., multivariate normality). This implies that the random coefficients (intercept and slope) and time-specific residuals are also normally distributed (Verbeke & Molenberghs, 2000). However, based on the statistical theory of finite normal mixture models, in a GMM the normality assumption is relaxed for the second-level equation (see Equations 7.2 and 7.3) in order to accommodate the mixture of several normal distributions corresponding to the different classes. These classes may have different mean structures and variance-covariance structures. That is, in a GMM, normality is only assumed within the classes (i.e., within-class normality), and heterogeneity is a characteristic of the residuals. If key variables are not included (like the distinction between two groups), the residuals may appear heterogeneous, but, in actuality, the apparent heterogeneity can be reduced or eliminated by incorporating information about the classes.

However, when the repeated measures data are not normally distributed, finite normal mixture models, such as a GMM, allow the estimation of latent class trajectories even without population heterogeneity. Particularly, there is the possibility of over-extracting latent trajectory classes in a homogeneous non-normal distribution (Bauer & Curran, 2003). Using a simulation study, Bauer and Curran (2003) showed that non-normality in the overall sample distribution leads to the over-extraction of classes. This over-extraction makes generalizing the finding of heterogeneous classes to the larger population doubtful.

Illustrative Example 7.3: A Non-Normal Distribution

Figure 7.7 is an example of a non-normal population distribution, which may lead to the incorrect identification of latent classes when estimating a GMM. Using mothers' hostility at Time 3 from the parent data set, Figure 7.7 shows non-normal data (skewness: 2.964). According to the model fit indices for a GMM utilizing these data, two classes were identified. These two classes most likely correspond to the two components of the distribution depicted by the broken lines with round dots and longer dashes. However, the second class only represents the long tail at

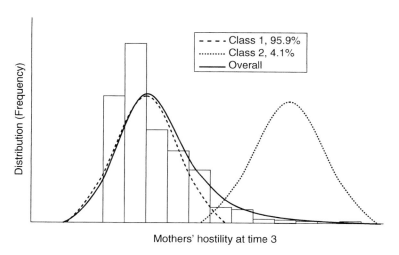

Mothers' hostility at time 3

FIGURE 7.7 Observed and Two-Class Normal Fitted Distribution of the Mothers' Hostility Measure at Time 3.
Note: Overlaid on the histogram are the probability distributions for a two-component normal mixture distribution (broken round dot and dash lines assume the GMM has a within-class normal distribution) and a skewed univariate distribution (unbroken line).

the end of the overall distribution (shown by the unbroken line). The existence of this class at the population level (with a normal distribution) is questionable.

In order to increase the accuracy of class identification (goodness of fit, Tofighi & Enders, 2007), we recommend checking the normality of the data in two ways before selecting the optimal class model. First, general normality indicators should be examined (i.e., multivariate normality; goodness of model fit). In order to assess multivariate normality, Hayduk (1987) suggested that each univariate distribution should have a normal distribution. Thus, the univariate skewness and kurtosis statistics should be investigated. Values less than ± 2.00 are generally accepted as a normal distribution (George & Mallery, 2010). Second, Muthén (2003) offered a statistical test known as the multivariate skewness/kurtosis test (SK test). This test compares the multivariate skewness and kurtosis values implied by the k-class model to those obtained from the observed data (Tofighi & Enders, 2007). Obtaining low (statistically significant) p-values for the SK test indicates that the skewness and kurtosis values of the k-class model differ from the corresponding sample distribution, which implies the existence of distinct modes (peaks). Thus, after selecting the optimal class model, a non-significant p-value is desired, because this suggests that the observed distribution meets the assumption of multivariate normality. This test is easily generated in M*plus* by specifying TECH13 in the OUTPUT command when the LISTWISE option of the DATA command is set to ON (see Figure 7.6).

Multiple model fit indices should be used to carefully select the optimal class model. Information criteria (IC) statistics have been conventionally used to select the optimal class model. However, Bauer and Curran (2003) reported that the IC statistics, especially the Akaike information criterion (AIC), often indicate that a multi-class model is superior to a single-class model despite the presence of a single class in the overall population. Thus, a more reliable approach to selecting the correct number of classes is to use multiple model fit indices. For example, Muthén (2003) recommended using the Lo-Mendell-Rubin likelihood ratio test (LMR-LRT, Lo et al., 2001) to take the correct distribution into account. Furthermore, Nylund, Asparouhov, and Muthén (2007) and Tofighi and Enders (2007) both showed that the sample size adjusted Bayesian information criterion (SSABIC) accurately identifies the number of latent classes. Most importantly, strong substantive theory and a meaningful interpretation of the sub-populations (classes) should support the empirical selection of classes. We return to these model fit indices in more detail in the next section (Step 5: Selecting the optimal class model).

POINT TO REMEMBER...

Non-normal growth mixture modeling

Muthén and Asparouhov (2015) recently proposed a non-normal GMM, which uses a skewed t-distribution to accommodate a skewed overall distribution. They contend that non-normal mixture models can fit non-normal data considerably better than normal mixtures. By appropriately selecting a more parsimonious model, these non-normal models reduce the risk of extracting spurious latent classes due to non-normality. According to this approach, three specific distributions can be specified: skew-normal (SKEWNORMAL), t (TDIST), and skew-t (SKEWT) distributions. The skew-normal distribution accounts for excessive skewness in the observed distribution. Excessive kurtosis can be accounted for using the t-distribution specification. The skew-t distribution accounts for *both* excessive skewness and kurtosis. Non-normal mixture modeling can be specified using the DISTRIBUTION = TDIST / SKEWNORMAL / SKEWT option in the ANALYSIS command. Below is the M*plus* syntax:

ANALYSIS: TYPE=MIXTURE;
DISTRIBUTION = SKEWT;

Estimation Problems Related to Local Maxima

A "local maxima" or "local solution" is obtained when the highest log-likelihood value is not replicated. (In the estimation process, the highest log-likelihood value should be replicated when the optimal class solution is achieved.) Thus, "local maxima" problems result in incorrect class solutions.

POINT TO REMEMBER…

Local maxima

With maximum likelihood estimation (MLE), the model solution is provided when the likelihood ratio reaches a replicable maximum value through an iterative optimization process. This iterative optimization process consists of two stages: (1) the initial stage, which is generated from the given starting values and (2) the final optimization stage (Myung, 2003). The iterative optimization process could stop prematurely and return a sub-optimal set of parameter values depending on the choice of the initial starting values (chosen either at random or by the researcher). This is called the "*local maxima*" problem.

Figure 7.8 depicts this problem. It is through this iterative optimization process that the starting log-likelihood parameter value at point *b* will lead to the optimal point B (called the global solution; the highest log-likelihood value). However, the starting value at point *a* may lead to point A, which is a local-optimal solution. Similarly, the starting log-likelihood parameter value at *c* may lead to another sub-optimal solution at point C. When analyzing a GMM, the iterative optimization with global maximum (point B) yields the proper number of classes, whereas the optimizations with local maxima (points A and C) yield an incorrect number of classes.

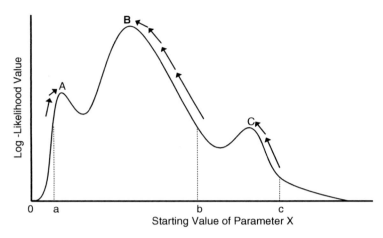

FIGURE 7.8 A Schematic Plot of the Log-Likelihood Functions for a One-Parameter Model.
Note: Point B is the global maximum (the highest log-likelihood value), whereas points A and C are two local maxima. The a, b, and c points represent starting values, and the arrows depict an iterative optimization process (adapted from Myung, 2003).

It is important to run the model with multiple random sets of starting values to guarantee that the global solution is obtained while avoiding "local maxima" (Hipp & Bauer, 2006). The global solution is the solution with the highest log-likelihood value (i.e., the best solution), which should be replicated at least twice in the final solution (Jung & Wickrama, 2008). When the highest log-likelihood value cannot be replicated in the final solution, it is possible that a local solution has been reached, and the results may not be trustworthy. If the model does not replicate the highest log-likelihood value (i.e., the best log-likelihood value), the following message is given in the M*plus* output:

WARNING: THE BEST LOG-LIKELIHOOD VALUE WAS NOT REPLICATED. THE SOLUTION MAY NOT BE TRUSTWORTHY DUE TO LOCAL MAXIMA. INCREASE THE NUMBER OF RANDOM STARTS.

POINT TO REMEMBER...

Determining if the model has a global solution

Recall that, in order to find the highest log-likelihood value in M*plus*, an iterative optimization process is carried out in two stages known as the initial stage and the final optimization stage. The ending values from the optimization with the highest log-likelihood value in the initial stage are used as the starting values in the final optimization stage (default = 4). The default M*plus* syntax for this maximum likelihood optimization is as follows:

ANALYSIS: TYPE=MIXTURE;

STARTS = 20 4;
STITERATIONS = 10;

where STARTS = 20 4 specifies 20 random sets of starting values and 4 final optimizations. STITERATIONS = 10 specifies 10 as the maximum number of iterations allowed in the initial stage. The estimation process history can be examined by specifying TECH8 in the output command.

Instead of using these default values, the user can choose to manually specify the number of random sets of starting values in the initial stage and the number of optimizations to use in the final stage. Jung and Wickrama (2008) suggested that between 100 and 500 random sets of starting values are necessary to locate the global solution (i.e., the highest log-likelihood value as opposed to a local solution).

We investigated whether our example model (using repeated composite measures of adolescents' internalizing symptoms (indicated by IS) from Time 1 to Time 5) has a local solution when estimating a GMM to assess a 4-class solution with a linear growth function. Below is the M*plus* syntax:

DATA: File is C:\example_ch_7.dat;
VARIABLE: NAMES are ID IS1-IS5;
USEVAR = IS1-IS5;
CLASSES = C (4);
MISSING = ALL (9);
ANALYSIS: TYPE=MIXTURE;
STARTS = 500 10;
STITERATIONS = 10;
MODEL: %OVERALL%
I S | IS1@0 IS2@1 IS3@3 IS4@4 IS5@6;
OUTPUT: TECH8;

The syntax is similar to the syntax shown in Figure 7.6 except that the variances and covariances of growth parameters across classes are now estimated by eliminating "I-S@0" from the syntax to change the model from an unconditional LCGA to a GMM. The M*plus* output on the next page provides the results from the 10 final optimizations including the log-likelihood values, seeds, and start numbers from the initial stage.

RANDOM STARTS RESULTS RANKED FROM THE BEST TO THE WORST LOG-LIKELIHOOD VALUES

Final stage log-likelihood values at local maxima, seeds, and initial stage start numbers:

-887.040	***850840***	232
-887.040	***283492***	435
-887.040	923437	398
-887.040	802779	122
-887.040	902278	21
-887.040	605358	321
-887.040	65651	214
-898.275	481835	57
-898.275	900268	327
-898.275	518828	432

If the model successfully replicates the highest log-likelihood value, the output will include the following message:

THE BEST LOG-LIKELIHOOD VALUE HAS BEEN REPLICATED. RERUN WITH AT LEAST TWICE THE RANDOM STARTS TO CHECK THAT THE BEST LOG-LIKELIHOOD IS STILL OBTAINED AND REPLICATED.

As the message indicates, even with successful replications of the highest log-likelihood value, it is necessary to determine whether this solution is the global solution. To assess this, the OPTSEED syntax can be utilized on the seed values from the highest log-likelihood values. (Seed values are random variables that determine the starting log-likelihood values.) This secondary analysis determines if the global solution has been found by attempting to replicate the same estimates again. M*plus* rank orders the log-likelikehood values from the best value to the worst. If the solution is replicated, the model parameters are estimated based on the global solution. The non-replication of the solution may be evidence of a local solution. In our example, as can be seen, most of the log-likelihood values are relatively close in value to one another. However, in order to replicate the same solution, the first two values (850840 and 283492) were used as seed values. Using the syntax below, we re-tested whether the estimation is replicated when specifying these seed values separately in ANALYSIS command:

ANALYSIS: TYPE = MIXTURE;
OPTSEED = 850840

After running this model, we re-ran the model using OPTSEED=283492 (the second highest seed value), and then compared the results of all three outputs. If the results from the original model are similar to the two models with the seed values from the highest log-likelihood values, then there is no evidence of a "local maxima" problem.

Estimation Problems Due to Model Non-Identification and Inappropriate Data

When the model is not identified (i.e., an under-identified model), the model does not converge (non-convergence), and standard errors (and related p-values) along with other meaningful estimates are to estimated. Non-identification is often due to model complexity; that is, models often fail to converge when too many parameters are simultaneously estimated in the model. In these instances, an error message is printed in the output, and the user needs to re-specify the hypothesized model, most commonly by constraining some parameters. Additional restrictions on parameters are needed to make the model identifiable (Muthén & Muthén, 2012). Non-convergence may also occur due to the use of inappropriate data, such as variables measured on different scales. In such instances of non-convergence, problems may be avoided by increasing the number of iterations and starting values (Muthén & Muthén, 2012).

Selecting the Optimal Class Model (Enumeration Indices) (Step Five)

After successful convergence and support for a global solution, the final step is to select the optimal class model. Conventionally, standard chi-square values have been recognized as the critical measuring tool for selecting the optimal model. However, in a mixture model, the chi-square values are not useful for determining the optimal model because the likelihood ratios between the k-class and k-1 class model do not follow a chi-square distribution (Feldman et al., 2009; Muthén, 2003).

As we suggested previously, it is necessary to investigate multiple model fit indices in order to select the final optimal model. So far, various statistical indices have been suggested for selecting the optimal class model. These indices fall into three categories (Feldman et al., 2009; Jung & Wickrama, 2008; Nylund et al., 2007):

(1) Information criteria (IC) statistics, such as the Bayesian information criterion (BIC) values and the sample-size adjusted BIC (SSABIC),
(2) Entropy values,
(3) Likelihood ratio tests (LRT), such as the Lo-Mendell-Rubin likelihood ratio test (LMR-LRT) and the bootstrapped LRT (BLRT) (Tein, Coxe, & Cham, 2013).

Information Criteria (IC) Statistics

IC statistics are generally computed using the deviance statistics (-2 log-likelihood values), the number of model parameters (p), and sample size (n). Thus, the number of parameters (as an indicator of high model complexity) and small sample size are taken into account as a parameterization penalty factor when IC statistics are estimated. In general, the BIC and SSABIC statistics have been reported as "better" performance indices than the AIC statistic (Peugh & Fan, 2012). While there are no cut-off points for these statistics, lower BIC and SSABIC values indicate a better model fit (Feldman et al., 2009).

Entropy and Average Posterior Probabilities

Entropy is a standardized index (i.e., ranging from 0 to 1) of model-based classification accuracy. Higher values indicate improved enumeration accuracy, which signals clear class separation (Feldman et al., 2009; Nagin, 2005). Entropy is estimated based on the average posterior probabilities (Muthén, 2004). Thus, an entropy value of zero indicates that the posterior probability of class membership does not vary across classes (no class separation), while an entropy value of one indicates that for any given individual, there is perfect posterior probability for membership in one class (a posterior probability value of 1 for that class) and a posterior probability value of 0 for membership in other classes (prefect class separation). According to Clark and Muthén (2009), entropy values of .40, .60, and .80 represent low, medium, and high class separation. Not only is entropy provided in the Mplus output, but the average posterior probability for all of the classes can be directly examined as well. The average probabilities of each class can be found under the "Model fit information" section (shown as "Average latent class probabilities for most likely latent class membership (row) by latent class (column)"). The average probabilities of each class are shown in matrix form. Diagonal components indicate the average posterior probabilities of an assigned class. Like the entropy statistic, in general, high probabilities indicate that the classes are distinct from each other. For example, a probability of .91 suggests that 91% of subjects in the assigned class fit that category, while 9% of the subjects in that given class are not accurately described by that category (Fanti & Henrich, 2010).

Likelihood Ratio Test (LRT): LMR-LRT and Bootstrapped LRT (BLRT)

Two LRTs, the Lo-Mendell-Rubin likelihood ratio test (LMR-LRT; Lo et al., 2001) and the bootstrapped likelihood ratio test (BLRT; McLachlan & Peel, 2000) are often used for model comparison when determining the optimal number of classes. The Lo-Mendell-Rubin LRT compares two adjacent class models (k-1 class model and k-class model) (Lo et al., 2001). Like the SSABIC, an adjusted LMR-LRT (Adj. LMR-LRT) statistic is provided in Mplus (Muthén, 2004) and accounts for sample size, which often impacts inference statistics (p-values).

McLachlan and Peel (2000) also developed another LRT known as the bootstrapped LRT (BLRT). Using this approach, the likelihood ratio test (LRT) between the k-1 and k-class models is conducted through a bootstrapped procedure (Asparouhov & Muthén, 2012). The bootstrap draw can be manually specified using the LRTBOOTSTRAP option in the ANALYSIS command. The default number of bootstraps varies from 2 to 100. For example, the following syntax specifies 50 bootstrap draws.

LRTBOOTSTRAP = 50;

One hundred bootstrap draws has been suggested by past research for the estimation of a precise p-value (McLachlan & Peel, 2000). In both tests (LMR-LRT and BLRT), the 2 time log-likelihood difference value (i.e., the difference of log-likelihood values) between the k-1 and k-class models is calculated and inference statistics are provided (p-values). Statistically significant p-values indicate that the k-class model (i.e., the current model) provides a better fit than the k-1 class model (i.e., a model with one less class).

In the M*plus* output, both LRTs (LMR-LRT and BLRT) provide the difference of log-likelihood values between k-1 and k-class models with p-values. A small p-value for both statistics indicates that the current k-class model is a significantly better fit to the data than a model with one less class. In most cases, the BLRT outperforms the LMR-LRT and other IC statistics (Nylund et al., 2007; Tein et al., 2013). However, the BLRT is more sensitive to a non-normal distribution and model complexity; all of which lead to convergence problems (i.e., local solutions; Nylund et al., 2007). If a local solution is detected in either the k-1 class or the k-class model (or both models), the difference of the log-likelihood values will be biased. In turn, this leads to biased inference statistics (p-values). In these instances, the results of the BLRT are not trustworthy even if the p-value is statistically significant. Consequently, the more robust LMR-LRT is the preferable statistic (Nylund et al., 2007).

In the "Model fit information" section of the M*plus* output, IC statistics and the entropy statistic are provided. However, the LMR-LRT and BLRT are only provided when TECH11 (for LMR-LRT) and TECH14 (for BLRT) are specified in the syntax.

Other Considerations

In addition to these three categories of model fit indices used for selecting the optimal number of classes, conventional model comparison approaches can also guide the selection of the optimal class model. For example, if similar information (e.g., fit indices) is provided for both k-1 and k-class models, the more parsimonious model, the model with fewer classes, is always preferred (Feldman et al., 2009). An additional consideration is the sample size of the smallest class. If the smallest class contains less than 5.0% of the sample and/or the n of the smallest class is less than 25, the model should only be retained as the optimal model if the researcher can accurately defend what is gained from this small class given the possibility of low power and a lack of statistical precision (Berlin, Williams,

& Parra, 2014; Lubke & Neale, 2006). Perhaps the most important other factor to consider is the interpretability of each class trajectory. For this purpose, plots often provide useful information when selecting the optimal class model. When viewing the plots, each class trajectory should be clearly distinct and separate from other classes. The classes should also be clearly interpretable and consistent with theory. All of these factors should be considered not only when determining the appropriate numbers of latent class trajectories, but also when determining which analytical approach to employ (GMM vs. LCGA) (Feldman et al., · 2009). Thus, enumerating the number of latent classes in a GMM or a LCGA is not a simple identification procedure based solely on model fit indices.

POINT TO REMEMBER...

Selecting the optimal class model involves considering more than model fit indices. When selecting the optimal class model, one must also take into account the theoretical expectations, substantive meaning and interpretability of each class solution, and the need for parsimony.

Illustrative Example 7.4: Identifying the Optimal Model

Using our example model (i.e., repeated composite measures of internalizing symptoms from Time 1 to Time 5; IS1 to IS5), we illustrate how to select the optimal model. We will discuss three incremental approaches to specifying a GMM. We label these approaches as a LCGA, GMM-CI, and GMM-CV. The most reduced approach is a LCGA, which we have discussed previously. The second approach is a GMM-CI, where variances and covariances are constrained to be the same across all classes (class-invariant variances and covariances). This is the default model in M*plus*. The other approach is a GMM-CV, where variances and covariances are freed to be estimated for all classes (class-varying variances and covariances). A GMM-CV may include some constrained parameters. In order to select the optimal class model, we first examined the model fit indices for a two-class model and compared those indices with indices from a model with three classes. This process continued incrementally by fitting models with a larger number of classes until model fit ceased to improve. Furthermore, in order to avoid the possibility of a local maxima, 500 random sets at the initial stage and 10 final optimizations were specified for all models (STARTS=500 10). Note that, for model simplicity, only a linear trajectory was assessed, but other trajectory forms can also be assessed. Table 7.1 provides model fit indices for both the LCGA and GMM.

As shown in Table 7.1, we tested all possible class model approaches, including a LCGA, a GMM-CI, and a GMM-CV. Below we review the results for each of these three models separately beginning with the LCGA.

TABLE 7.1 Fit Statistics for GMMs and LCGAs (total n = 436).

Fit statistics	2 Classes	3 Classes	4 Classes	5 Classes
LCGA				
LL (No. of Parameters)	−1100.208 (10)	−1014.793 (13)	−933.441 (16)	−890.484 (19)
BIC	2261.192	2108.595	1964.125	***1896.444***
SSABIC	2229.457	2067.340	1913.349	***1836.148***
Entropy	.940	.939	.932	.851
Adj. LMR−LRT (*p*)	453.812 (.24)	161.948 (.28)	***154.244 (.01)***	81.447 (.07)
BLRT (*p*)	478.701 (.00)	170.830 (.00)	162.703 (.00)	***85.914 (.00)**$_c$*
Group size (%) C1	48 (11.01%)	***23 (5.3%)***	5 (1.1%)	5 (1.1%)
C2	388 (88.99%)	***43 (9.9%)***	23 (5.3%)	22 (5.1%)
C3		***370 (84.8%)***	59 (13.5%)	28 (6.4%)
C4			349 (80.1%)	104 (23.9%)
C5				277 (63.5%)
GMM−CI				
LL (No. of Parameters)	−963.282 (13)	−918.234 (16)	−887.040 (19)	−853.867 (22)
BIC	2005.564	1933.711	1889.555	***1841.443***
SSABIC	1964.318	1882.935	1829.258	***1771.627***
Entropy	.974	.953	.957	.939
Adj. LMR−LRT (*p*)	***216.795 (.00)***	85.411 (.20)	58.615 (.12)	62.895 (.16)
BLRT (*p*)	228.685 (.00)	***90.096 (.00)***	61.830 (.00)$_c$	66.344 (.00)^c
Group size (%) C1	27 (6.2%)	***20 (4.6%)***	5 (1.1%)	6 (1.4%)
C2	409 (93.8%)	***38 (8.7%)***	21 (4.8%)	11 (2.5%)
C3		***378 (86.7%)***	33 (7.6%)	32 (7.4%)
C4			377 (86.5%)	34 (7.7%)
C5				353 (81.0%)
GMM− CV				
LL (No. of Parameters)	−894.590 (15)[b]	−838.735 (20)	−802.027 (25)[b]	−802.027 (30)[a][b]
BIC	1880.344	***1799.022***	−	−
SSABIC	1832.742	***1735.553***	−	−
Entropy	.714	*.767*	−	−
Adj. LMR−LRT (*p*)	354.408 (.00)	***108.151 (.00)***	−	−
BLRT (*p*)	366.071 (.00)	***111.710 (.00)***	−	−
Group size (%) C1	94 (21.6%)	***41 (9.4%)***	−	−
C2	342 (78.4%)	***68 (15.6%)***	−	−
C3		***327 (75.0%)***	−	−
C4			−	−
C5			−	−

Note: LCGA = Latent Class Growth Analysis. GMM-CI = Growth Mixture Model with class-invariant variances and covariances (M*plus* default model). GMM-CV = Growth Mixture Model with class-varying variances (constrained covariances to be equal across classes). LL = Log-Likelihood value. No. of Parameters = Number of estimated (freed) parameters. BIC = Bayesian Information Criteria. SSABIC = Sample Size Adjusted BIC. LMR-LRT = Lo-Mendell-Rubin Likelihood Ratio Test. Adj.LMR-LRT = Adjusted LMR. BLRT = Bootstrap Likelihood Ratio Test. *p* = *p*-value. [a] = No repeated log-likelihood value (i.e., local maxima). [b] = Inadmissible solution (i.e., negative variances). [c] = No replicated log-likelihood value (H0 log-likelihood value) of the k-1 class model. Bold and italic values indicate the best values for the model fit statistics in the set of models listed in the corresponding row.

Illustrative Example 7.4.A: Optimal LCGA: In the current example, no convergence problems or local maxima issues were encountered for the LCGAs. As mentioned previously, because a LCGA is more parsimonious than a GMM, difficulties with model convergence are less common for a LCGA than a GMM. However, a LCGA often identifies more classes than a GMM. As can be seen in Table 7.1, most of the model fit indices suggested that the 4- or 5-class model was the optimal model (e.g., lower BIC and SSABIC values, higher entropy values, and statistically significant p-values for the adj. LMR-LRT and BLRT). However, the results of the 4- and 5-class models were questionable because of the small class sizes. The smallest group size (n) in both models was 5. The existence of such a small class at the population level is doubtful. The 4-class model had better fit indices than the 3-class model, but the class sample sizes in the 3-class model contained reasonable proportions of the total sample. Also, compared to the 2-class model, the 3-class model had a lower BIC and SSABIC, a higher entropy value (with average posterior probabilities ranged from .897 to .982), and a statistically significant BLRT p-value. The entropy value of the 3-class model was also lower than those of the 4- and 5-class models. Taking all of this available information into consideration simultaneously, the 4- and 5-class models were rejected in favor of the 3-class LCGA model.

Illustrative Example 7.4.B: Optimal GMM-CI: No convergence problems (i.e., local maxima) were detected across the class models for the GMM-CI (class-invariant variances and covariances, which is the M*plus* default). The 5-class model had the lowest BIC and SSABIC values, which indicates that the 5-class model may be the preferred model. However, the 4- and 5-class models had classes with small group sizes. The smallest class size was 5 (1.1%) and 6 (1.4%) for the 4- and 5-class models, respectively. More importantly, given the non-replicated log-likelihood value of the k-1 class model, it is unlikely that the 4- and 5-class models exist in the general population. Instead, the 3-class model was selected as the optimal model because the model fit indices suggest that it fit better than the 2-class model (lower BIC and SSABIC values, higher entropy value [with average posterior probabilities ranging from .898 to .990], and a statistically significant p-value for the BLRT). While the p-value for the LMR-LRT was not statistically significant (indicating that the 3-class model does not improve the model fit compared to the 2-class model), Nylund et al. (2007) reported that the BLRT outperforms the LMR-LRT when a local solution is not detected.

Illustrative Example 7.4.C: Optimal GMM-CV: Recall that a GMM-CV allows for the estimation of variance-covariance parameters for all growth parameters in all classes (class-varying variances and covariances). However, in the GMM-CV, inadmissible solutions were consistently found (local solutions and negative variances of growth parameters within a class) for the 2-, 4-, and 5-class models. An exploratory approach with multiple remedies was employed to avoid these difficulties, including increasing the number of random starting values and constraining the negative variances to be zero (Grimm et al., 2009). However, inadmissible solutions remained for the 4-and 5-class models, so these models were excluded from all subsequent analyses (see Table 7.1). All of the model fit indices consistently supported the 3-class model over the 2-class model (lower BIC and SSABIC values, a higher entropy value [with average posterior probabilities ranging from .894 to .920], and statistically significant p-values for the adj. LMR-LRT and BLRT). All of the classes in these two models also contained a reasonable proportion of the sample. Thus, after considering all of the available information the 3-class model was selected as the optimal model. Growth parameters for each optimal model are shown in Table 7.2. As can be seen in Table 7.2, most growth parameters were statistically significant. Furthermore, statistically significant p-values for the equality test indicated that the variances and covariances of growth parameters were not equal across classes. This finding provides evidence supporting the use of a GMM-CV rather than a GMM-CI or a LCGA for assessing the current example data.

Plots of the observed trajectories for classes within a model can also provide important information for selecting the optimal class model (Muthén & Muthén, 2000). Figure 7.9 shows a comparison of the estimated mean trajectories generated by a LCGA (Panel A), a GMM-CI (Panel B), and a GMM-CV (Panel C).

Psychopathology literature has reported the existence of heterogeneity in internalizing symptom (IS) trajectories, such as symptoms of depression and anxiety in adolescence (Wickrama et al., 2008). Consistent with previous findings, Figure 7.9 clearly shows the existence of three general trajectory patterns across the three approaches. These three patterns include: (1) a consistently low IS trajectory, (2) an initially low or moderate and increasing trajectory, and (3) an initially high and decreasing trajectory. The entropy values and the average posterior probabilities are shown in Table 7.1. Figure 7.9 corroborates the model fit results by showing three clearly distinct trajectories in each approach. Interestingly, although these three approaches (LCGA, GMM-CI, and GMM-CV) seemingly produce similar patterns of trajectories, the class proportions among the three approaches are quite different (see Figure 7.9), especially for the initially low class and the moderate and increasing class. The proportion of individuals classified as having increasing IS over time was similar for the LCGA and the GMM-CI models, but the increasing IS class in the GMM-CV contained a noticeably larger proportion of the sample. These varying class proportions are a product of the different model parameter estimations across the approaches.

All of the growth parameters for each optimal model are provided in Table 7.2. As can be seen, most of the growth parameters were statistically significant. Furthermore, in order to investigate whether the estimated variances of the growth parameters were significantly different, we tested the equality of the growth parameters using a Wald chi-square test (see the equality test in Step 3). Statistically significant p-values for the growth parameter equality tests indicated that the variances and covariances of growth parameters were not equal across classes. This provides additional evidence for the value of utilizing a GMM-CV in this example. For

TABLE 7.2 Growth Parameters across the Three Optimal Models.

	Intercept		Linear Slope		Factor covariance	Growth parameter equality (Wald)		
	Mean	Variance	Mean	Variance		Intercept variance	Slope variance	Covariance
LCGA								
C1 (High and decreasing)	2.460***		−.165***					
C2 (Moderate and increasing)	1.722***		.133***					
C3 (Consistently low)	1.363***		−.022***					
GMM-CI								
C1 (High and decreasing)	2.651***	.050***	−.187***	.001*	−.003*			
C2 (Low and increasing)	1.432***	.050***	.181***	.001*	−.003*			
C3 (Consistently low)	1.374***	.050***	−.023***	.001*	−.003*			
GMM-CV								
C1 (High and decreasing)	2.556***	.140**	−.172***	.003*	−.003	4.829†[a, c]	23.631***[a, b, c]	Equally constrained
C2 (Low and increasing)	1.458***	.037**	.056**	.007***	−.003			
C3 (Consistently low)	1.346***	.034**	−.030***	.000	−.003			

Note: LCGA = Latent Class Growth Analysis. GMM-CI = Growth Mixture Model with class-invariant variances and covariances (Mplus default model). GMM-CV = Growth Mixture Model with class-varying variances (constrained covariances to be equal across classes). C = Class. Unstandardized coefficients are shown. Wald = Wald chi-square test (df = 2).

[a] High and decreasing vs. Low and increasing, $p < .05$.

[b] Low and increasing vs. Consistently low, $p < .05$.

[c] High and decreasing vs. Consistently low, $p < .05$.

† $p < .10$. * $p < .05$. ** $p < .01$. *** $p < .001$.

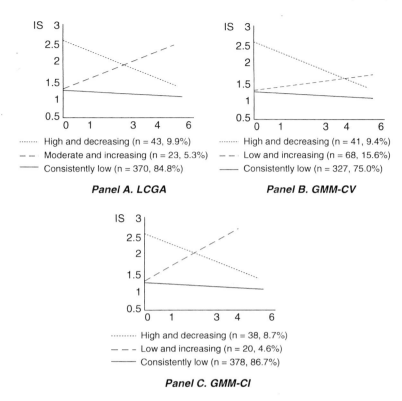

FIGURE 7.9 Predicted Mean Trajectories Over Time for a Three-Class LCGA, GMM-CI, and GMM-CV.
Note: LCGA = Latent Class Growth Analysis. GMM-CI = Growth Mixture Model with class-invariant variances and covariances. GMM-CV = Growth Mixture Model with class-varying variances (constrained covariances to be equal across classes). IS = Internalized Symptoms.

instance, the value of 4.829 indicates that the intercept variances among the three classes were significantly different. The slope variances were also found to be different across the three classes (Wald chi-square value (df) = 23.631(2)). Combining the information gathered from the model fit statistics, plots, and relevant theory, we selected the three-class GMM-CV model as the optimal class model.

Summary of a Model Building Strategy

1. Do latent classes really exist?

Latent classes can appear when they do not truly exist due to skewed/non-normal data (Bauer & Curran, 2003). Muthén (2003) suggests the following ways to increase the reliability of latent classes:

Suggestions:

- Use multiple measures of model fit and plots.
- Rely on substantive theory built over time.
- Include: antecedents (e.g., an intervention), concurrent events (e.g., time-varying covariates), and consequences (e.g., distal outcomes) in the model.

2. Non-convergence & local solution issues:

It is possible to select the wrong solution due to model complexity & model misspecification.

Suggestions:

- Vary the start values (using the STARTS and STITERATIONS options).
- Compare the results of competing solutions (LCGA vs. GMM).

3. What is the optimal number of classes and which fit index should be used?

Suggestions:

- Look for the smallest Bayesian Information Criteria (BIC) value.
- Look for a statistically significant Bootstrapped Likelihood Ratio Test (BLRT; specified by TECH 14 in the M*plus* output command).
- Look for a statistically significant Lo-Mendell-Rubin Test (LMR; specified by TECH 11 in the M*plus* output command).
- Consider the usefulness of the classes. Fewer classes with larger memberships (> 5%) are preferred over many excessively small classes.
- Look for classes with meaningful interpretations.

4. Rely on theory and multiple replications for clear conclusions:

Suggestions:

- Remember that latent trajectory classes are exploratory.
- Unobserved class trajectories should be confirmed from relevant theory.

5. The dependency of class membership:

Suggestion:

- The final solution depends on the variables included in the final model. Consider utilizing the 3-step approach introduced in the next chapter to avoid biased classes based on covariates included in the model.

6. Be familiar with the method, alternative ways to specify the best solution, and cautions about misspecification.

Chapter 7 Exercises

This hypothetical dataset consists of four composite (mean) scores of internalizing symptoms of fathers at T1~T4 (i.e., IS1F to IS4F). In the M*plus* syntax, indicate the variable order as follows:

DATA: File is exercise_Ch.7.dat;
VARIABLE: names are IS1F IS2F IS3F IS4F;
                    MISSSING ARE ALL (9);

Analyze a mixture model using the four repeated variables (IS1F–IS4F) and, using all available information, select the optimal class model. Follow the model building steps below.

1. Analyze a conventional linear growth curve model (the baseline LGCM with no autocorrelated errors, t = 0, 1, 2, and 3). Check whether the variances of the growth parameters are statistically significant.
2. Using the baseline LGCM (the model in Question 1),
    A. Analyze growth mixture models (GMM–CI and GMM–CV). Fill in Exercise Table 7.1 as you fit the models. In order to avoid local maxima problems, insert the M*plus* syntax below under the ANALYSIS command for all of the GMMs.

    ANALYSIS: TYPE=MIXTURE;
        STARTS = 200 10;
        STITERATIONS = 10;

    B. Check for estimation problems, such as negative variances of model parameters and model under-identification. If any problems are detected, report those in Exercise Table 7.1.
    C. Given all of the available information (model fit statistics and class parameters), select the optimal class model and provide reasons for your selection (including the model type [GMM-CI or GMM-CV] and the number of classes).
    D. Finally, create a plot of the optimal model by adding the M*plus* syntax below to the optimal class model, and check whether the plot shows clear class separation.

    PLOT: SERIES = IS1F-IS4F(S);TYPE=PLOT3;

    E. In order to confirm that the classes are clearly separated, test the equality of the growth mean parameters and report the results.

EXERCISE TABLE 7.1 *Fit Statistics for the GMM-CI and GMM-CV.*

Fit statistics	2 Classes	3 Classes	4 Classes
GMM-CI			
LL (No. of Parameters)			
BIC			
SSABIC			
Entropy			
Adj. LMR-LRT (*p*)			
BLRT (*p*)			
Group size (n, %)			
C1			
C2			
C3			
C4			
GMM- CV			
LL (No. of Parameters)			
BIC			
SSABIC			
Entropy			
Adj. LMR-LRT (*p*)			
BLRT (*p*)			
Group size (n, %)			
C1			
C2			
C3			
C4			

Notes: GMM-CI = Growth Mixture Model with class-invariant variances-covariances (M*plus* default model). GMM-CV = Growth Mixture Model with class-varying variances (constrained covariances to be equal across class). LL = Log-likelihood value. No. of Parameters = Number of estimated (free) parameters. BIC = Bayesian Information Criteria. SSABIC = Sample size adjusted BIC. LMR-LRT = Lo-Mendell-Rubin likelihood ratio test. Adj.LMR-LRT = Adjusted LMR. BLRT = Bootstrap likelihood ratio test. *p* = p-value. n = class sample size. C = Class. a = Inadmissible solution (i.e., negative variances of growth parameter) in classes. b = No replicated log-likelihood value (H0 log-likelihood value) for the k-1 class model.

References

Asparouhov, T., & Muthén, B. (2012). Using M*plus* tech11 and tech14 to test the number of latent classes. Retrieved from https://www.statmodel.com/examples/webnotes/webnote14.pdf. Accessed 4 December 2015.

Bauer, D. J., & Curran, P. J. (2003). Distributional assumptions of growth mixture models: Implications for overextraction of latent trajectory classes. *Psychological Methods*, 8(3), 338–363.

Bauer, D. J., & Curran, P. J. (2004). The integration of continuous and discrete latent variable models: Potential problems and promising opportunities. *Psychological Methods*, 9(1), 3–29.

Berlin, K. S., Williams, N. A., & Parra, G. R. (2014). An introduction to latent variable mixture modeling (Part 1): Cross sectional latent class and latent profile analyses. *Journal of Pediatric Psychology*, 39(2), 188–203.

Clark, S., & Muthén, B. (2009). Relating latent class analysis results to variables not included in the analysis. Retrieved from https://www.statmodel.com/download/relatinglca.pdf. Accessed 4 December 2015.

Collins, L. M., & Lanza, S. T. (2010). *Latent class and latent transition analysis: With applications in the social behavioral and health sciences.* Hoboken, NJ: John Wiley & Sons Inc.

Enders, C. K., & Tofighi, D. (2008). The impact of misspecifying class-specific residual variances in growth mixture models. *Structural Equation Modeling*, 15(1), 75–95.

Fanti, K. A., & Henrich, C. C. (2010). Trajectories of pure and co-occuring internalizing and externalizing problems from age 2 to age 12: Findings from the national institute of child health and human development study of early child care. *Developmental Psychology*, 46(5), 1159–1175.

Feldman, B., Masyn, K. E., & Conger, R. (2009). New approaches to studying problem behaviors: A comparison of methods for modeling longitudinal, categorical adolescent drinking data. *Developmental Psychology*, 45(3), 652–676.

George, D., & Mallery, M. (2010). *SPSS for Windows Step by Step: A Simple Guide and Reference, 16.0 update.* (10th Ed.). Boston: Pearson.

Hayduk, L. (1987). *Structural Equation Modeling with LISREL.* Baltimore: John Hopkins University Press.

Hipp, J. R., & Bauer, D. J. (2006). Local solutions in the estimation of growth mixture models. *Psychological Methods*, 11(1), 36–53.

Jung, T., & Wickrama, K. A. S. (2008). An introduction to latent class growth analysis and growth mixture modeling. *Social and Personality Psychology Compass*, 2(1), 302–317.

Kreuter, F. & Muthen, B. (2008). Longitudinal modeling of population heterogeneity: Methodological challenges to the analysis of empirically derived criminal trajectory profiles. In Hancock, G. R., & Samuelsen, K. M. (Eds.), *Advances in latent variable mixture models* (pp. 53–75). Charlotte, NC: Information Age Publishing, Inc.

Li, F., Duncan, T. E., Duncan, S. C., & Acock, A. (2001). Latent growth modeling of longitudinal data: A finite growth mixture modeling approach. *Structural Equation Modeling*, 8(4), 493–530.

Lo, Y., Mendell, N., & Rubin, D. (2001). Testing the number of components in a normal mixture. *Biometrika*, 88(3), 767–778.

Lubke, G. H., & Neale, M. C. (2006). Distinguishing between latent classes and continuous factors: Resolution by maximum likelihood. *Multivariate Behavioral Research*, 41(4), 499–532.

Magidson, J., & Vermunt, J. K. (2004). Latent class models. In D. Kaplan (Ed.), *Handbook of quantitative methodology for the social sciences* (pp. 175–198). Newbury Park, CA: Sage.

McLachlan, G., & Peel, D. (2000). *Finite mixture models.* New York: Wiley.

Morin, A. J. S., Maïano, C., Negengast, B., Marsh, H. W., Morizot, J., & Janosz, M. (2011). General growth mixture analysis of adolescents' developmental trajectories of anxiety: The impact of untested invariance assumptions on substantive interpretations. *Structural Equation Modeling*, 18(4), 613–648.

Muthén, B. (2002). Beyond SEM: General latent variable modeling. *Behaviormetrika*, 29(1), 81–117.

Muthén, B. (2003). Statistical and substantive checking in growth mixture modeling: Comment on Bauer and Curran (2003). *Psychological Methods*, 8(1), 369–377.

Muthén, B. (2004). Latent variable analysis: Growth mixture modeling and related techniques for longitudinal data. In D. Kaplan (Ed.), *Handbook of quantitative methodology for the social sciences* (pp. 345–368). Newbury Park, CA: Sage.

Muthén. B., & Asparouhov, T. (2015). Growth mixture modeling with non-normal distributions. Retrieved from https://www.statmodel.com/download/GMM.pdf. Accessed 4 December 2015.

Muthén, B., & Muthén, L. K. (2000). Integrating person-centered and variable-centered analyses: Growth mixture modeling with latent trajectory classes. *Alcoholism: Clinical and Experimental Research*, 24(6), 882–891.

Muthén, L. K., & Muthén, B. O. (1998–2012). *Mplus user's guide* (7th ed.). Los Angeles: Authors.

Muthén, B. O., & Shedden, K. (1999). Finite mixture modeling with mixture outcomes using the EM algorithm. *Biometrics*, 55(2), 463–469.

Myung, I. J. (2003). Tutorial on maximum likelihood estimation. *Journal of Mathematical Psychology*, 47(1), 90–100.

Nagin, D. S. (1999). Analyzing developmental trajectories: A semiparametric, group-based approach. *Psychological Methods*, 4(2), 139–157.

Nagin, D. S. (2005). *Group-based modeling of development.* Cambridge, MA: Harvard University Press.

Nylund, K. L., Asparouhov, T., & Muthén, B. (2007). Deciding on the number of classes in latent class analysis and growth mixture modeling: A Monte Carlo simulation study. *Structural Equation Modeling*, 14(4), 535–569.

Peugh, J., & Fan, X. (2012). How well does growth mixture modeling identify heterogeneous growth trajectories? A simulation study examining GMM's performance characteristics. *Structural Equation Modeling*, 19(2), 204–226.

Shiyko, M. P., Ram, N., & Grimm, K. J. (2012). An overview of growth mixture modeling: A simple nonlinear application in OpenMx. In R. H. Hoyle (Ed.), *Handbook of Structural Equation Modeling* (pp. 532–546). New York: Guilford Press.

Tein, J-Y., Coxe, S., & Cham, H. (2013). Statistical power to detect the correct number of classes in latent profile analysis. *Structural Equation Modeling*, 20(4), 640–657.

Tofighi, D., & Enders, C.K. (2007). Identifying the correct number of classes in growth mixture models. In G. R. Hancock & K. M. Samuelsen (Eds.), *Advances in latent variable mixture models* (pp. 317–341). Greenwhich, CT: Information Age.

Verbeke, G. & Molenberghs, G. (2000). *Linear Mixed Models for Longitudinal Data.* New York: Springer.

Vermunt, J. K. & Magidson, J. (2002). Latent class cluster analysis. In J.A. Hagenaars & A.L. McCutcheon (Eds.), *Advances in Latent Class Analysis* (pp. 89–106). Cambridge University Press.

Wickrama, K.A.S., Conger, R. D., & Abraham, W. T. (2008). Early family adversity, youth depressive symptoms trajectories, and young adult socioeconomic attainment: A latent trajectory class analysis. *Advances in Life Course Research*, 13, 161–192.

Wurpts, I. C., & Geiser, C. (2014). Is adding more indicators to a latent class analysis beneficial or detrimental? Results of a Monte-Carlo study. *Frontiers in Psychology*, 5, 920–934.

8

ESTIMATING A CONDITIONAL GROWTH MIXTURE MODEL (GMM)

Introduction

Growth mixture modeling (GMM) employs a person-centered approach to analysis because it explicitly focuses on class membership. The identification of heterogeneous classes of individuals with qualitatively different trajectories of depressive symptoms allows us to focus not only on class membership but also to examine covariates, including (a) antecedents, or predictors (e.g., preceding socioeconomic conditions and life experiences) and (b) consequences, or distal outcomes (e.g., subsequent socioeconomic attainment and life experiences), of class membership. This person-centered approach is particularly informative when examining the association between heterogeneity in a time-varying attribute (e.g., health) and its predictors and outcomes.

This chapter illustrates the incorporation of predictors and outcomes (covariates) into growth mixture models (GMMs). Like other modeling approaches, growth mixture models (including latent class growth analyses, or LCGA) can incorporate multiple covariates (predictors and distal outcomes). Two approaches can be used in M*plus* to add covariates into a growth mixture model: (1) the direct specification approach (i.e., 1-step approach) and (2) the 3-step approach. It is important to note that the direct specification approach may result in a different class classification than that identified in an unconditional GMM (see Chapter 7), while the 3-step approach preserves the class classification from the unconditional GMM. We will return to the 3-step approach later in this chapter. First, we introduce how to utilize the 1-step approach for testing the influence of covariates in a growth mixture model. This chapter provides M*plus* syntax for the incorporation of predictors and outcomes into a GMM using both approaches and guidance for interpreting the results.

Growth Mixture Models: Predictors and Distal Outcomes

Researchers have argued that heterogeneity among the trajectories of an attribute may stem from background characteristics and preceding life experiences. For example, consistent with sociological and developmental theories, early stressful experiences stratify individuals' trajectories of internalizing symptoms (IS), resulting in sub-populations (or "clusters" of individuals) with varying patterns of IS growth over time. Returning to the sub-populations we identified in the example unconditional GMM in Chapter 7, recall that we expect the majority of youth who face relatively few stressful experiences to exhibit IS trajectories characterized by relatively low and stable levels of IS. In contrast, we expect that the IS trajectories of youth who experience many stressful life experiences will be non-normative. That is, these individuals should be more likely to experience an escalation in their internalizing symptoms than youth with few preceding stressful experiences. Some adolescents who initially experience high levels of IS may experience positive life events (e.g., entering into a successful close relationship), which serve as "turning points" to alter the course of their IS trajectories for the better. This sub-population reflects a recovering group, in which the IS trajectory is "turned" and the unhealthy trajectory is "slowed down," or deviated, resulting in a reduction in IS over time. In contrast, some adolescents who initially experience low levels of IS may be redirected by central negative life experiences (e.g., their parents' divorce). These individuals may comprise an escalating group in which the IS trajectory accelerates over time.

The background socioeconomic conditions and preceding life experiences that contribute to the development of specific trajectories provide important prognostic information for modifying personal attribute patterns. For instance, learning more about the life experiences that place individuals at-risk for an escalating trajectory of IS across adolescence is helpful for identifying youth who should be targeted for early intervention and prevention efforts (Feldman, Masyn, & Conger, 2009; Wickrama, Conger, & Abraham, 2008).

This person-centered approach can be applied to various developmental and social science research areas. Time-varying attributes, such as verbal development, physical growth, body mass index (BMI), educational performance, health behaviors, income, work satisfaction, and relationship quality are just a few possible candidates for this person-centered approach due to their potential for heterogeneous growth. For example, youth's career development trajectories may be clustered into heterogeneous groups, with a "fast achieving or excelling" group, a "slow or normative achieving" group, and a "stagnant or under-achieving" group. An examination of the individuals comprising these trajectory groups with unique career development growth patterns may identify variations in family background characteristics (antecedents) and subsequent socioeconomic outcomes in adulthood across the clusters.

The One-Step Approach to Incorporating Covariates into a GMM

A 1-step analysis examines the association between latent class variables and observed variables by directly specifying the covariates in the GMM (e.g., Muthén & Shedden, 1999). This direct inclusion approach is known to result in a more precise estimation of the covariates' effects (Bolck, Croon, & Hagenaars, 2004; Clark & Muthén, 2009). Furthermore, like other types of modeling in a SEM framework, the 1-step approach allows *latent variables* to be used as covariates (both predictors and distal outcomes). Muthén (2004) introduced three types of paths that can be used to directly incorporate covariates in a GMM using the 1-step approach: (1) incorporating predictors of latent classes (multinomial regression), (2) incorporating predictors of latent growth factors, and (3) incorporating distal outcomes of the latent classes. For our illustrative purpose, this chapter demonstrates each type of path individually within the GMM-CV (a GMM with within-class variances) selected as the optimal model in Chapter 7. However, assuming the model is not overly complex, all three of these types of paths can be modeled simultaneously. Furthermore, while latent variables can be used as covariates, for the purpose of our illustration we used a manifest predictor variable (mothers' hostility at Time 1).

Predictors of Latent Classes (Multinomial Regression)

If predictor X is included in a GMM, a multinomial regression will be performed. Multinomial regression analyses allow researchers to investigate the influence of a predictor on between-class variation. Thus, regressing latent class C on X estimates the influence of predictor X on class membership (C). Let us assume that the reference class is the last class (K) where class k = 1, 2, 3, ... K. The logistic equation for individual i being in class k; that is, $\log [P(C_i = k \mid X_i) / (C_i = K \mid X_i)]$ is written as:

$$\text{Logistic of being in class k (logit)} = \beta_{k0} + \beta_{k1} X_i \tag{8.1}$$

where β represents a logistic coefficient in the multinomial regression (i.e., β_{k0} = the threshold [intercept] of a specific k class and β_{k1} = the logistic coefficient for class k). Logistic coefficient β_{k1} can be interpreted as the increase in the log-odds (i.e., logit) of being in the class k versus the reference class K for a one-unit increase in X. The odds-ratio (Exp (β_{k1})) can be used for an alternative interpretation. "Exp" indicates an exponential function (equal to 2.71828). The odds-ratio can be interpreted as the percent (%) change in the odds of being in class k for a unit increase in X (odds-ratio is approximately equal to 100 × (Exp (β_{k1}) − 1). The solid line representing the path from the predictor X to the class variable C in Figure 8.1 indicates the β_{k1} parameter. The probability (P) of individual i being in latent class C = k (1, 2, 3, ..., K) is:

$$P = \frac{\text{Exp}\left(\beta_{k0} + \beta_{k1} X_i\right)}{\sum_{s=1}^{K} \text{Exp}\left(\beta_{s0} + \beta_{s1} X_i\right)} \tag{8.2}$$

These probabilities (Ps) are referred to as **posterior probabilities**. For each individual, the posterior probabilities sum to 1 and are based on that individual's observed outcome responses and covariate values. The individual trajectory of the outcome variable is determined by the probability of class membership and the class-specific means and covariance structures as well as covariates (Feldman et al., 2009).

Illustrative Example 8.1: Incorporating a Time-Invariant Predictor into a GMM

Figure 8.1 shows a conditional growth mixture model (GMM) with a time-invariant predictor of between-class differences where the influence of the predictor is estimated using a multinomial logit regression. Using mothers' hostility at Time 1 as a manifest predictor variable (i.e., MH1), we tested a conditional GMM-CV (a GMM with within-class variances). The bold syntax can be incorporated in the unconditional GMM-CV from Chapter 7 to add this predictor.

MODEL:
%OVERALL%
I S | IS1@0 IS2@1 IS3@3 IS4@4 IS5@6;
C ON MH1;
%C#1%   !Freeing within-class variability
I-S;
%C#2%
I-S;
%C#3%
I-S;

Predictors of Latent Growth Factors Within Classes

The main advantage of a GMM (both the GMM-CV and GMM-CI approaches) is that covariates can be specified to predict both within-class variations and between-class variations simultaneously. Thus, a GMM with within-class variation (GMM-CV) is appropriate for this analysis because a GMM-CV already estimates both within-class and between-class variation. To investigate whether predictor X

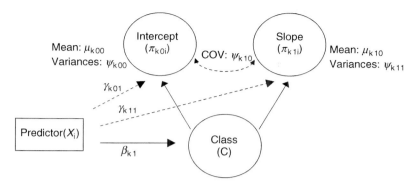

FIGURE 8.1   Parameter Specification of a Conditional Growth Mixture Model (GMM) and a LCGA (Latent Class Growth Analysis) with a Manifest Predictor. Note: Subscript k represents a categorical class variable k (k = 1, 2, 3, ..., K). Solid lines are parameters present in both a GMM and a LCGA, and dashed lines represent parameters that define the model as a GMM. The measurement model is not shown. Adapted from Feldman et al. (2009).

is associated with within-class variation, a mixture regression analysis can be used. The second-level equation (i.e., the structural model) for a GMM for individual i is indexed by class k:

$$\pi_{k0i} = \mu_{k00} + \zeta_{k0i}, \qquad \text{where } \zeta_{k0i} \sim \text{NID } (0, \Psi_{k00}) \tag{8.3}$$

$$\pi_{k1i} = \mu_{k10} + \zeta_{k1i}, \qquad \text{where } \zeta_{k1i} \sim \text{NID } (0, \Psi_{k11}) \tag{8.4}$$

$$\Psi_{k} = \begin{bmatrix} \Psi_{k00} & \\ \Psi_{k10} & \Psi_{k11} \end{bmatrix} \tag{8.5}$$

where μ_{k00} and μ_{k10} represent the average intercept and slope for latent trajectory class k; ζ_{k0i} and ζ_{k1i} indicate the disturbances (i.e., errors) reflecting the variability of the estimated intercepts and slopes across individuals within the latent trajectory class. Then, predictor X can be specified to explain within-class variation. The second-level equation of a GMM can be specified as:

$$\pi_{k0i} = \mu_{k00} + \gamma_{k01} \times X_i + \zeta_{k0i} \tag{8.6}$$

$$\pi_{k1i} = \mu_{k10} + \gamma_{k11} \times X_i + \zeta_{k1i} \tag{8.7}$$

where γ_{k01} and γ_{k11} represent regression coefficients for k class. In a GMM-CV, the covariate can be specified at both the within-class and between-class levels.

Illustrative Example 8.2: Adding Within-Class Effects of Predictors to a GMM

Using the same predictor, the bold syntax below can be added to estimate a conditional class-varying GMM (GMM–CV):

MODEL:
%OVERALL%
I S | IS1@0 IS2@1 IS3@3 IS4@4 IS5@6;
C ON MH1;
I S ON MH1;
%C#1%
I S ON MH1;   !Estimating the within–class effect of the predictor
%C#2%
I S ON MH1;
%C#3%
I S ON MH1;

The upper section of Table 8.1 presents results of the multinomial regression (estimating between-class variation), while the lower section of the table presents results of the mixture regression (estimating within-class variation). In interpreting the *multinomial* coefficients, a statistically significant logit coefficient was detected in the comparison between Class 1 and Class 3. The logistic value of .517 indicates that the log-odds of being in Class 1 (the high and decreasing IS class) in comparison to the log-odds of being in Class 3 (the consistently low IS class) increased by .517 for every one unit increase in mothers' hostility. Alternatively, this logit coefficient can be interpreted using the odds-ratio 1.677 (= Exp (.517)). This odds-ratio represents the percent (%) change in odds for a one unit increase in mothers' hostility. For example, for a one unit increase in mothers' hostility, the odds of being a member of Class 1 are 67.7% (= 100 (Exp (.517) − 1)) *higher* than the odds of being a member of Class 3. Another statistically significant logit coefficient is .796. This coefficient indicates that the odds of being a member of Class 1 (the high and decreasing IS class) are 121.7% (= 100 (Exp (.796) − 1)) *higher* than the odds of being a member of Class 2 (the low and increasing IS class) for every one unit increase in mothers' hostility.

The *mixture regression* coefficients can be interpreted as standard regression coefficients reflecting the influence of mothers' hostility on within-class growth parameters. As can be seen in Table 8.1, most of coefficients were statistically significant in the expected direction with varying effect sizes. Overall, these coefficients indicate that mothers' hostility differentially influences latent classes of IS trajectories, suggesting a heterogeneous regression. In general, mothers' hostility adversely influenced almost all within-class growth factors of IS (except for the slope of IS in the consistently low class).

TABLE 8.1. The Influence of Mothers' Hostility on Class Membership (total n = 412).

	Est.	OR
Between-class (multinomial regression coefficients)		
High and decreasing vs Consistently low [a]	.517*	1.678
Low and increasing vs Consistently low [a]	−.278	.757
High and decreasing vs Low and increasing [b]	.796*	2.217
Within-class (mixture regression coefficients)		
C1 (High and decreasing IS)		
Intercept	.013	
Slope	.124***	
C2 (Low and increasing IS)		
Intercept	−.113	
Slope	.068*	
C3 (Consistently low IS)		
Intercept	.165***	
Slope	−.020***	

Note: Logistic coefficients are shown. GMM-CV= Growth Mixture Model with class-varying variances. C = Class. Est. = Estimate. OR = Odds Ratio. IS = Internalizing Symptoms.
[a] Consistently low is the reference class.
[b] Low and increasing is the reference class.
*p < .05. ***p < .001.

For the sake of brevity, we do not present similar analyses incorporating predictors for the other two unconditional mixture model approaches introduced in Chapter 7 (LCGA and GMM-CI). However, for a similar analysis (incorporating predictors) with a class-invariant GMM (i.e., GMM-CI) in M*plus*, the following syntax can be incorporated into the unconditional GMM-CI.

```
MODEL:
%OVERALL%
I S | IS1@0 IS2@1 IS3@3 IS4@4 IS5@6;
C ON X;
I S ON X;
```

If the researcher selects a LCGA as the optimal model, it is only possible to add predictors explaining the between-class variation. That is, when estimating a conditional LCGA, the covariate can only be used for predicting between-class variation because a LCGA assumes no within-class variation exists. The M*plus* syntax below is utilized for adding predictors of between-class variation in a LCGA.

MODEL:
%OVERALL%
I S | IS1@0 IS2@1 IS3@3 IS4@4 IS5@6;
C ON X;
%C#1%
I-S@0;
%C#2%
I-S@0;
%C#3%
I-S@0;

Adding Distal Outcomes of Latent Classes (Categorical and Continuous)

The specification of a conditional LCGA or GMM with distal outcomes varies depending on whether the outcome is binary or continuous. First, we discuss incorporating binary (or categorical) outcomes into a 1-step GMM, followed by a discussion of how to incorporate continuous outcomes.

For a binary (or categorical) outcome, logistic coefficients are estimated. With the example of a binary distal outcome U (scored 0 or 1) and the latent class variable C as a covariate, the probability (P) of individual i having a value of U_i that equals 1 is:

$$P\ (U_i = 1\ |\ C_i = k) = \frac{1}{1 + Exp\left(\tau_k\right)} \tag{8.8}$$

where the main effect of C is captured by the class-varying threshold τ_k (an intercept of the logistic regression equation with its sign reversed). By specifying a binary outcome, M*plus* provides a threshold of the outcome variable for each class. A threshold value above 0 indicates that, within the class, the proportion of individuals with the lower outcome value (0) is larger than the proportion of individuals with the higher outcome value, while a threshold value below 0 indicates the opposite. That is, values below 0 indicate that the proportion of individuals with the higher outcome value (1) is larger than that of the lower outcome value. A threshold value of 0 indicates that the proportion of individuals with low and high outcome values in the class is equal. Muthén and Muthén (2008) provide additional guidelines for interpreting and using a threshold.

Illustrative Example 8.3: Incorporating a Binary Distal Outcome into a GMM

We created a binary variable named NEE5_B by splitting the negative economic events variable (NEE5) at the median, and we examined the influence of class membership on this binary distal outcome in a GMM-CV. M*plus* syntax for this logistic regression predicting binary outcome U is shown in Figure 8.2:

```
DATA: FILE=example_ch_8.dat;
VARIABLE: NAMES = ID IS1 IS2 IS3 IS4 IS5 MALE MH1
                  NLE1 NSE1 NEE5 RV5 PC5 NEE5_B RV5_B PC5_B;
          USEV = IS1-IS5 NEE5_B;
          CLASSES = C(3);
          CATEGORICAL = NEE5_B;
          MISSING ARE ALL (9.00);

ANALYSIS: TYPE=MIXTURE;
          STARTS = 500 10;
          STITERATIONS = 10;

MODEL:
%OVERALL%
      I S | IS1@0 IS2@1 IS3@3 IS4@4 IS5@6;
      [NEE5_B $1];

%C#1%
      I S; [NEE5_B$1](T1);      !Estimating within class variances and thresholds

%C#2%
      I S; [NEE5_B$1](T2);

%C#3%
      I S; [NEE5_B$1](T3);

MODEL TEST:
      T1=T2; T2=T3;
```

FIGURE 8.2 M*plus* Syntax for the 1-Step Approach Incorporating a Categorical Distal Outcome.
Note: I = Initial Level. S = Slope. IS = Internalizing Symptoms. NEE5_B = Negative Economic Events (binary variable) at Time 5.

Note that in order to use a binary outcome variable, the CATEGORICAL option should be specified under the VARIABLE command. Although square brackets in an M*plus* syntax usually refer to mean structures, for binary and ordered categorical indicators, the square brackets can also be used to refer to threshold structures. The number of thresholds for a categorical variable is equal to the number of categories minus one. Thus, this binary variable has one threshold. Thresholds are referenced by adding a $ followed by the threshold number to the variable names (e.g., [NEE5_B$1]).

Note that the regression specification for this type of model is different from other regression models in a SEM framework using M*plus*. Whereas investigating the influence of an exogenous variable on an endogenous variable in a regression model using a SEM framework typically employs the "ON" syntax, this syntax is *not* used to specify X on C. Instead, threshold structures (like mean structures) are specified for each class with assigned labels, such as (T1), (T2), and (T3). These thresholds and labels are used for threshold equality tests with the MODEL TEST option. By specifying "T1=T2;" and "T2=T3;" in the syntax, the output provides the overall Wald chi-square value (similar to the overall F-value an ANOVA) along with its p-value. For instance, the statistically significant p-value in Table 8.2 indicates that the thresholds among the three classes are statistically different. This threshold equality test is equivalent to a simple regression test, which is commonly used to test a causal effect.

TABLE 8.2 The Results Pertaining to Distal Outcomes in the GMM-CV (total n = 438).

Outcomes	High and decreasing	Low and increasing	Consistently low	Wald chi-square equality test (df)
Binary outcomes (Thresholds)				
Negative economic events	.305	.661*	1.030***	3.284 (2)
Continuous outcomes (Means)				
Negative economic events	1.778**	1.364***	1.024***	3.444 (2)

Note: Unstandardized coefficients are shown. df = Degrees of Freedom. The 3-class GMM-CV was used.
*p < .05. **p < .01. ***p < .001.

As shown in Table 8.2, in our example, the threshold value of the "consistently low IS class" is 1.030. This threshold can be converted into the probability $(= \frac{1}{1+EXP(1.030)})$ of individuals in this class having an outcome value of 1 (U = 1). By carrying out this calculation, we see that the probability is .263. If this same threshold value was negative (-1.030), this threshold would be converted into the probability $(= \frac{1}{1+EXP(-1.030)})$ of individuals in this class having an outcome value of 0 (U = 0). Carrying out this calculation leads to a probability of .737. In the M*plus* output, these probabilities are automatically calculated and provided in the "Result in probability scale" section. Table 8.3 presents within-class probabilities of each category for the binary distal outcomes.

Note that for each latent class, Category 1 indicates the probability of having an outcome value of 0 (U = 0), and Category 2 indicates the probability of having an outcome value of 1 (U = 1). The second column shows the probabilities of each class, and the third, fourth, and fifth columns represent the standard error, t-values, and p-values, respectively. The overall Wald chi-test was not statistically significant (3.284(2), p = .193), which indicates that, in general, the probability of experiencing high or low negative economic events was similar across the three classes of IS trajectories.

Next, the inclusion of continuous distal outcomes in a LCGA or GMM is presented. When incorporating continuous outcomes, class means are estimated and the statistical significance of the differences in these means are analyzed using a mean equality test (similar to an ANOVA).

Illustrative Example 8.4: Incorporating a Continuous Distal Outcome into a GMM

To demonstrate the incorporation of a continuous distal outcome into a GMM, we used the continuous negative economic events (NEE5) variable. In order to

TABLE 8.3 M*plus* Output for a Binary Distal Outcome: Within-Class Probabilities for Each Category of the Binary Outcomes.

RESULTS IN PROBABILITY SCALE

Latent Class 1				
NEE5_B				
Category 1	0.659	0.072	9.218	0.000
Category 2	0.341	0.072	4.760	0.000
Latent Class 2				
NEE5_B				
Category 1	0.576	0.098	5.867	0.000
Category 2	0.424	0.098	4.327	0.000
Latent Class 3				
NEE5_B				
Category 1	0.737	0.029	25.605	0.000
Category 2	0.263	0.029	9.139	0.000

analyze the continuous distal outcome model, the bold text below can be added to the M*plus* syntax for the GMM-CV:

MODEL:
%OVERALL%
   I S | IS1@0 IS2@1 IS3@3 IS4@4 IS5@6;
%C#1%
   I S; **[NEE5](M1);**
%C#2%
   I S; **[NEE5](M2);**
%C#3%
   I S; **[NEE5](M3);**
MODEL TEST:
M1=M2; M2=M3;

There are three differences between this M*plus* syntax and the syntax for including a distal binary outcome shown previously. First, the variable name (NEE5) was changed because this example uses the continuous negative economic events variable. Second, the CATEGORICAL option in the VARIABLE command was deleted, and third, the "$" and numbers in square brackets were deleted. Note that in this syntax we assigned three labels (i.e., M1, M2, and M3) to allow for testing mean equality using a Wald chi-square test. The continuous distal outcome results are shown in Table 8.2.

Uncertainty of Latent Class Membership With the Addition of Covariates

As previously mentioned, although this 1-step approach is easy to implement (because covariates to the GMM are directly specified in the model) this approach sometimes leads to substantive problems. When the covariates are specified directly in the model, they can influence the formation of the latent classes. The influence of covariates is particularly likely when the entropy value is low (Asparouhov & Muthén, 2013). Therefore, although this 1-step approach, which utilizes maximum likelihood estimation, is the typical method for conducting such an analysis with predictors and outcomes, this approach can be flawed if the researcher expects to identify latent classes based *solely* on the indicator variables (without considering covariates for class identification) because the model with covariates may affect the latent class formation. Consequently, the meaning of the latent classes may no longer be based exclusively on the indicator variables. For example, in a latent class model with predictors, if the observed predictor variables that are intended to be assessed as antecedents of the latent classes have a direct effect on one of the indicator variables, the inclusion of the predictor variables can result in a substantial change to the latent classes, and the latent class variable will lose its intended meaning (Vermunt, 2010).

Thus, when estimating a GMM, specific procedures have been developed to allow for the incorporation of predictors and/or outcomes, while still protecting the class formation from the potential influence of covariates (Muthén, 2003). This procedure has been termed the "3-step" approach for model estimation (Vermunt, 2010). This "3-step" procedure can be implemented through three separate analytical steps carried out by the researcher (also known as the "manual" approach) or using the recently introduced, and more convenient, "auxiliary variable" option. In the following section, we discuss both the 3-step manual procedure and the alternative auxiliary option.

The Three-Step Approach: The "Manual" Method

The 3-step "manual" approach (Vermunt, 2010) includes the following steps:

Step One
    Estimate the unconditional latent class model (i.e., unconditional GMM).
Step Two
    Use the variable identifying the class to which each individual most likely belongs (S) from the latent class posterior distribution to specify the latent class indicator variable. M*plus* automatically estimates the misclassification error rates (i.e., uncertainty rate) through the logit approach. This step allows for the estimation of the effects of covariates *after* taking into account class uncertainty rates.
Step Three
    Include covariates (i.e., predictors and distal outcomes) directly in the GMM using the class variable (S) and the misclassification error rates.

Illustrative Example 8.5: The Three-Step Procedure for Incorporating Predictor(s)

Using our example GMM-CV, we demonstrate how to implement these steps in M*plus*. First, in order to identify the most likely class membership for each individual in the sample (S), we estimated a GMM-CV using the M*plus* syntax shown in Figure 8.3.

Note that we used the SAVEDATA command with the IDVARIABLE option to obtain information on class membership (i.e., the most likely class, S). Using this syntax, a text file (CLASS.txt) was generated. Note that this file is automatically saved in the same computer location as the model data file (i.e., example_ch_8.dat in this instance). Files with the .txt extension can be opened in other statistical packages, including SPSS (we used SPSS Version 22). Figure 8.4 shows a sample .txt file opened in SPSS.

In Figure 8.4, the first five variables correspond to the manifest variables from the observed dataset that were used to identify the classes (IS1 to IS5). The ID variable is then shown (V6 in the figure) followed by each individual's posterior probability for each of the three classes. More specifically, V7, V8, and V9 correspond to individual posterior probabilities for Classes 1, 2, and 3, respectively. The final column provides the membership information; that is, it indicates the latent class number assigned to

```
DATA:FILE=example_ch_8.dat;

VARIABLE:NAMES = ID IS1 IS2 IS3 IS4 IS5 MALE MH1
                 NLE1 NSE1 NEE5 RV5 PC5 NEE5_B RV5_B PC5_B;
         USEV = IS1-IS5;
         CLASSES = C(3);
         IDVARIABLE = ID;
         MISSING ARE ALL (9.00);

SAVEDATA: FILE=CLASS.txt;
          SAVE=CPROB;

ANALYSIS: TYPE=MIXTURE;
          STARTS = 500 10;
          STITERATIONS = 10;

MODEL: %OVERALL%
       I S | IS1@0 IS2@1 IS3@3 IS4@4 IS5@6;
%C#1%
       I S;

%C#2%
       I S;

%C#3%
       I S;
```

FIGURE 8.3 M*plus* Syntax for the First Step of the Manual 3-Step Approach. Note: I = Initial Level. S = Slope. IS = Internalizing Symptoms.

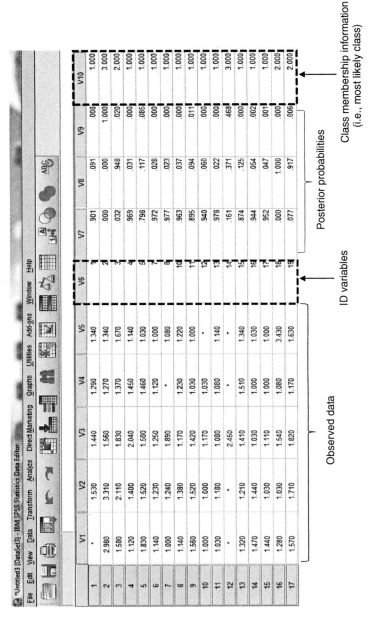

FIGURE 8.4 SPSS Screenshot of an Imported .txt Class File Obtained from the First Step of the 3-Step Procedure.

Note: Asterisks in the dataset correspond to missing values.

TABLE 8.4 M*plus* Output for the Logit Values of Classification Uncertainty Rates for the Most Likely Class Variable.

Average Latent Class Probabilities for Most Likely Latent Class Membership (Row) by Latent Class (Column)

	1	2	3
1	.920	.065	.015
2	.082	.894	.024
3	.067	.039	.894

Classification Probabilities for the Most Likely Latent Class Membership (Column) by Latent Class (Row)

	1	2	3
1	.973	.018	.009
2	.255	.726	.019
3	.113	.038	.849

Logits for the Classification Probabilities for the Most Likely Latent Class Membership (Column) by Latent Class (Row)

	1	2	3
1	4.695	.710	.000
2	2.586	3.631	.000
3	−2.017	−3.120	.000

the individual by M*plus* as the most likely class, or the class for which the individual had the highest posterior probability. The current GMM has three classes. Thus, the values of this column range from one to three. The information in the final column is essential for implementing the 3-step approach. Another essential component is the logit values for misclassification error rates (i.e., the uncertainty rates of classification). These logit values can be found in the M*plus* output (see Table 8.4) under the heading "Logits for the classification probabilities for the most likely latent class membership (column) by latent class (row)."

The M*plus* output detailing "Average latent class probabilities for most likely latent class membership (row) by latent class (column)" is presented in Table 8.4. Each column represents classification probabilities (averages of the individual probabilities in each class, corresponding to variables V7 to V9 in Figure 8.4) for the estimated GMM, and each row represents the classification probability for the most likely class (corresponding to variable V10 in Figure 8.4) (Asparouhov & Muthén, 2014). Thus, if the classes are perfectly separated, the diagonal components of the matrix should be 1.00, and the off-diagonal components should be 0 (i.e., identity matrix). This can be interpreted as if no uncertainty exists in the classifications (i.e., misclassification rates).

However, such a scenario is unlikely as there is usually some degree of uncertainty among the classifications. As can be seen in the top portion of Table 8.4, in the current example, each off-diagonal component has values ranging from .015 to .082, which indicates some degree of uncertainty exists in the classifications in this example. If these off-diagonal values (i.e., uncertainty rates) are high compared to the diagonal values, the classification is poor (i.e., the model has low entropy). Poor classification may change the class membership or even cause the GMM to fail to run when covariates are specified (Asparouhov & Muthén, 2014). However, if the model has high entropy, the class classification is good and is not likely to change significantly when covariates are included.

These uncertainty rates (misclassification rates) can be manually fixed in M*plus* so that when covariates are added the formation of classes is not influenced by the covariates. In the middle portion of Table 8.4, the classification probabilities after taking the uncertainty rates into account are shown, and the bottom portion of Table 8.4 shows these probabilities converted into logit values. Note that in the bottom portion of the table all logit values are fixed to 0 in the column representing the last class, which indicates that the last class was treated as the reference group. Using these logit values, each class can be protected (i.e., class classification in the unconditional GMM is held constant) so that the class membership solution will not be influenced by the subsequent addition of covariates.

Each column corresponds to each class's logit values for uncertainty rates. These logit values are used with the membership information identifying the most likely class of each individual in the final step. Using the ID variable (ID), the class membership variable, and the seven covariate variables, a new M*plus* file (.dat) was created (name: 3step.dat). The covariates include: MALE, MH1 (mothers' hostility), NSE1 (negative school environment), NLE1 (negative life events), NEE5 (negative economic events), RV5 (romantic violence), and PC5 (physical complaints).

Now, an unconditional GMM can be estimated using this 3-step approach. The corresponding M*plus* syntax is shown in Figure 8.5.

As can be seen, in the VARIABLE command, the class membership information (N) is directly specified as a nominal (unordered) categorical variable using the NOMINAL option. This class membership variable (N) is now recognized as a latent class variable. In the MODEL command of M*plus*, the categories of a nominal variable are referenced by adding the number sign (#) and the category number to the name of the unordered categorical variable. In our example model, N#1 refers to the first class of N, N#2 refers to the second class of N. N#3 does not have to be specified in the model, because N#3 is treated as the reference class. Square brackets refer to the mean structures of each class.

As can be seen, the means of each class are now fixed using the logit values (i.e., uncertainty rates for the most likely class of each individual). These logit values were obtained from the results of the original GMM (see Table 8.4). This is the central idea of the 3-step approach; by protecting the class membership solution using the uncertainty rates (i.e., logit values), class membership will not be influenced by the covariates. In the final step, we added four predictors: MALE, MH1, NSE1, and NLE1. This procedure is similar to the 1-step approach with direct specification. The bold text in the M*plus* syntax below can be added to the unconditional GMM specified in Figure 8.5 to incorporate predictors using the manual 3-step approach:

DATA: FILE=3step.dat;
VARIABLE: NAMES = N MALE MH1 NLE1 NSE1 NEE5 RV5 PC5;
   USEV = N **MALE MH1 NSE1 NLE1;**
   NOMINAL = N;
   CLASSES = C(3);
   MISSING ARE ALL (9.00);
ANALYSIS: TYPE=MIXTURE;
MODEL: %OVERALL%
C ON MALE MH1 NSE1 NLE1;
⋮

The results are provided in Table 8.5. Note that when using this 3-step approach M*plus* does not allow the inclusion of cases with missing data for exogenous predictors. Consequently, for this example, the program automatically deleted 29 cases with missing data.

Illustrative Example 8.6: The Three-Step Procedure for Incorporating Distal Outcome(s)

For this example, we used negative economic events (NEE5), romantic violence (RV5), and physical complaints (PC5) as distal outcomes of the latent classes. The

```
DATA:FILE=3step.dat;

VARIABLE: NAMES = N MALE MH1 NLE1 NSE1 NEE5 RV5 PC5;

        USEV = N;
        NOMINAL = N;
        CLASSES = C(3);
        MISSING ARE ALL (9.00);

ANALYSIS: TYPE=MIXTURE;

MODEL:
%OVERALL%

%C#1%
     [N#1@4.695]; [N#2@.710];

%C#2%
     [N#1@2.586]; [N#2@3.631];

%C#3%
     [N#1@-2.017]; [N#2@-3.120];

OUTPUT: SAMPSTAT TECH7;
```

FIGURE 8.5 M*plus* Syntax for the Third Step of the Manual 3-Step Approach Testing an Unconditional GMM.
Note: I = Initial Level. S = Slope. IS = Internalizing Symptoms.

TABLE 8.5 The Logit Coefficients of Predictors from the Manual 3-Step and Auxiliary 3-Step Approaches (total n = 407).

	Manual 3-Step approach					
	High and decreasing[a]		*Low and increasing*[a]		*High and decreasing*[b]	
Predictors	*Est.*	*OR*	*Est.*	*OR*	*Est.*	*OR*
Male (vs. Female)	**−1.910***	.148	−.477	.620	−1.433	.238
Mothers' hostility (MH)	**.931***	2.537	.333	1.395	**.598***	1.818
Negative school environment (NSE)	**1.225****	3.404	**.971****	2.640	.255	1.290
Negative life events (NLE)	.103	1.108	.026	1.026	.077	1.080

	AUXILLARY = (R3STEP) option					
	High and decreasing[a]		*Low and increasing*[a]		*High and decreasing*[b]	
Predictors	*Est.*	*OR*	*Est.*	*OR*	*Est.*	*OR*
Male (vs. Female)	**−1.910***	.148	−.477	.620	−1.433	.238
Mothers' hostility (MH)	**.931***	2.537	.333	1.395	**.598***	1.818
Negative school environment (NSE)	**1.226****	3.407	**.971***	2.640	.255	1.290
Negative life events (NLE)	.103	1.108	.026	1.026	.077	1.080

Note: Unstandardized coefficients are shown. Est. = Estimate. OR = Odds Ratio. The 3-class GMM-CV was used.

[a] Consistently low is the reference class.

[b] Low and increasing is the reference class.

*p < .05. **p < .01. ***p < .001.

```
DATA: FILE = 3step.dat;

VARIABLE:NAMES = N MALE MH1 NLE1 NSE1 NEE5 RV5 PC5;
        USEV = N NEE5 RV5 PC5;
        NOMINAL = N;
        CLASSES = C(3);
        MISSING ARE ALL (9.00);

ANALYSIS: TYPE = MIXTURE;

MODEL:
%OVERALL%
        [NEE5 RV5 PC5];

%C#1%
        [N#1@1.045]; [N#2@-2.586];
        [RV5](M1); [NEE5] (M2); [PC5] (M3);

%C#2%
        [N#1@-1.102]; [N#2@2.017];
        [RV5](M4); [NEE5] (M5); [PC5] (M6);

%C#3%
        [N#1@-3.985]; [N#2@-4.695];
        [RV5](M7); [NEE55] (M8); [PC5] (M9);

MODEL TEST:
        M3 = M6; M3 = M9;
```

FIGURE 8.6 M*plus* Syntax for the Manual 3-Step Approach Incorporating a Continuous Distal Outcome.
Note: NEE5 = Negative Economic Events at Time 5. RV5 = Romantic Violence at Time 5. PC5 = Physical Complaints at Time 5.

M*plus* syntax for incorporating these distal outcomes using the manual 3-step procedure is shown in Figure 8.6.

By specifying the mean structure in square brackets for each distal outcome in each class, M*plus* estimates the means of each distal outcome separately for each class. To test mean equality using a Wald chi-square test, a label can be assigned to each mean, and the significance of mean differences can be examined using the MODEL TEST option (see Illustrative Example 8.4 for more information on how to test mean equality in a distal outcome model). It is important to note that in this model with distal outcomes the total sample size (n = 436) was retained as M*plus* allows for missing data when assessing the impact of class membership on distal outcomes (unlike when assessing exogenous/predictor variables). The results are provided in Table 8.6.

TABLE 8.6 The Means of Distal Outcomes for Two 3-Step Approaches (Manual 3-Step and Auxiliary 3-Step Approaches) (total n = 436).

	Manual 3-step approach (Mean)			
Outcomes	High and decreasing	Low and increasing	Consistently low	Overall Chi-square test
Romantic violence (RV)	2.844	2.475	1.962	27.766*** [b, c]
Negative economic events (NEE)	1.742	1.908	.841	8.554*
Physical complaints (PC)	2.890	4.889	2.461	6.999* [a, b]

	AUXILLARY = (DE3STEP) option (Mean)			
Outcomes	High and decreasing	Low and increasing	Consistently low	Overall Chi-square test
Romantic violence (RV)	2.847	2.556	1.935	33.726*** [b, c]
Negative economic events (NEE)	1.917	1.349	.985	4.612
Physical complaints (PC)	2.890	4.305	2.644	7.790* [a, b]

Note: Unstandardized coefficients are shown. The 3-class GMM-CV was used.
[a] High and decreasing vs. Low and increasing, $p < .05$.
[b] Low and increasing vs. Consistently low, $p < .05$.
[c] High and decreasing vs. Consistently low, $p < .05$.
*$p < .05$. ***$p < .001$.

AUXILIARY Option for the Three-Step Approach

The 3-step approach can easily be implemented in M*plus* using the AUXILIARY option in the VARIABLE command. When an auxiliary variable (i.e., covariate) is identified as "(R3STEP)" all three steps are implemented automatically. One of the advantages to using this auxiliary option is that class membership does not change with the addition of covariates. Like the manual 3-step approach, in this auxiliary option approach, missing cases are removed for computing multinomial logit regression coefficients, but the auxiliary option in M*plus* allows the assessment of distal outcomes with the inclusion of cases with missing values.

If the "(DU3STEP)" option is specified for the auxiliary variable, the 3-step method will be used, and the auxiliary variable will be analyzed as a distal outcome with unequal means and *unequal* variances across classes. Specifying the "(DE3STEP)" option for auxiliary variables indicates that the auxiliary variables are to be treated as distal outcomes with unequal means and *equal* variances across classes. This equal variance estimation is useful for models with small classes, which can cause convergence problems when utilizing distal outcome estimation with unequal variance due to near-zero variances within the classes (i.e., a binary distal outcome; Asparouhov & Muthén, 2013).

However, previous studies have suggested that, in general, the equal variance option should not be used because it may lead to biased estimates and standard errors if the equal variance assumption is violated (Asparouhov & Muthén, 2013). In order to demonstrate how to utilize the auxiliary option of the 3-step approach, we specified the R3STEP option for all predictors and the DE3STEP option for all distal outcomes in the VARIABLE command. Note that M*plus* cannot recognize options for *both* predictors (i.e., R3STEP) and distal outcomes (i.e., DE3STEP) simultaneously. Thus, all auxiliary 3-step test commands listed in the model must be tested separately. In other words, two separate models are analyzed with one model testing the influence of all predictor variables simultaneously and a second model testing the effect of class membership on the distal outcome variables.

Illustrative Example 8.7: Utilizing the Auxiliary Option with the 3-Step Approach

Using the same covariates as in Illustrative Examples 8.5 and 8.6, we tested multiple predictors and multiple outcomes in a GMM. The M*plus* syntax is provided in Figure 8.7.

In the M*plus* output, the results of the predictor estimation can be found under the heading "Tests of categorical latent variable multinomial logistic regressions using the 3-step procedure." The results for the distal outcomes are shown under the heading "Equality tests of means across classes using the 3-step procedure with 2 degree(s) of freedom for the overall test." (Note that these headings are

DATA:FILE = example_ch_8.dat;
VARIABLE: NAMES = ID IS1 IS2 IS3 IS4 IS5 **MALE MH1 NLE1**
    NSE1 NEE5 RV5 PC5 NEE5_B RV5_B PC5_B;

    USEV = IS1-IS5;
    CLASSES = C(3);

    AUXILIARY variables are not specified in the
    USEVARIABLES (USEV) option.

    AUXILIARY = (R3STEP) MALE MH1 NSE1 NLE1;
    AUXILIARY = (DE3STEP) NEE5 RV5 PC5;

    The **AUXILIARY** option is used in conjunction with the
    TYPE=MIXTURE option. Auxiliary variables are not
    used to estimate the analysis model. Instead, the auxiliary
    variables are used to estimate coefficients for predictors
    or distal outcomes of the latent classes.

    MISSING ARE ALL (9.00);

ANALYSIS: TYPE = MIXTURE;
    STARTS = 500 10;
    STITERATIONS = 10;

    AUXILIARY = (R3STEP) is used to specify the list of predictors.
    AUXILIARY = (DE3STEP) is used to specify the list of
    continuous or binary distal outcomes.

MODEL: %OVERALL%
    I S | IS1@0 IS2@1 IS3@3 IS4@4 IS5@6;
%C#1%
    I S;
%C#2%
    I S;
%C#3%
    I S;

    Although shown together here, when analyzing a GMM, only one
    AUXILIARY option should be specified (for either a predictor model
    or a distal outcome model).
    If more than one auxiliary option is specified simultaneously,
    the model will not be estimated.

OUTPUT: TECH 7;

FIGURE 8.7 *Mplus* Syntax for Auxiliary Options Utilizing the Auxiliary (3-Step) Approach.

Note: MH1 = Mothers' Hostility at Time 1. NSE1 = Negative School Environment at Time 1. NLE1 = Negative Life Events at Time 1. NEE5 = Negative Economic Events at Time 5. RV5 = Romantic Violence at Time 5. PC5 = Physical Complaints at Time 5. I = Initial Level. S = Slope. IS = Internalizing Symptoms. The 3-class GMM-CV (the optimal class model) was utilized as the unconditional (base) GMM.

not included in the tables.) For the predictor model (see Table 8.5), the results are identical to those from the manual 3-step approach. Comparing the distal outcome results across the manual and auxiliary 3-step approaches indicates that the results are very similar, although not identical (see Table 8.6).

Illustrative Example 8.8: Utilizing the Auxiliary Option with "Lanza Commands"

Recently, a new method for the estimation of auxiliary distal outcomes has been proposed by Lanza, Tan, & Bray (2013) with a more flexible model-based approach. This method allows users to take into account the variance (i.e., distribution) of a distal outcome variable when predicting a distal outcome. Thus, this approach can be used for both continuous and categorical distal outcomes (i.e., nominal outcomes). The Lanza et al. (2013) approach assumes that there are no correlations between the latent class indicators and the distal outcomes. That is, conditional independence is assumed between the latent class indicators and the distal outcomes. Thus, when estimating a conditional GMM using this approach, the distal outcomes do not influence the GMM classification (Lanza et al., 2013).

In M*plus*, specifying an auxiliary variable as "(DCON)" tells the program to use the Lanza et al. (2013) approach and treat the specified variable as a distal continuous outcome. In the M*plus* output, the mean of each distal outcome is provided for all classes. If an auxiliary variable is specified as "(DCAT)," the Lanza et al. (2013) method will be used, and the variable will be treated as a distal categorical outcome (this is the appropriate specification for either a binary outcome or a categorical unordered outcome, such as a nominal variable). Note that unlike the previously discussed 3-step approach, both DCON and DCAT specifications can be incorporated into a GMM simultaneously. M*plus* syntax for the Lanza et al. (2013) approach is provided in Figure 8.8.

In the M*plus* output under the heading "Equality tests of means/probabilities across classes," the probabilities of each distal outcome for each class are provided. When we used both the DCON (for continuous variables) and DCAT (for categorical variables) estimations, we encountered difficulties due to missing cases which led to the exclusion of all missing cases from the distal outcome analysis (i.e., mean and probability equality test). The results are provided in Table 8.7. Note that, in order to demonstrate the model with categorical variables, we created a new file (name: lanza.dat), which included three binary distal outcomes (i.e., NEE5_B, RV5_B, and PC5_B) using the median splitting method.

In this chapter, we have provided many options for incorporating covariates in a GMM. We have included a summary of these options in Table 8.8 to allow readers to draw some comparisons between the options.

```
DATA:FILE = example_ch_8.dat;
VARIABLE: NAMES = ID IS1 IS2 IS3 IS4 IS5 MALE MH1 NLE1
                 NSE1 NEE5 RV5 PC5 NEE5_B RV5_B PC5_B;
          USEV = IS1-IS5;
          CLASSES = C(3);

AUXILIARY = RV5 (DCON) NEE5 (DCON) PC5 (DCON) RV5_B (DCAT) NEE5_B (DCAT) PC5_B (DCAT);
```

AUXILIARY = (DCON) is used to specify continuous or binary distal outcomes.

AUXILIARY = (DCAT) is used to specify categorical (binary or nominal) distal outcomes.

```
          MISSING ARE ALL (9.00);

ANALYSIS: TYPE = MIXTURE;
          STARTS = 500 10;
          STITERATIONS = 10;

MODEL: %OVERALL%
       I S | IS1@0 IS2@1 IS3@3 IS4@4 IS5@6;
%C#1%
       I S;
%C#2%
       I S;
%C#3%
       I S;

OUTPUT: TECH7;
```

When analyzing a GMM, *both DCON and DCAT options can be specified simultaneously*. By doing so, the associations with both continuous and categorical distal outcomes can be estimated simultaneously.

FIGURE 8.8 *Mplus* Syntax for Auxiliary Options with Distal Outcomes Using the Lanza et al. (2013) Approach. Note: MH1 = Mothers' Hostility at Time 1. NSE1 = Negative School Environment at Time 1. NLE1 = Negative Life Events at Time 1. NEE5 = Negative Economic Events at Time 5. RV5 = Romantic Violence at Time 5. PC5 = Physical Complaints at Time 5. I = Initial Level. S = Slope. IS = Internalizing Symptoms. The 3-class GMM-CV (the optimal class model) was utilized as the unconditional (base) GMM.

TABLE 8.7 The Mean (Probability) Structures of Continuous and Categorical Distal Outcomes Using Auxiliary DCON and DCAT Options.

Continuous Outcomes	AUXILLARY = (DCON) option (Mean)			
	High and decreasing	Low and increasing	Consistently low	Overall Chi-square test
Romantic violence (RV)	2.934	2.430	1.926	82.016*** [a, b, c]
Negative economic events (NEE)	1.726	1.417	1.015	5.469
Physical complaints (PC)	2.974	4.312	2.667	26.693*** [a, b]

Binary Outcomes	AUXILLARY = (DCAT) option (Probability)			
	High and decreasing	Low and increasing	Consistently low	Overall Chi-square test
Romantic violence (RV)	.904	.609	.368	42.503*** [a, b, c]
Negative economic events (NEE)	.434	.355	.261	3.032
Physical complaints (PC)	.418	.619	.302	15.112*** [b]

Note: Unstandardized coefficients are shown. For the binary distal outcomes (0, 1), probabilities for the "1" response (i.e., Category 2 in the M*plus* output) are provided. The 3-class GMM-CV was used. The number of missing cases varied depending on the distal outcome of interest.

[a] High and decreasing vs. Low and increasing, $p < .05$.

[b] Low and increasing vs. Consistently low, $p < .05$.

[c] High and decreasing vs. Consistently low, $p < .05$.

***$p < .001$.

TABLE 8.8 A Summary of Auxiliary Options.

Options	Continuous vs. Categorical	Predictor(s) vs. Distal outcome(s)	Specification with other auxiliary options	Assumptions	Estimation issues for predictors and distal outcomes
3-Step Approach					
(R3STEP)	Considered as Continuous	Predictor(s)	No		Missing cases of predictors [a]
(DU3STEP)	Continuous or Binary	Distal outcome(s)	No	Unequal means and variances across classes.	
(DE3STEP)	Continuous or Binary	Distal outcome(s)	No	Unequal means and equal variances across classes (Good for small class sizes or binary distal outcomes).	
Lanza et al. (2013) Approach					
(DCON)	Continuous or Binary	Distal outcome(s)	Yes (with DCAT option)	Latent class indicators (X) are not correlated with distal outcomes (U) given a latent class variable (C) (conditional independence).	Missing cases of distal outcomes [a]
(DCAT)	Only Categorical (Binary or nominal)	Distal outcome(s)	Yes (with DCON option)		Missing cases of distal outcomes [a]

[a] The class memberships are retained, but the coefficients may be biased depending on how many missing cases exist for the covariates (predictors or distal outcomes).

Chapter 8 Exercises

This exercise is an extension of the exercise in Chapter 7. The covariates (i.e., predictors and distal outcomes) can be incorporated into the identified class model in Chapter 7 to form a conditional GMM. Three additional variables were included in the dataset used in the Chapter 7 Exercises. The M*plus* syntax below indicates the variable order:

DATA: File is exercise_Ch.8.dat;

VARIABLE: NAMES are IS1F IS2F IS3F IS4F FEP1 GHP4F GHP4F_B;

MISSSING ARE ALL (9);

Note that IS1F~IS4F indicate composite (mean) scores for internalizing symptoms of fathers. FEP1 represents family economic hardship at Time 1. GHP4F represents global health problems reported by fathers at Time 4. GHP4F_B represents a binary variable of GHP4F created by taking a median split of the continuous variable.

In order to avoid local maxima problems, insert the M*plus* syntax below under the ANALYSIS command for this exercise.

ANALYSIS: TYPE = MIXTURE;

STARTS = 200 10;

STITERATIONS = 10;

1. Investigate the influence of a manifest predictor (FEP1) on the latent class variable (C):
    A. Using the 1-step approach,
        1) Investigate the influence of a predictor (FEP1) on the between-class variation. (Specify a path between the predictor and the latent class variable using multinomial logistic regression.) Interpret all statistically significant logistic coefficients (β) using the percent change in the odds ($100 (\text{Exp} (\beta) - 1)$).
        2) Add paths between the predictor (FEP1) and latent growth factors of IS (specify regression paths between the predictor and the latent growth factors (π)) in the above model. Recall that these coefficients represent effects that exist across classes (the effects are common to all classes). Interpret all statistically significant regression coefficients (γ).

EXERCISE TABLE 8.1 The Mean (or Probability) Difference for Several Approaches (1-Step, 3-Step, and Lanza's Approaches).

Outcomes	Specify each class name here	Wald chi-square equality test (df)
One-step approach		
Global health problems (Continuous)		
Three-step approach (DE3STEP)		
Global health problems (Continuous)		
Lanza's approach (DCON)		
Global health problems (Continuous)		
Lanza's approach (DCAT)		
Global health problems (Binary)		

Note: Unstandardized coefficients are shown. df = Degrees of Freedom.
***p < .001.

    B. Using the 3-step approach (R3STEP), estimate the logistic regression coefficients (β).

    C. Compare the logistic regression coefficients (β) between the 1-step approach (Question 1.A.1) and 3-step approach (Question 1.B). Are the results similar or different? Why?

2. Investigate the influence of the latent class variable (C) on a distal outcome (GHP4F).

    A. Test mean equality using the 1-step approach and report the results using Exercise Table 8.1.

    B. Test mean equality using the auxiliary 3-step approach (with the DE3STEP option) and report the results using Exercise Table 8.1.

    C. Test mean equality using Lanza's approach (with the DCON option) and report the results using Exercise Table 8.1.

    D. Compare the results of the mean equality tests across the three approaches (i.e., 1-step, 3-step, and Lanza's 3-step auxiliary approach utilizing the DCON syntax). Are the results similar or different? Why?

    E. Using the DCAT option, estimate the effect of class membership on a categorical distal outcome (GHP4F_B) and interpret the results using the probability of each class.

References

Asparouhov, T., & Muthén, B. O. (2013). Auxiliary variables in mixture modeling: A 3-step approach using Mplus [Online web notes]. Retrieved from http://statmodel.com/examples/webnotes/AuxMixture_submitted_corrected_webnote.pdf. Accessed 6 December 2015.

Asparouhov, T., & Muthén. B. O. (2014). Auxiliary variables in mixture modeling: Three-step approaches using Mplus. *Structural Equation Modeling*, 21(3), 329–341.

Bolck, A., Croon, M., & Hagenaars, J. (2004). Estimating latent structure models with categorical variables: One-step versus three-step estimators. *Political Analysis*, 12(1), 3–27.

Clark, S., & Muthén, B. (2009). Relating latent class analysis results to variables not included in the analysis. Retrieved from https://www.statmodel.com/download/relatinglca.pdf. Accessed 6 December 2015.

Feldman, B., Masyn, K. E., & Conger, R. D. (2009). New approaches to studying problem behaviors: A comparison of methods for modeling longitudinal, categorical adolescent drinking data. *Developmental Psychology*, 45(3), 652–676.

Lanza, S. T., Tan, X., & Bray, B. C. (2013). Latent class analysis with distal outcomes: A flexible model-based approach. *Structural Equation Modeling*, 20(1), 1–26.

Muthén, B. O. (2003). Statistical and substantive checking in growth mixture modeling: Comment on Bauer and Curran (2003). *Psychological Methods*, 8(1), 369–377.

Muthén, B. O. (2004). Latent variable analysis: Growth mixture modeling and related techniques for longitudinal data. In D. Kaplan (Ed.), *Handbook of quantitative methodology for the social sciences* (pp. 345–368). Newbury Park, CA: Sage.

Muthén, L. K., & Muthén, B. O. (2008). *Mplus user's guide* (5th ed.). Los Angeles: Muthén & Muthén.

Muthén, B. O., & Shedden, K. (1999). Finite mixture modeling with mixture outcomes using the EM algorithm. *Biometrics*, 55(2), 463–469.

Vermunt, J. K. (2010). Latent class modeling with covariates: Two improved three-step approaches. *Political Analysis*, 18(4), 450–469.

Wickrama, K. A. S., Conger, R. D., & Abraham, W. (2008). Early family adversity, youth depressive symptom trajectories, and young adult socioeconomic attainment: A latent trajectory class analysis. *Advances in life course research*, 161–192.

9

SECOND-ORDER GROWTH MIXTURE MODELS (SOGMMS)

Introduction

In Chapters 3 through 6, we examined second-order growth trajectories of a global domain as a curve-of-factors model (CFM) (Chapters 3 and 4) and a factor-of-curves model (FCM) (Chapters 5 and 6). We considered internalizing symptoms (IS) as the global domain comprised of three subdomains (symptoms of depression [DEP], anxiety [ANX], and hostility [HOS]). The main purpose of this chapter is to introduce an approach to investigating potential heterogeneity in higher-order trajectories of global domains (both as a CFM and a FCM). This approach involves combining second-order growth curve modeling (Chapters 3 through 6) and growth mixture modeling (Chapters 7 and 8) procedures.

Accordingly, in this chapter, two extensions to a GMM involving second-order growth curves (known as second-order growth mixture models, SOGMMs) are discussed. First, this chapter discusses a GMM extension of a curve-of-factors model (CFM). Recall from Chapters 3 and 4 that a CFM estimates second-order trajectories of a global domain and, therefore, is able to account for measurement error in the manifest indicators of confirmatory factors (at the population level). Consequently, a CFM may provide a more precise test for population heterogeneity compared to a LGCM with composite measures. This chapter also discusses a GMM extension of a factor-of-curves model (FCM). Because a FCM estimates second-order trajectories after accounting for measurement error in the manifest indicators of primary growth factors (see Chapters 5 and 6), a FCM, like a CFM, may also provide a more precise test of the heterogeneity that exists within the sample population compared to a LGCM with composite measures.

Research suggests that understanding underlying global domains and their trajectories is important because the developmental course, as well as the antecedents and consequences, of these global domains may differ from those of the constituent subdomains. Trajectories of global domains are also substantively important. As previously discussed, examples of global domains that underlie different subdomains include: internalizing symptoms (IS), healthy lifestyle behaviors, socioeconomic status, work quality, and marital quality. Theory often posits that heterogeneity exists in these global domains, and recent studies have also begun to demonstrate the existence of heterogeneity in higher-order trajectories of global domains (Kirves, Kinnunen, de Cuyper, & Mäkikangas, 2014; Vaske, Ward, Boisvert, & Wright, 2012).

Furthermore, in order to investigate potential heterogeneity among several growth trajectories specific to individual subdomains, this chapter also introduces a multidimensional growth mixture model (MGMM). Typically, this model does not include second-order growth factors. Instead, primary growth factors of multiple subdomains are used together as indicators of latent classes. This chapter provides illustrative examples of various types of GMM applications using longitudinal data in M*plus*. Figures, equations, and M*plus* syntax are provided for all of the examples.

Also, it should be noted that for illustrative purposes we utilize linear trajectories for our SOGMMs in this chapter, but non-linear global trajectories (e.g., quadratic slopes) may also exist. It is possible to use the same analytical approach that will be discussed in this chapter to estimate SOGMMs with non-linear global trajectories.

Estimating a Second-Order Growth Mixture Model: A Curve-of-Factors Model (SOGMM of a CFM)

A growth mixture model (GMM) can be extended to a second-order growth mixture model (SOGMM) to investigate latent classes of second-order trajectories. For example, latent classes from a curve-of-factors model (CFM) can be investigated using a GMM approach. For the measurement model of a CFM, the equation can be written as in the case of a longitudinal confirmatory factor analysis (LCFA). The first-order measurement model (measurement factor model) at each time point, t, is defined by:

- the measurement intercept values (τ_{jt}) for each observed indicator (y_{jti}),
- the factor loading for each indicator variable (λ_{jt}),
- a latent factor variable (η_{ti}),
- and the indicator-specific variance (ε_{jti})

where the subscript j refers to a particular indicator of the first-order factor model and the subscript i refers to individuals. Thus, the equation for observed indicator y_{jti} can be written as:

$$y_{jti} = \tau_{jti} + \lambda_{jt}\, \eta_{ti} + \varepsilon_{jti}, \quad \varepsilon_{jti} \sim \text{NID}\,(0, \sigma^2_{jt}) \tag{9.1}$$

The growth curve equation can be written to incorporate second-order growth parameters (intercept and slope). For a second-order growth model (the structural model), the first-order latent factors (η) can be used as the indicators of second-order factors. In this model specification, the loadings for the intercept factor, π_{0i}, are set to 1.0, and loadings for the slope factor, π_{1i}, are set to correspond to the measurement time intervals (commonly, 0, 1, 2, 3, ... for equally spaced intervals). Like a first-order growth curve model, for each intercept and slope factor, a mean (μ) and variance (ψ) value is estimated. Thus, the equations of a CFM are as follows (see Figure 3.2 in Chapter 3):

$$\eta_{ti} = \pi_{0i} + \lambda \times \pi_{1i} + \zeta_{it}, \quad \zeta_{it} \sim \text{NID}\,(0, \psi_t) \tag{9.2}$$

$$\pi_{0i} = \mu_{00} + \zeta_{0i}, \quad \zeta_{0i} \sim \text{NID}(0, \psi_{00}) \tag{9.3}$$

$$\pi_{1i} = \mu_{10} + \zeta_{1i}, \quad \zeta_{1i} \sim \text{NID}\,(0, \psi_{11}) \tag{9.4}$$

$$\Psi = \begin{bmatrix} \psi_{00} & \\ \psi_{10} & \psi_{11} \end{bmatrix} \tag{9.5}$$

where μ_{00} represents the average baseline values of the second-order trajectories if time increments 0, 1, 2, and 3 are used, and μ_{10} represents the average slope. Residuals are assumed to be normally and independently distributed (NID) with a mean of zero and variance-covariance structure (ψ).

As discussed in the previous chapter, based on a latent class analysis (LCA), a GMM assumes that a categorical latent variable, C, specifies the most likely category of which each individual is a member. In other words, individuals are assigned to one of the distinct second-order growth pattern categories (with a second-order intercept and second-order slope) according to their own growth pattern based on their "posterior probabilities." As in the case of the GMM introduced in Chapter 7, a mixture model of a second-order growth curve (SOGMM) uses a categorical latent variable to represent *a mixture of classes indicating sub-populations of second-order trajectories* (hereafter, second-order latent classes) based on posterior probabilities (which are similar to factor scores). Posterior probabilities of each person for all classes are inferred from the data. (See Chapter 7 for more information on posterior probabilities).

POINT TO REMEMBER...

A SOGMM estimates the individual probabilities associated with each second-order latent class for all individuals, and the person is typically classified into the class for which he/she has the highest posterior probability.

The second-level equation (i.e., the structural model) for individual i is indexed by the latent class category (q = 1, 2, 3, ..., Q):

$$\pi_{q0i} = \mu_{q00} + \zeta_{q0i}, \qquad \zeta_{q0i} \sim \text{NID } (0, \psi_{q00}) \qquad\qquad (9.6)$$

$$\pi_{q1i} = \mu_{q10} + \zeta_{q1i}, \qquad \zeta_{q1i} \sim \text{NID } (0, \psi_{q11}) \qquad\qquad (9.7)$$

$$\Psi_q = \begin{bmatrix} \psi_{q00} & \\ \psi_{q10} & \psi_{q11} \end{bmatrix} \qquad\qquad (9.8)$$

where μ_{q00} and μ_{q10} represent the average intercept and slope in latent trajectory class q, and ζ_{q0i} and ζ_{q1i} are the disturbances (i.e., errors) reflecting the variability of the estimated intercepts and slopes across individuals within each latent trajectory class. These disturbances have a variance-covariance structure represented by ψ_q. ψ_{q00} and ψ_{q11} are the error variances of the estimated intercept and slope factors, respectively, and ψ_{q10} is the covariance between intercept and slope factors within the latent trajectory class q. The q subscript indicates that most parameters are allowed to vary between estimated latent trajectory classes. Thus, each second-order latent class (or group) could be defined by its own CFM based on class-specific parameters consisting of the variance and covariance structure (ψ_q) and means (i.e., μ_{q00} and μ_{q10}) of growth. Note that the q subscript can also be used to index other model parameters, such as factor loadings (λ), intercepts (τ), residual variances of manifest indicator variables (σ^2 and ψ), and first-order latent variables (if those parameters vary across classes). However, the additional parameters increase model complexity, which often leads to convergence problems (e.g., no available standard errors or an inadmissible solution). Therefore, to estimate a simple and straightforward model, we assume that the variances of manifest indicator variables and first-order latent variables are invariant across classes. The SOGMM-CF specification is shown in Figure 9.1.

Illustrative Example 9.1: A Second-Order Growth Mixture Model of a CFM (SOGMM-CF)

In order to select the optimal class model using a SOGMM-CF, we used the same CFM that we estimated as the baseline model in Chapter 4 (see Figure 4.13). Recall that our example CFM met the assumption of longitudinal invariance (i.e., partial strong invariance of the indicators). It is important that the longitudinal invariance of indicators is confirmed (at least partial strong invariance) before estimating a SOGMM-CF in order to ensure that the same confirmatory latent factor has been assessed over time.

In order to select the optimal class model, we followed the same model selection process we introduced in Chapter 7. First, several SOGMM-CFs were analyzed using the following mixture models based on different assumptions: (a) a

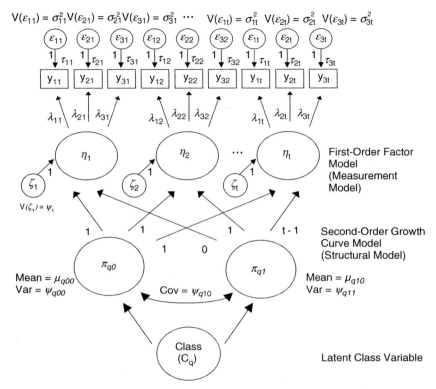

FIGURE 9.1 Second-Order Growth Mixture Model – Curve-of-Factors (SOGMM-CF).

Note: Cov = Covariance. Var = Variance. Subscript q is a categorical class variable q (q = 1, 2, 3, ..., Q).

LCGA, (b) a GMM-CI, and (c) a GMM-CV. Below is a brief review of the varying assumptions of these models. See Chapter 7 for a detailed discussion of these three approaches.

- A LCGA fixes variances of global growth factors to zero across classes.
- A GMM-CI constrains the variances of the global growth factors (i.e., π_{q0i} and π_{q1i}) across classes to be equal (homogeneous classes).
- A GMM-CV freely estimates all variances of the global growth factors across classes.
- A LCGA is a reduced GMM-CI, which is a reduced GMM-CV. The researcher may prefer one over the other based on theoretical expectations.

TROUBLESHOOTING

Researchers often experience convergence problems with second-order growth mixture models, particularly a GMM-CV (e.g., negative variances) because it is the most "freed" model in comparison to the other two mixture models. *In order to address these convergence problems, it may be necessary to follow an exploratory approach by constraining and/or equalizing some of the parameters in a GMM-CV,* which may produce a model with a combination of elements from all three types of models (Grimm & Ram, 2009). Furthermore, this combined model requires increased computational time. By specifying PROCESSORS=8 under the ANALYSIS command, the computational speed can be increased.

The optimal model was selected given all available information, such as the model fit indices, plot, and interpretability of latent classes of trajectories as well as theoretical considerations. In order to avoid local maxima, we applied the "STARTS = 500 10;" and "STITERATIONS = 10;" options to all class models. The M*plus* syntax for our example SOGMM-CF using a GMM-CI (the M*plus* default model for a GMM) is provided in Figure 9.2.

Although the M*plus* syntax for a SOGMM-CF looks quite long and complex, a closer inspection reveals that this syntax is simply a combination of the CFM and GMM syntaxes, and most of this syntax has already been described in Chapter 4 (CFM) and Chapter 7 (GMM). One important aspect of a SOGMM-CF specification is related to the within-class mean structure (μ_{q00} and μ_{q10}). Unless the mean structures of each class are specified (see the last part of the syntax), the fixed factor scale setting approach is automatically used as the default approach for the SOGMM-CF identification (i.e., the intercept mean will be fixed to zero for the last class in the GMM). In order to avoid this fixed factor approach and apply the marker variable approach, as seen in Figure 9.2, we fixed the intercept of each depressive symptoms manifest variable to 0 using the syntax [DEP1-DEP5@0];. Then, by specifying the means of the intercept and slope factors across classes, a SOGMM-CF can estimate means (μs) of the global intercept factors for all classes.

To utilize the latent class growth analysis (LCGA) approach to a SOGMM-CF, the M*plus* syntax is as follows:

```
MODEL: %OVERALL%
:
I_IS S_IS | IS1@0 IS2@1 IS3@3 IS4@4 IS5@6;
[I_IS S_IS]; I_IS-S_IS@0;
%C#1%
[I_IS S_IS];
%C#2%
```

[I_IS S_IS];
To utilize the growth mixture model with class-varying variances approach (GMM-CV) to a SOGMM-CF, the M*plus* syntax is as follows:
MODEL: %OVERALL%
⋮

I_IS S_IS | IS1@0 IS2@1 IS3@3 IS4@4 IS5@6;
[I_IS S_IS]; I_IS–S_IS;
⋮

%C#1%
[I_IS S_IS]; I_IS–S_IS;
%C#2%
[I_IS S_IS]; I_IS–S_IS;

Note that for a SOGMM-CF utilizing the GMM-CI approach, the variances of global growth factors, I_GIS–S_GIS, must be set to be equal across classes by specifying those parameters only under the %OVERALL% statement (see Figure 9.2).

However, if certain parameters need to be released in a specific class, specify those parameters under the appropriate class model (e.g., %C#1% for Class 1). In the same manner, the GMM-CV approach also releases the covariance structures across classes by specifying "I_IS WITH S_IS;" in both the %OVERALL% statement and again under each class model (i.e., %C#1% and %C#2%). It is important to note that releasing covariances across classes releases *all* class parameters (i.e., μ_{q00}, μ_{q10}, ψ_{q00}, ψ_{q11}, and ψ_{q10}), which may lead to convergence problems. More importantly, estimating these parameters may change the class classification, model fit indices, and growth parameters. Thus, the class model parameters should be specified carefully and in a meaningful manner. Because we detected many convergence problems (e.g., negative variances) when estimating our example SOGMM-CF, we excluded the fully released SOGMM-CF from subsequent analyses.

The initial model fit results are shown in Table 9.1. As can be seen, all model solutions were inadmissible based on the warning messages in the M*plus* output. Inadmissible solutions (i.e., negative variances) were not identified in the curve-of-factors model (CFM). As previously noted, an inadmissible solution suggests the need for modifications to the specified SOGMM-CF using an exploratory approach (Grimm & Ram, 2009). In our experience, we have observed that convergence problems (e.g., negative variances) are common when the CFM portion of the SOGMM is particularly complex.

This modification process is not always simple or straightforward because the modification process itself may lead to other convergence issues, such as local solutions. In order to avoid these problems, we propose the following exploratory approach. Although this is an exploratory approach (merely a modification to the model based on empirical evidence), it is important that maximum consideration be given to theoretical expectations and the interpretability of results in this process.

```
Title:SOGMM of a CFM
DATA: File is C:\example_ch_9.dat;
VARIABLE: NAMES are ANX1-ANX5 DEP1-DEP5 HOS1-HOS5;
        USEV = ANX1-ANX5 DEP1-DEP5 HOS1-HOS5;
        CLASSES = C(2);
        MISSING = ALL(9);

ANALYSIS: TYPE = MIXTURE;
        STARTS = 500 10;
        STITERATIONS = 10;
        PROCESSORS = 8;
```

The **PROCESSORS** option is used to *increase computational speed* by specifying the number of processors to be used for parallel computing.

```
MODEL:%OVERALL%
    IS1 by DEP1
        ANX1 HOS1 (1-2);
    IS2 by DEP2
        ANX2 HOS2 (1-2);
    IS3 by DEP3
        ANX3 HOS3 (1-2);
    IS4 by DEP4
        ANX4 HOS4 (1-2);
    IS5 by DEP5
        ANX5 HOS5 (1-2);

[DEP1-DEP5@0]; [ANX2-ANX5] (3); [HOS1-HOS5] (4);
```

By assigning labels to the model parameters, the parameters are constrained (*longitudinal invariance*). See how to test this in Chapter 4.

```
I_GIS S_GIS | IS1@0 IS2@1 IS3@3 IS4@4 IS5@6;
[I_GIS S_GIS]; I_GIS-S_GIS;

    DEP1 WITH DEP2 DEP3 DEP4 DEP5;
    DEP2 WITH DEP3 DEP4 DEP5;
    DEP3 WITH DEP4 DEP5; DEP4 WITH DEP5;
    ANX1 WITH ANX2 ANX3 ANX4 ANX5;
    ANX2 WITH ANX3 ANX4 ANX5;
    ANX3 WITH ANX4 ANX5; ANX4 WITH ANX5;
    HOS1 WITH HOS2 HOS3 HOS4 HOS5;
    HOS2 WITH HOS3 HOS4 HOS5;
    HOS3 WITH HOS4 HOS5; HOS4 WITH HOS5;
    HOS5 WITH DEP3 ANX1 ANX5;
    HOS4 WITH DEP1;
```

Autocorrelated errors are employed to avoid model misspecification (see how to specify this structure in Chapter 4).

```
    %C#1%
    [I_GIS S_GIS];
    %C#2%
    [I_GIS S_GIS];
```

By specifying [I_GIS S_GIS]; for each class except the reference group, *mean structures* of the intercept and slope factors are freely estimated across classes. All other model parameters are constrained to be equal across classes.

```
OUTPUT: SAMPSTAT STANDARDIZED TECH7 TECH11 TECH14;
```

FIGURE 9.2 M*plus* Syntax of the Two-Class SOGMM-CF using the GMM-CI approach.

Note: DEP = Depressive Symptoms. ANX = Anxiety Symptoms. HOS = Hostility Symptoms. IS = Internalizing Symptoms. I_GIS = Intercept of IS. S_GIS = Slope of IS. GMM-CI = Growth Mixture Model with a class–invariant variance-covariance model. The marker variable scale setting approach was used.

TABLE 9.1 Fit Statistics for the Second-Order Growth Mixture Models – Curve-of-Factors (SOGMM-CF, total n = 436).

Fit statistics	2 Classes	3 Classes	4 Classes
LCGA			
LL (No. of Parameters)	-2546.751 (64)[b]	-2438.296 (67) [b,c]	-2379.527 (70) [b,c]
BIC	5482.471	5283.794	5184.488
SSABIC	5279.368	5071.171	4962.345
Entropy	.988	.952	.953
Adj. LMR-LRT (*p*)	430.363 (.074)	205.632 (.460)	131.465 (.266)
BLRT (*p*)	453.967 (.000)	216.910 (.000)	138.676 (.000)
Group size (%) C1	25 (5.7%)	18 (4.1%)	5 (1.2%)
C2	411 (94.3%)	42 (9.6%)	18 (4.1%)
C3		376 (862%)	43 (9.8%)
C4			370 (84.9%)
GMM-CI			
LL (No. of Parameters)	-2432.887 (67)[b]	-2357.981 (70)[b]	-2315.674 (73)[b]
BIC	5272.975	5141.396	5075.015
SSABIC	5060.352	4919.253	4843.351
Entropy	.984	.954	.936
Adj. LMR-LRT (*p*)	249.688 (.04)	142.023 (.139)	122.071 (.171)
BLRT (*p*)	263.383 (.00)	149.812 (.000)	128.767 (.000)
Group size (%) C1	22 (5.0%)	16 (3.7%)	13 (3.0%)
C2	414 (95.0%)	42 (9.6%)	33 (7.5%)
C3		378 (86.7%)	37 (8.5%)
C4			353 (81.0%)
GMM- CV			
LL (No. of Parameters)	-2333.305 (69)[b]	-2339.804 (74) [b,c]	-2210.022 (79) [a,b]
BIC	5091.967	5129.353	—
SSABIC	4872.997	4894.516	—
Entropy	.720	.835	—
Adj. LMR-LRT (*p*)	442.001 (.436)	—	—
BLRT (*p*)	456.547 (.000)	—	—
Group size (%) C1	123 (28.2%)	0 (0.0%)	—
C2	313 (71.8%)	93 (21.3%)	—
C3		343 (78.6%)	—
C4			—

Notes. LCGA = Latent Class Growth Analysis. GMM-CI = Growth Mixture Model with class-invariant variances-covariances. GMM-CV = Growth Mixture Model with class-varying variances (constrained covariances to be equal across class). LL = Log-Likelihood value. No. of Parameters = Number of estimated (free) parameters. BIC = Bayesian Information Criteria. SSABIC = Sample Size Adjusted BIC. LMR-LRT = Lo-Mendell-Rubin Likelihood Ratio Test. Adj.LMR-LRT = Adjusted LMR. BLRT = Bootstrapped Likelihood Ratio Test.
[a] = no repeated log-likelihood values (i.e., local maxima). [b] = inadmissible solution (i.e., negative variances of growth parameter) in classes. [c] = no replicated log-likelihood values (H0 log-likelihood value) for the k-1 class model.

The following modification sequence can be applied to fitting a SOGMM-CF as well as other types of SOGMMs (such as a SOGMM-FC).

(1) First, try increasing the number of classes in the SOGMM-CF to see if the same parameters consistently produce the negative variances. If the parameters that initially produced negative variances continue to produce negative variances after increasing the number of classes, these parameters are potential candidates for modification. However, if *different* parameters produce negative variances when the number of classes is increased, it may be difficult to locate the proper parameters to modify. Thus, a SOGMM-CF may not fit the data well.

(2) Determine whether the negative variances are estimated in parameters of the measurement model (i.e., the first-order factor model) or the structural model (i.e., the second-order growth curve model) of the SOGMM-CF (see Figure 9.1).

   (a) If the residual variances (σ^2) of manifest indicators or first-order latent factors (ψ_t) are negative (which are generally constrained to be zero or equal across classes for a class-invariant model), parameters in the baseline CFM portion of the model can be modified and then the SOGMM-CF can be re-analyzed. The consistent detection of negative variances in the LCGA, GMM-CI, and GMM-CV even after modifying the CFM portion of the model indicates that a SOGMM-CF specification may not fit the data well.

   (b) If the negative variances are found in parameters of the CFM portion of the model (i.e., the second-order growth parameters), we recommend directly modifying structural model parameters, including the variances of global growth factors (ψ_{q00} and ψ_{q11}) for the SOGMM-CF.

(3) It is also possible that negative variances may exist in both the measurement (first-order factor model) and structural (second-order growth curve model) models of the SOGMM-CF. In this instance, we recommend modifying the measurement model first, and then the structural model can be modified, if needed. Because the structural model is built on the measurement model, the negative variances found in parameters from the structural model often disappear after modifying the measurement model.

This exploratory approach, depicted in Figure 9.3 in detail, may produce a SOGMM-CF that is a combination of elements from several different GMM approaches (i.e., LCGA, GMM-CI, and GMM-CV).

Illustrative Example 9.2: Avoiding Convergence Problems

In order to select the optimal model, we applied our proposed exploratory approach in Figure 9.3 to all three models separately (i.e., LCGA, GMM-CI,

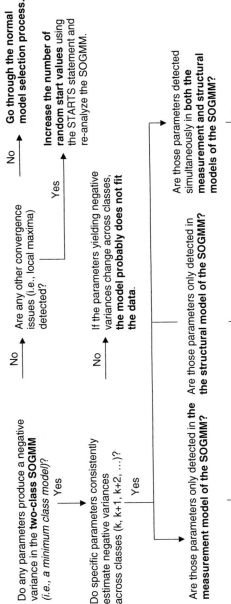

FIGURE 9.3 An Exploratory Modification Suggestion for Correcting Negative Variances in a Second-Order Growth Mixture Model (SOGMM).

and GMM-CV). First, in order to see whether negative variances are consistently estimated across classes, we increased the number of classes for all three models. In both the LCGA and the GMM-CI, we detected that the parameters with negative variances changed when the numbers of classes increased. Thus, we excluded those two models from subsequent analyses. For the GMM-CV, we consistently detected negative variances across the classes for anxiety symptoms at Time 5 (i.e., ANX5) and the slope of the global growth factor. Thus, those two parameters were selected as candidates for modification. The end result was an imposed equality constraint for ANX1, ANX4, and ANX5, and global slope factor variance fixed to zero. Using this exploratory approach, we reached a best-fitting model, which was a reduced version of the GMM-CV with three classes (see the syntax below). This model can also be obtained by adding the bold and italic M*plus* syntax to the two-class GMM-CI (recall that M*plus* considers a GMM-CI as the default model).

MODEL: **%OVERALL%**
IS1 by DEP1
        ANX1 HOS1 (1-2);
IS2 by DEP2
        ANX2 HOS2 (1-2);
IS3 by DEP3
        ANX3 HOS3 (1-2);
IS4 by DEP4
        ANX4 HOS4 (1-2);
IS5 by DEP5
        ANX5 HOS5 (1-2);
[DEP1-DEP5@0];
[ANX-ANX5] (3); ***ANX1(V1); ANX4(V1); ANX5(V1);***
[HOS1-HOS5] (4);
I_IS S_IS | IS1@0 IS2@1 IS3@3 IS4@4 IS5@6;
[I_IS S_IS]; ***I_IS;***
%C#1%
[I_IS S_IS]; ***I_IS; S_IS@0;***
%C#2%
[I_IS S_IS]; ***I_IS; S_IS@0;***
⋮

Table 9.2 provides the model fit indices for the modified SOGMM-CF from the two-class model to the five-class model. Given the model fit indices (i.e., SSABIC and entropy) and the statistically significant BLRT value (BLRT = 236.660, p = .000), the 3-class model of the modified SOGMM-CF (SSABIC = 4913.211, entropy = .922) was the preferred model compared to the 2-class model (SSABIC = 5138.254, entropy = .888). The model did not produce a proper solution for

TABLE 9.2 Fit Statistics for the Modified Second-Order Growth Mixture Model – Curve-of-Factors (SOGMM-CF; total n = 436).

Fit statistics	2 Classes	3 Classes	4 Classes	5 Classes
Modified GMM-CV				
LL (No. of Parameters)	−2476.194 (64)	−2357.864 (68)	−2270.534 (72) [b]	−2289.610 (76) [a]
BIC	5341.356	5129.007	4978.658	—
SSABIC	5138.254	4913.211	4750.168	—
Entropy	.888	.922	.830	—
Adj. LMR–LRT (p)	224.756 (.042)	227.309 (.120)	167.759 (.174)	—
BLRT (p)	234.001 (.000)	236.660 (.000)	174.660 (.000)	—
Group size (%) C1	45 (10.3%)	27 (6.2%)	22 (5.0%)	—
C2	391 (89.7%)	45 (10.3%)	46 (10.6%)	—
C3		364 (83.5%)	107 (24.5%)	—
C4			261 (59.9%)	—

Notes. GMM-CV = Growth Mixture Model with class-varying variances (constrained covariances to be equal across class).
LL = Log-Likelihood value. No. of Parameters = Number of estimated (free) parameters. BIC = Bayesian Information Criteria.
SSABIC = Sample Size Adjusted BIC. Adj.LMR–LRT = Adjusted Lo–Mendell–Rubin Likelihood Ratio Test.
BLRT = Bootstrapped Likelihood Ratio Test.

[a] = no repeated log-likelihood values (i.e., local maxima). [b] = inadmissible solution (i.e., negative variances of growth parameter) in classes.

TABLE 9.3 Class-Specific Global Growth Parameters of the Second-Order Growth Mixture Models (SOGMMs) and the Subdimensional Growth Parameters for the Multidimensional Growth Mixture Model (MGMM).

	Initial level		Slope level		Factor covariance
	Mean	Variance	Mean	Variance	
SOGMM-CF(GMM-CV) [a]					
High and decreasing	2.614***	.069*	-.186***	.000	.000
Moderate and increasing	1.660***	.276**	.154***	.000	.000
Consistently low	1.455***	.018***	-.022***	.000	.000
SOGMM-FC (GMM-CI)					
High and decreasing	2.779***	.061***	-.213**	.001	-0.004*
Moderate and increasing	1.886***	.061***	.203***	.001	-0.004*
Consistently low	1.441***	.061***	-.007	.001	-0.004*

TABLE 9.3 (cont.)

	Initial level		Slope level		Factor covariance
	Mean	Variance	Mean	Variance	
MGMM (GMM-CI) [b]					
High and decreasing					
Depressive symptoms	2.574***	.118***	-.152	.001	Ranged -.015*** (H_I and H_S) to .053* (H_I and D_I)
Anxiety symptoms	2.478***	.044***	-.206***	.000	
Hostility symptoms	2.547***	.113***	-.179***	.003***	
Moderate and increasing					
Depressive symptoms	1.727***	.118***	.138**	.001	Ranged -.015*** (H_I and H_S) to .053* (H_I and D_I)
Anxiety symptoms	1.525***	.044***	.137***	.000	
Hostility symptoms	1.644***	.113***	.176***	.003***	
Consistently low					
Depressive symptoms	1.452***	.118***	-.009	.001	Ranged -.015*** (H_I and H_S) to .053* (H_I and D_I)
Anxiety symptoms	1.299***	.044***	-.024***	.000	
Hostility symptoms	1.392***	.113***	-.025***	.003***	

Note: Unstandardized coefficients are shown. GMM-CV = Growth Mixture Model with class-varying variances. GMM-CI = Growth Mixture Model with class-invariant variances-covariances. Subscript I and S represent intercept and slope, respectively. D = Depressive Symptoms. H = Hostility Symptoms. [a] = The residual variances of three manifest indicators (ANX1, ANX2, ANX4, ANX5) were constrained to be equal, and the slope variance of the global growth factors was fixed to zero. [b] = The variance of the anxiety slope factor was fixed to zero. Total n = 436.

*p < .05. **p < .01. ***p < .001.

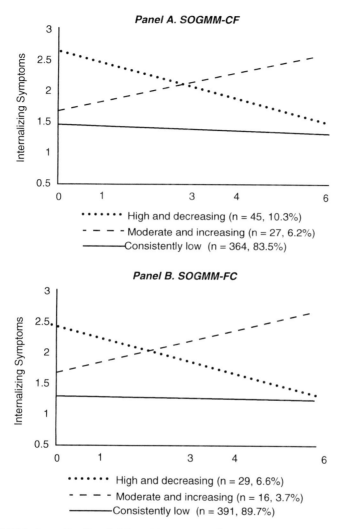

FIGURE 9.4 Predicted Mean Trajectories for Two SOGMM Approaches (CFM vs. FCM).
Note: SOGMM = Second-Order Growth Mixture Model. CF = Curve-of-Factors. FC = Factor-of-Curves.

the 4- and 5-class models. Also, as introduced in Chapter 7, examining the plot can provide useful evidence when attempting to select the optimal class model. Thus, plots are provided in Figure 9.4. Furthermore, class-specific growth parameters are presented in the top part of Table 9.3.

Overall, the three class trajectories appear to be clearly separated and represent substantively meaningful IS trajectories. These classes represent trajectories that are "initially high and decreasing IS," "initially moderate and increasing IS," and "consistently low IS." All of the class sizes were acceptable. Given all of this available information from the parameter values, plots, and model fit indices, the 3-class SOGMM-CF was selected as the optimal SOGMM-CF.

Estimating a Second-Order Growth Mixture Model: A Factor-of-Curves Model (SOGMM of a FCM)

The equation for the measurement model of a SOGMM incorporating a FCM is similar to the measurement model introduced in Chapter 5 for estimating trajectories that are specific to each subdomain (a first-order LGCM, which is the basis of a FCM). A first-order LGCM for a specific subdomain k is defined as:

$$y_{ikt} = \pi_{0ik} + \lambda_{kt} \times \pi_{1ik} + \varepsilon_{ikt}, \qquad \varepsilon_{ikt} \sim \text{NID}(0, \sigma_{kt}^2) \tag{9.9}$$

where $k = 1, 2, ..., K$ represents the subdomain, $i = 1, 2, ... , n$ indicates the respondent, and, ε_{it} is the error term for the ith individual at Time t ($t = 1, 2, 3, ...$). For a linear change with four equally-spaced measurements t can be 0, 1, 2, and 3. Residuals at each point in time are assumed to be normally and independently distributed (NID) with means of zero.

In a FCM, the primary growth curve factors of different subdomains along with their intercepts (π_{0ik}) and slopes (π_{1ik}) are used as multiple indicators for the second-order factors (η_{0i} and η_{1i}) in order to estimate the global domain. (In our example, the global domain is internalizing symptoms, or IS.) For primary latent growth factor structures for any subdomain k ($k = 1, 2, ..., K$), the equations and variance-covariance matrix can be written as follows:

$$\pi_{0ik} = \mu_{00k} + \lambda_{0k0} \times \eta_{0i} + \zeta_{0ik}, \qquad \zeta_{0ik} \sim \text{NID}(0, \psi_{00kk}) \tag{9.10}$$

$$\pi_{1ik} = \mu_{10k} + \gamma_{1k1} \times \eta_{1i} + \zeta_{1ik}, \qquad \zeta_{1ik} \sim \text{NID}(0, \psi_{11kk}) \tag{9.11}$$

$$\Psi_{kk} = \begin{bmatrix} \Psi_{0011} & & \\ \vdots & \ddots & \\ \Psi_{10k1} & \cdots & \Psi_{11kk} \end{bmatrix} \tag{9.12}$$

As in a SOGMM-CF, a SOGMM-FC uses a categorical latent variable *to represent a mixture of sub-populations of second-order trajectories* based on posterior probabilities. For each person, posterior probabilities for all the classes are inferred from the data. That is, individual probabilities associated with each latent class of second-order trajectories are estimated for all individuals. For each person, these posterior probabilities are calculated for each class (or group) of second-order trajectories (similar to factor scores), and the person is typically classified into the class for which he/she has the highest posterior probability. Thus, the latent trajectory class q can then be incorporated into the second-order global factor structure (η_{0i} and η_{1i}) using the following equations:

$$\eta_{q0i} = \alpha_{q00} + \zeta_{q0i}, \qquad \zeta_{q0i} \sim \text{NID} \ (0, \psi_{q00}) \tag{9.13}$$

$$\eta_{q1i} = \alpha_{q10} + \zeta_{q1i}, \qquad \zeta_{q1i} \sim \text{NID} \ (0, \psi_{q11}) \tag{9.14}$$

$$\Psi_q = \begin{bmatrix} \psi_{q00} & \\ \psi_{q10} & \psi_{q11} \end{bmatrix} \tag{9.15}$$

where α_{q00} and α_{q10} represent the average intercept and slope in latent trajectory class q; ζ_{q0i} and ζ_{q1i} are the disturbances (i.e., errors) reflecting the variability of the estimated intercepts and slopes, respectively, across individuals within each latent trajectory class. These disturbances have a variance-covariance structure represented by ψ_q. ψ_{q00} and ψ_{q11} are the error variances of the estimated intercept and slope factors, respectively, and ψ_{q10} is the covariance between the intercept and slope factors within the latent trajectory class q. The q subscript indicates that most parameters are allowed to vary across the estimated latent trajectory classes. Thus, technically each latent trajectory class could be defined by its own FCM based on its class-specific parameters consisting of the variance and covariance structure (ψ_q) and the means (i.e., α_{q00} and α_{q10}) of growth.

Note that, as in the case of a SOGMM-CF, the subscript q can also be specified into other model parameters, such as the mean (μ) of primary growth factors and the residual variances (σ^2 and ψ of manifest variables and first-order primary growth factors, respectively). However, the model complexity will be dramatically increased by estimating additional parameters, and increased model complexity often leads to convergence problems (e.g., no available standard errors). For simplicity, and for easy model convergence, we assume that the variances of manifest variables and first-order latent variables are constant (i.e., invariant) across classes. The SOGMM-FC specification is shown in Figure 9.5.

Illustrative Example 9.3: A Second-Order Growth Mixture Model of a FCM (SOGMM-FC)

In order to select the optimal SOGMM-FC, we used the FCM that we estimated in Chapter 6 as the baseline FCM (see Table 6.3). This FCM has already shown "loading equality," which was assessed by the tau-equivalent test introduced in Chapter 5.

In order to select the optimal SOGMM-FC, we followed the same model selection process used to select the optimal SOGMM-CF. First, several SOGMM-FCs were analyzed using different modeling approaches, including a LCFA, a GMM-CI, and a GMM-CV. The optimal model was selected after taking all available information into consideration, such as the model fit indices and plot, as well as interpretability of the latent class trajectories and theoretical considerations. In order to avoid local maxima, we applied the "STARTS = 500 10; STITERATIONS = 10;" options across all of the class models. The M*plus* syntax for a SOGMM-FC using a GMM-CI (the default GMM in M*plus*) is shown in Figure 9.6.

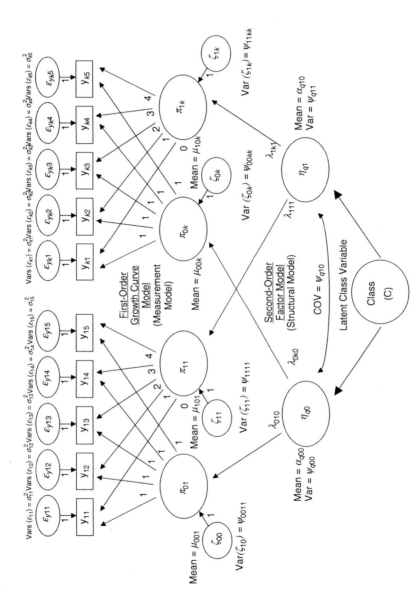

FIGURE 9.5 Second-Order Growth Mixture Model – Factor-of-Curves (SOGMM-FC).
Note: Subscript k represents a sub-domain (k = 1, 2, 3, ..., K). C = Categorical class variable.

```
Title: SOGMM of a FCM
DATA: File is C:\example_ch_9.dat;
        ⋮
ANALYSIS: TYPE = MIXTURE;
          STARTS = 500 10;
          STITERATIONS = 10;
          PROCESSORS = 8;
```

The **PROCESSORS** option is used to *__increase computational speed__* by specifying the number of processors to be used for parallel computing.

```
MODEL: %OVERALL%

  I_DEP S_DEP|DEP1@0 DEP2@1 DEP3@3 DEP4@4 DEP5@6;
  I_ANX S_ANX|ANX1@0 ANX2@1 ANX3@3 ANX4@4 ANX5@6;
  I_HOS S_HOS|HOS1@0 HOS2@1 HOS3@3 HOS4@4 HOS5@6;

I BY I_DEP@1
     I_ANX (L1)
     I_HOS (L2);

S BY S_DEP@1
     S_ANX (L1)
     S_HOS (L2);

I_DEP I_ANX I_HOS PWITH S_DEP S_ANX S_HOS;
[I_DEP-S_DEP@0];
[I_ANX I_HOS];
[S_ANX S_HOS];
[I S]; I-S;

DEP1 WITH ANX1 HOS1; ANX1 WITH HOS1;
DEP2 WITH ANX2 HOS2; ANX2 WITH HOS2;
DEP3 WITH ANX3 HOS3; ANX3 WITH HOS3;
DEP4 WITH ANX4 HOS4; ANX4 WITH HOS4;
DEP5 WITH ANX5 HOS5; ANX5 WITH HOS5;
DEP3 WITH DEP4; HOS3 WITH HOS4;
```

__Autocorrelated errors__ are specified to avoid model misspecification. See how to specify this structure in Chapter 4.

```
%C#1%
  [I S];
  [I_DEP-S_DEP@0];
  [I_ANX I_HOS] (M1-M2);
  [S_ANX S_HOS] (M3-M4);

%C#2%
  [I S];
  [I_DEP-S_DEP@0];
  [I_ANX I_HOS] (M1-M2);
  [S_ANX S_HOS] (M3-M4);
```

In order to constrain all mean structures of primary growth factors (μ) to be equal across classes, the *__same mean structures should be specified for each within-class model specification__* by assigning the same labels across classes.

```
OUTPUT: SAMPSTAT STANDARDIZED TECH7 TECH11 TECH14;
```

FIGURE 9.6 M*plus* Syntax for the Two-Class SOGMM-FC using the GMM-CI Approach.

Note: DEP = Depressive Symptoms. ANX = Anxiety Symptoms. HOS = Hostility Symptoms. I = Intercept of Internalizing Symptoms. S = Slope of Internalizing Symptoms. GMM-CI = Growth Mixture Model with a class-invariant variance-covariance model. The marker variable scale setting approach was used.

Like the SOGMM-CF, this SOGMM-FC is also a complex model. Thus, we specified the PROCESSORS=8 option under the ANALYSIS command to increase the program's computational speed. Across the classes, a SOGMM-FC includes many mean parameters (i.e., the means of global factors [α_{q0} and α_{q1}] and the means of primary growth factors [μs]). Thus, in order to avoid convergence problems, most mean parameters were constrained to be equal. Only certain mean parameters were released across classes. Furthermore, because our primary interest is the global growth factors, we constrained the mean parameters (μs) of all primary growth factors and released the mean parameters for the global factors (α_{q0} and α_{q1}). In terms of variance parameters, in order to be consistent with the GMM-CI approach, we constrained all variances of primary and global growth factors to be equal across classes.

It is also possible to utilize the LCGA and GMM-CV approaches for fitting a SOGMM-FC. The syntax for these two approaches is provided below.

M*plus* syntax for the LCGA version of a SOGMM-FC:

MODEL: **%OVERALL%**
⋮
I_DEP I_ANX I_HOS PWITH S_DEP S_ANX S_HOS;
[I_DEP-S_DEP@0];
[I_ANX I_HOS];
[S_ANX S_HOS];
[I S]; *I-S@0;*
⋮
%C#1%
[I S];
[I_DEP-S_DEP@0];
[I_ANX I_HOS] (M1-M2);
[S_ANX S_HOS] (M3-M4);
%C#2%
[I S];
[I_DEP-S_DEP@0];
[I_ANX I_HOS] (M1-M2);
[S_ANX S_HOS] (M3-M4);

Note that if the constraint "@0" is removed from the "*I-S@0;*" syntax, this syntax will analyze a SOGMM using the GMM-CI approach.

M*plus* syntax for the GMM-CV version of this SOGMM-FC:

MODEL: **%OVERALL%**
I_DEP I_ANX I_HOS PWITH S_DEP S_ANX S_HOS;
[I_DEP-S_DEP@0];
[I_ANX I_HOS];

[S_ANX S_HOS];
[I S]; *I-S;*
⋮
%C#1%
[I S]; *I-S;*
[I_DEP-S_DEP@0];
[I_ANX I_HOS] (M1-M2);
[S_ANX S_HOS] (M3-M4);
%C#2%
[I S]; *I-S;*
[I_DEP-S_DEP@0];
[I_ANX I_HOS] (M1-M2);
[S_ANX S_HOS] (M3-M4);

Note that in both of the above syntaxes (LCGA and GMM-CV), we did not release the covariance structures of the primary and global growth factors across classes in order to be consistent with the SOGMM-CF illustration.

The model fit results are shown in Table 9.4. Like the results of our SOGMM-CF, when utilizing the LCGA and GMM-CV approaches, the SOGMM-FC solutions were inadmissible due to growth parameters with negative variances. More specifically, the SOGMM-FC utilizing the LCGA approach consistently produced negative variances for the 2-, 3-, and 4-class models even after several modifications. Convergence problems were not detected for the SOGMM-FC utilizing the GMM-CI approach. Thus, we selected the GMM-CI as the optimal SOGMM-FC.

As can be seen in Table 9.4, the SSABIC and BIC values indicated the 4-class model as the optimal model. However, the class sizes were not acceptable (n of smallest class = 2, .4%). Consequently, the 3-class model was selected as the optimal model. This model had lower SSABIC (4947.171) and BIC (5131.233) values compared to the 2-class model (5048.822 and 5223.363 for SSABIC and BIC, respectively). All of the GMM-CI class models evaluated had good entropy values (ranged from .95 to .99). The plots indicated that distinct and meaningful classes were derived from the three-class model. The estimated parameters of the SOGMM-FC (using the GMM-CI approach) are shown in the middle portion of Table 9.3, and the plots are provided in Figure 9.4. Overall, the three classes of trajectories are clearly separated and represent trajectories of initially high and decreasing IS, initially moderate and increasing IS, and consistently low IS.

Comparison of Classification Between a First-Order GMM With Composite Measures and Second-Order GMMs

In general, across our examples, both first-order growth mixture models using a composite measure of IS and second-order growth mixture models identified

TABLE 9.4 Fit Statistics for the Second-Order Growth Mixture Models – Factor-of-Curves (SOGMM-FC, total n = 436).

Fit statistics	2 Classes	3 Classes	4 Classes
LCGA			
LL (No. of Parameters)	-2522.680 (53) [b,c]	-2418.104 (56) [b, c]	-2366.921 (59) [b,c]
BIC	5367.475	5176.555	5092.424
SSABIC	5199.280	4998.840	4905.189
Entropy	.937	.953	.961
Adj. LMR–LRT (p)	311.475 (.029)	232.735 (.185)	97.042 (.133)
BLRT (p)	328.559 (.000)	245.500 (.000)	102.364 (.000)
Group size (%) C1	42 (9.6%)	16 (3.7%)	2 (.5%)
C2	394 (90.4%)	37 (8.5%)	13 (3.0%)
C3		383 (87.8%)	42 (9.6%)
C4			379 (86.9%)
GMM-CI			
LL (No. of Parameters)	-2444.546 (55)	-2389.365 (58)	-2350.116 (61)
BIC	5223.363	5131.233	5070.969
SSABIC	5048.822	4947.171	4877.387
Entropy	.990	.954	.964
Adj. LMR–LRT (p)	263.052 (.027)	219.244 (.199)	74.416 (.329)
BLRT (p)	277.479 (.000)	231.268 (.000)	78.497 (.000)
Group size (%) C1	18 (4.1%)	16 (3.7%)	2 (.4%)
C2	418 (95.9%)	29 (6.6%)	14 (3.2%)
C3		391 (89.7%)	35 (8.0%)
C4			385 (88.4%)
GMM- CV			
LL (No. of Parameters)	-2297.761 (57) [b]	-2246.929 (62) [a,b]	-2151.747 (67) [a, b]
BIC	4941.948	–	–
SSABIC	4761.060	–	–
Entropy	.798	–	–
Adj. LMR–LRT (p)	552.856 (.018)	–	–
BLRT (p)	571.049 (.000)	–	–
Group size (%) C1	179 (41.1%)	–	–
C2	257 (.58.9%)	–	–
C3		–	–
C4			–

Notes. LCGA = Latent Class Growth Analysis. GMM-CI = Growth Mixture Model with class-invariant variances-covariances. GMM-CV = Growth Mixture Model with class-varying variances (constrained covariances to be equal across class). LL = Log-Likelihood value. No. of Parameters = Number of estimated (free) parameters. BIC = Bayesian Information Criteria. SSABIC = Sample Size Adjusted BIC. Adj.LMR-LRT = Adjusted Lo-Mendell-Rubin Likelihood Ratio Test. BLRT = Bootstrapped Likelihood Ratio Test.

[a] = no repeated log-likelihood (i.e., local maxima). [b] = inadmissible solution (i.e., negative variances of growth parameter) in classes. [c] = no replicated log-likelihood values (H0 log-likelihood value) for the k-1 class model.

the 3-class model as the optimal classification. However, the composition of these classes varied between the first-order and second-order GMMs. For example, in the first-order GMM (more specifically, the GMM-CV), the "consistently low IS" class was the largest class (n = 327, 75.0%), followed by the "low and increasing IS" class (n = 68, 15.6%), and then the "high and decreasing IS" class (n = 41, 9.4%). In both the SOGMM-CF and the SOGMM-FC, we also detected that the largest group was the "consistently low IS" class (n = 364, 83.5% and n = 391, 89.7% for the CFM and the FCM, respectively). However, the second largest group in both SOGMMs was the "high and decreasing IS" class (n = 45, 10.3% and n = 29, 6.6% for the CFM and the FCM, respectively), while those with a "moderate and increasing" trajectory comprised the smallest class (n = 27, 6.2% and n = 16, 3.7% for the CFM and the FCM, respectively).

The classification similarities in the first-order GMM, which utilized composite measures of IS, and the second-order GMMs, which utilized subdomain symptoms, may be attributed to (a) high reliabilities of the subdomain manifest indicator variables and (b) high correlations among the subdomain manifest indicator variables. However, compared to the results of the first-order GMM using composite measures of IS, both of the SOGMMs (CFM and FCM) provided more distinct class trajectories. For example, compared to the "low and increasing IS" trajectory class from the first-order GMM (see Panel C of Figure 7.9), the slope means of the corresponding classes in both of the SOGMMs (i.e., the initially moderate and increasing class) were steeper. This difference suggests that the classification is clearer in the second-order GMMs compared to the first-order GMM.

Estimating a Conditional Model (Conditional SOGMM)

As in the case of a first-order GMM (Chapter 8), covariates can be incorporated into a SOGMM using two approaches: the one-step approach (with direct specification) and the three-step approach (using the AUXILIARY option in M*plus*). The main advantage of the 1-step approach is that it allows researchers to investigate how covariates are associated with between-class variation and within-class variation *simultaneously* by directly specifying covariates in the SOGMM. However, the 1-step approach can be problematic because class membership can vary when adding covariates in this approach. Asparouhov and Muthén (2014) reported that the 1-step approach has the same efficiency as the 3-step approach *if the model has a clear classification* (i.e., entropy ≥ .60). Because the 1-step approach for a SOGMM with the direct specification of covariates is straightforward and is performed almost exactly like the 1-step approach for a first-order GMM, we move on to discuss how the 3-step approach can be applied to estimating a SOGMM with covariates.

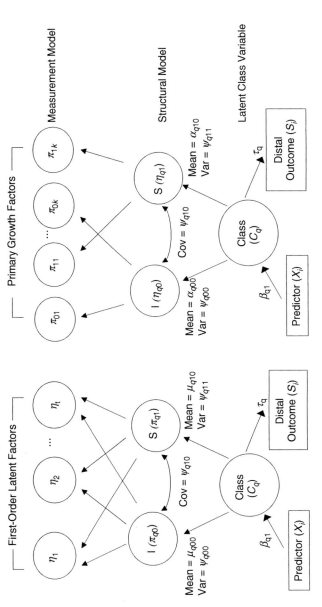

FIGURE 9.7 Covariate Specifications Using the Three-Step Approach for Conditional Second-Order Growth Mixture Models (SOGMMs).

Note: I = Intercept. S = Slope. Measurement models are not shown. η = A latent factor. π = A primary growth factor. β_{q1} = A logistic coefficient for q class. τ_q = A threshold (intercept) of distal outcomes for q class. Subscripts i, t, and q indicate an individual, a time, a sub-domain, and a categorical class variable, respectively.

Panel A. A SOGMM-CF with covariates   Panel B. A SOGMM-FC with covariates

The Three-Step Approach (Using the AUXILIARY Option) to Add Predictors of Second-Order Trajectory Classes

As discussed in Chapter 8 for a first-order GMM, it is possible to add predictors to both a SOGMM-CF and a SOGMM-FC. Figure 9.7 specifies how to add predictors to a SOGMM.

The bottom portion of the model illustrates the associations between the latent classes and the covariates. This portion of the SOGMM is similar to that of a first-order GMM. Thus, Equation 8.1 of Chapter 8 is applicable here. β_{q1} parameters are logistic coefficients from a multinomial regression (i.e., a logistic coefficient to class q).

Illustrative Example 9.4: Estimating a Conditional SOGMM with Predictors

The main advantage of the 3-step approach is that class membership remains stable regardless of the covariates specified in the model. Using the optimal SOGMM previously selected (i.e., the 3-class SOGMM-CF and SOGMM-FC), we demonstrate how to incorporate predictors into a SOGMM. Here we use the same four predictors that we incorporated into the conditional first-order GMM in Chapter 8 (i.e., Sex [Male], mothers' hostility [MH1], negative school environment [NSE1], and negative life events [NLE1]). The M*plus* syntax for the example SOGMM-CF and SOGMM-FC is shown in Figures 9.8 and 9.9, respectively.

As discussed in Chapter 8, many methods can be specified with the AUXILIARY option in order to estimate the effects of covariates in a GMM (see Table 8.5 of Chapter 8). For estimating predictor effects, the R3STEP option is employed in both types of SOGMMs (-CF and -FC). The results are shown in the top part of Table 9.5. Overall, the SOGMM-CF results are similar to those of the first-order GMM, which utilized a composite measure of IS. However, compared to the first-order GMM results (see Table 8.5 in Chapter 8), the conditional SOGMM-CF revealed more statistically significant predictors of classes. For example, in the first-order GMM, the logistic coefficient of negative life events for the low and increasing class was .026, which was not statistically significant. For a clearer interpretation, this coefficient can be converted into the "percent change" (i.e., $100 (e_\beta - 1)$), which is interpreted as the odds of being classified in the low and increasing IS group compared to the reference group. In this instance, the reference class was the consistently low class. Therefore, the odds of being in the low and increasing IS class were only 2.63% (= $100 (\text{Exp} (.026) - 1)$) *higher* than the odds of being in the consistently low IS class for every one unit increase in negative life events. However, in the SOGMM-CF, the corresponding coefficient was statistically significant ($\beta = .124$). For every one unit increase in negative life events, the odds of being in the moderate and increasing IS class were 13.20% (= $100 (\text{Exp} (.124) - 1)$) *higher* than the odds of being in the consistently low

Title: A conditional SOGMM of a CFM (3-STEP)
DATA: File is C:\example_ch_9_1.dat;
VARIABLE: NAMES are ID ANX1-ANX5 DEP1-DEP5 HOS1-HOS5 **MALE MH1 NLE1**
                NSE1 RV5 NEE5 PC5 RV5_B NEE5_B PC5_B;
        USEV= ANX1-ANX5 DEP1-DEP5 HOS1-HOS5;
        CLASSES = C(3);
        MISSING = ALL(9);

        AUXILIARY: (R3STEP) MALE MH1 NSE1 NLE1;

ANALYSIS: TYPE = MIXTURE;
        STARTS = 500 10;   The 3-step approach is
        STITERATIONS = 10;   used to estimate predictors'
        PROCESSORS = 8;   effects with the
                                            AUXILIARY option

MODEL: %OVERALL%

    IS1 by DEP1
        ANX1 HOS1 (1-2);
    IS2 by DEP2
        ANX2 HOS2 (1-2);
    IS3 by DEP3
        ANX3 HOS3 (1-2);
    IS4 by DEP4
        ANX4 HOS4 (1-2);
    IS5 by DEP5
        ANX5 HOS5 (1-2);
[DEP1-DEP5@0];
[ANX2-ANX5] (3); ANX1(V1); ANX4(V1); ANX5(V1);
[HOS1-HOS5] (4);

I_GIS S_GIS | IS1@0 IS2@1 IS3@3 IS4@4 IS5@6;
[I_GIS S_GIS]; I_GIS;S_GIS@0;

    DEP1 WITH DEP2 DEP3 DEP4 DEP5;
    DEP2 WITH DEP3 DEP4 DEP5;
    DEP3 WITH DEP4 DEP5; DEP4 WITH DEP5;
    ANX1 WITH ANX2 ANX3 ANX4 ANX5;
    ANX2 WITH ANX3 ANX4 ANX5;
    ANX3 WITH ANX4 ANX5; ANX4 WITH ANX5;
    HOS1 WITH HOS2 HOS3 HOS4 HOS5;
    HOS2 WITH HOS3 HOS4 HOS5;
    HOS3 WITH HOS4 HOS5; HOS4 WITH HOS5;
    HOS5 WITH DEP3 ANX1 ANX5;
    HOS4 WITH DEP1;

  %C#1%
  [I_GIS S_GIS]; I_GIS;
   %C#2%
  [I_GIS S_GIS]; I_GIS;
   %C#3%
  [I_GIS S_GIS]; I_GIS;

FIGURE 9.8 M*plus* Syntax for Specifying Predictors in a SOGMM–CF (GMM–CI)
Using the 3-Step Approach.
Note: DEP = Depressive Symptoms. ANX = Anxiety Symptoms. HOS = Hostility
Symptoms.
IS = Internalizing Symptoms. I_GIS = Intercept of IS. S_GIS = Slope of IS. The
marker variable scale setting approach was used.

Title: A conditional SOGMM of a FCM (3-STEP)
DATA: File is C:\example_ch_9_1.dat;
        ⋮
        CLASSES = C(3);

AUXILIARY: (R3STEP) MALE MH1 NSE1 NLE1;

ANALYSIS:TYPE = MIXTURE;
        STARTS = 500 10;
        STITERATIONS = 10;
        PROCESSORS = 8;

> The 3-step approach is used to estimate the effect of predictors using the AUXILIARY option in M*plus*.

MODEL: %OVERALL%
    I_DEP S_DEP | DEP1@0 DEP2@1 DEP3@3 DEP4@4 DEP5@6;
    I_ANX S_ANX | ANX1@0 ANX2@1 ANX3@3 ANX4@4 ANX5@6;
    I_HOS S_HOS | HOS1@0 HOS2@1 HOS3@3 HOS4@4 HOS5@6;

    I BY I_DEP@1
        I_ANX (L1)
        I_HOS (L2);

    S BY S_DEP@1
        S_ANX (L1)
        S_HOS (L2);

    I_DEP I_ANX I_HOS PWITH S_DEP S_ANX S_HOS;
    [I_DEP-S_DEP@0];
    [I_ANX I_HOS];
    [S_ANX S_HOS];
    [I S]; I-S;

    DEP1 WITH ANX1 HOS1; ANX1 WITH HOS1;
    DEP2 WITH ANX2 HOS2; ANX2 WITH HOS2;
    DEP3 WITH ANX3 HOS3; ANX3 WITH HOS3;
    DEP4 WITH ANX4 HOS4; ANX4 WITH HOS4;
    DEP5 WITH ANX5 HOS5; ANX5 WITH HOS5;
    DEP3 WITH DEP4; HOS3 WITH HOS4;
%C#1%
    [I S];
    [I_DEP-S_DEP@0];
    [I_ANX I_HOS] (M1-M2);
    [S_ANX S_HOS] (M3-M4);
%C#2%
    [I S];
    [I_DEP-S_DEP@0];
    [I_ANX I_HOS] (M1-M2);
    [S_ANX S_HOS] (M3-M4);
%C#3%
    [I S];
    [I_DEP-S_DEP@0];
    [I_ANX I_HOS] (M1-M2);
    [S_ANX S_HOS] (M3-M4);

FIGURE 9.9 M*plus* Syntax for Specifying Predictors in a SOGMM-FC Using the 3-Step Approach.
Note: DEP = Depressive Symptoms. ANX = Anxiety Symptoms. HOS = Hostility Symptoms. I = Intercept of internalizing symptoms. S = Slope of internalizing symptoms. The marker variable scale setting approach was used.

TABLE 9.5 The Logit Coefficients of Predictors for Two SOGMMs (i.e., CFM vs. FCM) Using the AUXILIARY Option (3-Step Approach).

Curve-of-Factors Model (CFM)

Predictors	High and decreasing[a]		Moderate and increasing[a]		High and decreasing[b]	
	Est.	OR	Est.	OR	Est.	OR
Male (vs. Female)	**-1.271***	.281	-.083	.920	-1.188	3.281
Mothers' hostility	**.749****	2.115	.021	1.021	**.728***	2.071
Negative school environment	**.977****	2.656	**1.073***	2.924	-.096	.908
Negative life events	.096	1.101	**.124***	1.132	-.028	.972

Factor-of-Curves Model (FCM)

Predictors	High and decreasing[a]		Moderate and increasing[a]		High and decreasing[b]	
	Est.	OR	Est.	OR	Est.	OR
Male (vs. Female)	**-1.449***	.235	-.271	.763	-1.178	.308
Mothers' hostility	**.959****	2.609	.300	1.350	**.659†**	1.933
Negative school environment	**.870***	2.387	**1.510****	4.527	-.640	.527
Negative life events	.112	1.119	.111	1.117	.001	1.001

Note: Unstandardized coefficients are shown. Est. = Estimated coefficients. OR = Odds Ratio. Bold and italic values are statistically significant. Total n = 407.
[a] Consistently low is the reference class.
[b] Moderate and increasing is the reference class.
† p < .10. *p < .05. **p < .01.

IS class. For more on interpreting logistic regression coefficients see "Predictors of latent classes" in Chapter 8. The coefficient for negative life events explaining membership in the moderate and increasing IS class was statistically significant, which indicates that negative life events occurred at a higher rate for youth in the moderate and increasing IS class compared to the consistently low IS class.

Regarding the SOGMM-FC results, the effects of predictors were generally similar to the effects found in the SOGMM-CF. However, compared to the SOGMM-CF results, there were fewer statistically significant predictors. For example, the coefficient for negative life events was not statistically significant for the moderate and increasing class, which indicates that youth in the moderate and increasing IS class and the consistently low IS class experienced negative life events at similar rates.

The Three-Step Approach (Using the AUXILIARY Option) to Add Outcomes of second-Order Trajectory Classes

As discussed in Chapter 8, the AUXILIARY option in M*plus* also allows users to specify a GMM with distal outcomes (as either binary or continuous outcomes). In this section, we demonstrate how to apply the AUXILIARY option to estimate the effects of distal outcomes in two example SOGMMs (-CF and -FC). τ_q indicates a threshold (intercept) of a distal outcome for class q, which is shown in Figure 9.7. The full model equation and a detailed interpretation of distal outcomes in a GMM are provided in Chapter 8 (see "Adding distal outcomes of latent classes"). We now demonstrate how to apply the AUXILARY option with a conditional SOGMM for the distal outcomes.

Illustrative Example 9.5: Estimating a Conditional SOGMM with Outcomes

In order to demonstrate a conditional SOGMM with distal outcomes, we employed the same distal outcomes utilized in our example first-order GMM in Chapter 8. These outcomes at Time 5 included the three continuous distal outcomes (negative economic events [NEE5], romantic violence [RV5], and physical complaints [PC5] for continuous outcomes) and three binary versions of the same outcomes (NEE5_B, RV5_B, and PC5_B) used in previous examples. The model specification in M*plus* is similar to that of a first-order GMM. The syntax below should be specified under the VARIABLE command:

AUXILIARY = (DU3STEP) NEE5 RV5 PC5 NEE5_B RV5_B PC5_B;

In order to estimate the effects of distal outcomes, we applied the DU3STEP option (i.e., a distal outcome with unequal means and unequal variances) for all of the continuous and binary distal outcomes. (Other approaches, including DE3STEP, DCON, and DCAT, failed to estimate the effects of the distal outcomes for certain classes.) The results are shown in Table 9.6.

TABLE 9.6 The Mean and Probabilities for Continuous and Binary Distal Outcomes.

	Curve-of-Factors Model (CFM)			
	High and decreasing	*Moderate and increasing*	*Consistently low*	*Overall chi-test*
Continuous Outcomes				
Romantic violence	2.753	2.728	2.033	22.115***[b,c]
Negative economic events	2.755	1.229	.925	7.431*[a,b]
Physical complaints	2.407	4.151	2.968	5.233
Binary Outcomes				
Romantic violence	.781	.752	.417	9.497**[b,c]
Negative economic events	.496	.368	.261	5.993*[c]
Physical complaints	.306	.630	.357	6.822*[a,b]

	Factor-of-Curves Model (FCM)			
	High and decreasing	*Moderate and increasing*	*Consistently low*	*Overall chi-test*
Continuous Outcomes				
Romantic violence	2.815	2.688	2.082	20.888***[b,c]
Negative economic events	2.433	1.961	.997	5.308
Physical complaints	2.768	3.270	3.009	.466
Binary Outcomes				
Romantic violence	Estimation Problems			
Negative economic events	.460	.375	.276	2.876
Physical complaints	.322	.573	.367	2.523

Notes. Unstandardized coefficients are shown. For the binary distal outcomes (0, 1), probabilities for having a "1" response are provided. The auxiliary DU3STEP option was used. The optimal class of each SOGMM was used. Missing cases varied depending on the distal outcome of interest. [a] High and decreasing vs. Moderate and increasing, $p < .05$. [b] Moderate and increasing vs. Consistently low, $p < .05$. [c] High and decreasing vs. Consistently low, $p < .05$.
*$p < .05$. ***$p < .001$.

For the continuous distal outcomes, the results of the SOGMM-CF showed that the mean level of romantic violence and negative economic events varied significantly across classes. This is mostly consistent with the corresponding results from the first-order GMM with the composite IS measures (see Table 8.6 of Chapter 8), where the mean levels of romantic violence and physical complaints were also found to be significantly different across classes. However, for the binary distal outcomes, all three outcomes were significantly predicted by class membership in the SOGMM-CF (i.e., romantic violence, negative economic events, and physical complaints), while only two of these distal outcomes (i.e., romantic violence and physical complaints) were significantly predicted by class membership in the first-order GMM using composite measures of IS. In general, there were more statistically significant associations for the SOGMM-CF (including predictors and distal outcomes) than the first-order GMM.

Also, as can be seen in Table 9.6, in the SOGMM-FC, the mean level of romantic violence varied significantly across the classes. However, for both the distal continuous and binary outcomes, the other two distal outcomes were not significantly different across the classes, indicating that the mean levels and probabilities (or thresholds) of these distal outcomes were similar across the classes. Importantly, for the binary outcomes, we also found that the probability (threshold) difference of romantic violence was not estimated. This may be related to the small class size of one class in our SOGMM-FC example.

Estimating a Multidimensional Growth Mixture Model (MGMM)

A multidimensional growth mixture model (MGMM) can be thought of as an extension of a multivariate first-order GMM. An advantage of this modeling approach is the simultaneous estimation of heterogeneity in trajectories specific to the subdomains. This modeling approach allows added flexibility over second-order growth curves (including a CFM and a FCM) because non-significant or weak correlations among the levels and slopes in a first-order LGCM make it difficult to create second-order growth factors (Duncan, Duncan, & Strucler, 2006). However, the first-order trajectories of each subdomain can combine over a longitudinal context to form heterogeneous classes that are substantively meaningful. These classes represent the heterogeneity in first-order growth parameters of different subdomains taken together without estimating second-order growth parameters. For example, Wickrama, Mancini, Kwag, and Kwon (2013) identified latent classes of trajectories of overall health among older adults using growth trajectories of different health subdomains as indicators (depressive symptoms, physical illness, and physical functioning). These latent trajectory classes of *overall health* were more strongly associated with early socioeconomic adversity than the latent trajectory classes for the *individual health subdomains*.

Thus, the MGMM approach can be employed to identify latent classes using all first-order growth parameters (e.g., primary intercepts and slopes of

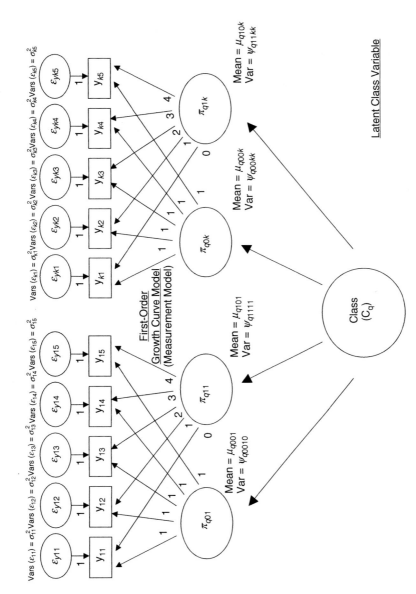

FIGURE 9.10 A Multidimensional Growth Mixture Model (MGMM).

Note: Subscript k represents a sub-domain (k = 1, 2, 3, ..., K). Subscript q represents the categorical class variable q (q = 1, 2, 3, ..., Q). The covariance structures of the primary growth factors (πs) are not shown.

all subdomains) as classification indictors of an overall domain. The equation is identical to that of the parallel process growth curve model (PPM) except for the class-specific parameters. A MGMM can be defined by the subscript q denoting class-specific parameters. An SEM figure for this MGMM is shown in Figure 9.10.

Illustrative Example 9.6: Estimating a Multidimensional Growth Mixture Model

We demonstrate a MGMM by building on the parallel process latent growth curve model (PPM) we estimated in Chapter 5 (see Figure 5.1). M*plus* syntax for the two-class MGMM (using the GMM-CI approach because it is the default M*plus* model) is shown in Figure 9.11.

As can be seen in Figure 9.11, most of the syntax is similar to that of a SOGMM-FC. However, the syntax of a MGMM is simpler, because a MGMM

```
Title: MGMM
DATA: File is C:\example_ch_9.dat;
         ⋮
         CLASSES = C(2);


ANALYSIS: TYPE = MIXTURE;
          STARTS = 500 10;
          STITERATIONS = 10;
          PROCESSORS = 8;

MODEL: %OVERALL%

    I_DEP S_DEP | DEP1@0 DEP2@1 DEP3@3 DEP4@4 DEP5@6;
    I_ANX S_ANX | ANX1@0 ANX2@1 ANX3@3 ANX4@4 ANX5@6;
    I_HOS S_HOS | HOS1@0 HOS2@1 HOS3@3 HOS4@4 HOS5@6;


    DEP1 WITH ANX1 HOS1; ANX1 WITH HOS1;
    DEP2 WITH ANX2 HOS2; ANX2 WITH HOS2;
    DEP3 WITH ANX3 HOS3; ANX3 WITH HOS3;
    DEP4 WITH ANX4 HOS4; ANX4 WITH HOS4;
    DEP5 WITH ANX5 HOS5; ANX5 WITH HOS5;
    DEP3 WITH DEP4; HOS3 WITH HOS4;

OUTPUT: SAMPSTAT STANDARDIZED MOD(3.84) TECH7 TECH11 TECH14;
```

FIGURE 9.11. M*plus* Syntax for the Two-Class Multidimensional Growth Mixture Model (MGMM) Using the GMM-CI Approach.
Note: DEP = Depressive Symptoms. ANX = Anxiety Symptoms. HOS = Hostility Symptoms. I = Intercept. S = Slope. GMM-CI = Growth Mixture Model with a class-invariant variance-covariance model. The marker variable scale setting approach was used.

does not contain two global intercept and slope factors (i.e., I and S at the second-order level). Instead, all of the primary growth factors (i.e., I_DEP, S_DEP, I_ANX, S_ANX, I_HOS, and S_HOS) are specified under the %OVERALL% option. This simplification may help to avoid convergence problems. In order to select the optimal class model, we conducted a model selection process using all three approaches (i.e., LCGA, GMM-CI, and GMM-CV). The syntax in Figure 9.11 can be used to estimate the MGMM using the GMM-CI approach. To estimate a MGMM using the LCGA approach, the bold and italic syntax below can be directly added to the syntax in Figure 9.11:

MODEL:%OVERALL%
I_DEP S_DEP | DEP1@0 DEP2@1 DEP3@3 DEP4@4 DEP5@6;
I_ANX S_ANX | ANX1@0 ANX2@1 ANX3@3 ANX4@4 ANX5@6;
I_HOS S_HOS | HOS1@0 HOS2@1 HOS3@3 HOS4@4 HOS5@6;
⋮
I_DEP-S_HOS@0;

To estimate a MGMM using the GMM-CV approach, add the bold and italic syntax below to the syntax in Figure 9.11.

MODEL:%OVERALL%
I_DEP S_DEP | DEP1@0 DEP2@1 DEP3@3 DEP4@4 DEP5@6;
I_ANX S_ANX | ANX1@0 ANX2@1 ANX3@3 ANX4@4 ANX5@6;
I_HOS S_HOS | HOS1@0 HOS2@1 HOS3@3 HOS4@4 HOS5@6;
⋮
%C#1%
I_DEP-S_HOS;
%C#2%
I_DEP-S_HOS;

The covariance structures of the six primary growth factors (i.e., I_DEP, S_DEP, I_ANX, S_ANX, I_HOS, and S_HOS) are estimated in these syntaxes, but the covariances among these growth factors are constrained to be equal across the classes. It is possible to estimate these covariances, but in doing so, the model may encounter convergence problems because of the large number of parameters released simultaneously. To estimate these covariance structures, add the following bold and italic syntax:

MODEL:%OVERALL%
I_DEP S_DEP | DEP1@0 DEP2@1 DEP3@3 DEP4@4 DEP5@6;
I_ANX S_ANX | ANX1@0 ANX2@1 ANX3@3 ANX4@4 ANX5@6;
I_HOS S_HOS | HOS1@0 HOS2@1 HOS3@3 HOS4@4 HOS5@6;
⋮
%C#1%
I_DEP WITH S_DEP I_ANX S_ANX I_HOS S_HOS;

S_DEP WITH I_ANX S_ANX I_HOS S_HOS;
I_ANX WITH S_ANX I_HOS S_HOS;
S_ANX WITH I_HOS S_HOS;
I_HOS WITH S_HOS;
%C#2%
I_DEP WITH S_DEP I_ANX S_ANX I_HOS S_HOS;
S_DEP WITH I_ANX S_ANX I_HOS S_HOS;
I_ANX WITH S_ANX I_HOS S_HOS;
S_ANX WITH I_HOS S_HOS;
I_HOS WITH S_HOS;

Based on the model fit indices and model modification process (which we introduced in relation to a SOGMM-CF), we selected the 3-class MGMM utilizing the GMM-CI approach (with fixed slope variance for the anxiety growth factors [S_ANX@0]) as the optimal model. The results are shown in the bottom panel of Table 9.3. As can be seen in Figure 9.12, the plots depict estimated mean trajectories of each subdomain separately for each class. For example, in the high and decreasing IS class (see Panel A of Figure 9.12), all three of the subdomain trajectories (i.e., depressive, anxiety, and hostility symptoms) decreased over time. Although the class size of the moderate and increasing group (n = 19, 4.4%) is relatively small, we selected this class model as the optimal model because all three classes are clearly separated and interpretable.

Conclusion

In this chapter, two extensions to a GMM involving second-order growth curves (known as second-order growth mixture models, SOGMMs) were discussed. These included a GMM extension of a curve-of-factors model (CFM) and an extension of a factor-of-curves model (FCM). Both SOGMM extensions reveal the heterogeneity of second-order global trajectories of internalizing symptoms. The heterogeneity of global trajectories is substantively important and may provide different information compared to analyses of subdomains. For example, classes of internalizing symptoms may be significantly different from classes of depressive, anxiety, and hostility symptom trajectories, and may have different antecedents and consequences. As previously discussed, SOGMMs may be a useful tool for analyzing global phenomena, such as healthy lifestyle behaviors, socioeconomic status, work quality, and marital quality, that underlie different subdomains. Another advantage of a SOGMM is that it accounts for measurement error in the manifest indicators at the lowest level of the model, with or without autocorrelations. This may provide a more precise test for population heterogeneity and associated parameters compared to a LGCM with composite measures.

However, as shown by the example models in this chapter, there may often be convergence problems in SOGMM estimations. Thus, following previous

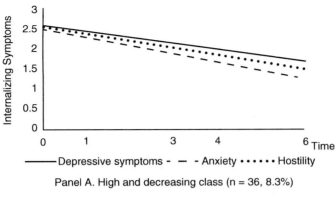

Panel A. High and decreasing class (n = 36, 8.3%)

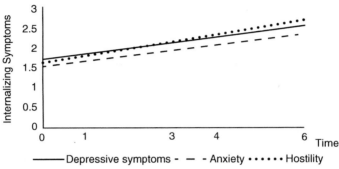

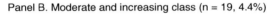

Panel B. Moderate and increasing class (n = 19, 4.4%)

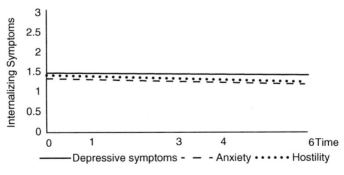

Panel C. Consistently low class (n = 381, 87.3%)

FIGURE 9.12 Predicted Mean Trajectories for a Multidimensional Growth Mixture Model (MGMM).

researchers, we proposed an exploratory approach to aid in identifying the best fitting model. As we have previously indicated, exploratory methods should be treated with caution as they do not always arrive at a solution that is consistent with existing theory. Thus, it is important that the decisions about identification of plausible models be based on theory, interpretability, prior research, *and* practical considerations.

Chapter 9 Exercises

This sample dataset consists of three internalizing symptoms (IS) scores of mothers at T1~T4 and multiple covariates. In the M*plus* syntax, indicate the variable order as follows:

DATA: File is exercise_Ch.9.dat;
VARIABLE: names are DEP1M DEP2M DEP3M DEP4M ANX1M ANX2M ANX3M ANX4M HOS1M HOS2M HOS3M HOS4M FC1M FWC1M JD1M PS1M MD4M TAB4M GHP4M;
  MISSSING ARE ALL (9);

Note that DEP1M~DEP4M indicate mothers' depressive symptoms at T1~T4. ANX1M~ANX4M indicate mothers' anxiety symptoms at T1~T4. HOS1M~HOS4M indicate mothers' symptoms of hostility at T1~T4. Four variables to be used as predictors of class membership are also included. More specifically, FC1M represents financial cutbacks reported by the mother at Time 1. FWC1M represents family-work conflict reported by the mother at Time 1. JD1M represents job dissatisfaction reported by the mother at Time 1. PS1M represents mothers' reports of fathers' support at Time 1. Three variables to be used as distal outcomes of class membership are also included. More specifically, MD4M represents marital dissatisfaction reported by mothers at Time 4. TAB4M represents children's antisocial behavior as reported by the child at Time 4. GHP4M represents global health problems reported by mothers at Time 4.

1. Investigate the optimal class model using two second-order growth mixture models (SOGMM-CF and -FC). Build both models with the following features:
   - Specify linear time functions in all growth curves (t = 0, 1, 2, and 3).
   - Specify autocorrelated errors among manifest indicators.
   - Use the anxiety manifest indicators (or growth factors) as marker variables to estimate the global growth factors.

- Assume that both models meet the requirements for measurement invariance (i.e., longitudinal invariance for a CFM or loading invariance for a FCM).
- Use the following M*plus* syntax under the ANALYSIS command to avoid local maxima.

  ANALYSIS: TYPE=MIXTURE;
  STARTS = 500 10;
  STITERATIONS = 10;

A. Investigate the optimal class model of a SOGMM-FC using the GMM-CI approach. Complete Exercise Table 9.1 by filling in the results. Note any local maxima problems, inadmissible solutions (i.e., negative variances), and/or convergence issues (e.g., under-identified model) in the table.

EXERCISE TABLE 9.1 Fit Statistics for a Second-Order Growth Mixture Model – Factor-of-Curves Model (SOGMM-FC).

Fit statistics	2 Classes	3 Classes	4 Classes
GMM-CI			
LL (No. of Parameters)			
BIC			
SSABIC			
Entropy			
Adj. LMR-LRT (*p*)			
BLRT (*p*)			
Group size (%) C1			
C2			
C3			
C4			

Notes. GMM-CI = Growth Mixture Model with class-invariant variances and covariances (M*plus* default model). LL = Log-likelihood value. No. of Parameters = Numbers of estimated (free) parameters. BIC = Bayesian Information Criteria. SSABIC = Sample size adjusted BIC. LMR-LRT = Lo-Mendell-Rubin likelihood ratio test. Adj.LMR-LRT = Adjusted LMR. BLRT = Bootstrap likelihood ratio test.

a = no repeated log-likelihood (i.e., local maxima). b = Inadmissible solution (i.e., negative variances of growth parameter) in classes. c = No replicated log-likelihood value (H0 log-likelihood value) for the k-1 class model.

B. Investigate the optimal class model of a SOGMM-CF using the GMM-CI approach. Complete Exercise Table 9.2 by filling in the results. Note any local maxima problems, inadmissible solutions (i.e., negative variances), and/or convergence issues (e.g., under-identified model) in the table.

C. Using all available information (model fit statistics, class sizes, plots), select the optimal SOGMM-FC and SOGMM-CF class models. Why are the selected models the optimal models?

EXERCISE TABLE 9.2 Fit Statistics for a Second-order Growth Mixture Model –
Curve-of-Factors (SOGMM-CF).

Fit statistics	2 Classes	3 Classes	4 Classes
GMM-CI			
LL (No. of Parameters)			
BIC			
SSABIC			
Entropy			
Adj. LMR–LRT (p)			
BLRT (p)			
Group size (%) C1			
C2			
C3			
C4			

Notes. GMM-CI = Growth Mixture Model with class-invariant variances and covariances
(M*plus* default model). LL = Log-likelihood value. No. of Parameters = Numbers of estimated
(free) parameters. BIC = Bayesian Information Criteria. SSABIC = Sample size adjusted BIC.
LMR–LRT = Lo-Mendell-Rubin likelihood ratio test. Adj.LMR–LRT = Adjusted LMR.
BLRT = Bootstrap likelihood ratio test.

a = no repeated log-likelihood (i.e., local maxima). b = Inadmissible solution (i.e., negative variances of growth parameter) in classes. c = No replicated log-likelihood value (H0 log-likelihood
value) for the k-1 class model.

D. Draw the plot of each optimal model and label each class using the class
mean parameters of growth factors.

E. Using plots and estimated parameters, compare the classification between
the two SOGMMs (-CF and -FC).

1) Do you think the two SOGMMs estimate a similar or different
classification?

2) Do they have similar growth parameters (i.e., means and variances)?

2. Next, analyze conditional SOGMMs (i.e., -CF and -FC) using the two optimal models you selected in Question 1.

A. The conditional SOGMM-FC:

1) Using the 3-step approach (R3STEP option), estimate the influence
of multiple predictors (i.e., FC1M, FWC1M, JD1M, and PS1M) on
the latent class variable, and interpret all statistically significant coefficients (use the $100 \times (Exp\ (\beta) - 1)$ formula for interpretation).

2) Using the 3-step approach (DU3STEP option), estimate the influence of the latent class variable on multiple distal outcomes (i.e.,
MD4M, TAB4M, and GHP4M), and interpret all statistically significant coefficients.

B. The conditional SOGMM-CF:
1) Using the 3-step approach (R3STEP option), estimate the influence of multiple predictors (i.e., FC1M, FWC1M, JD1M, and PS1M) on the latent class variable, and interpret all statistically significant coefficients.
2) Using the 3-step approach (DU3STEP option), estimate the influence of the latent class variable on multiple distal outcomes (i.e., MD4M, TAB4M, and GHP4M), and interpret all statistically significant coefficients.

References

Asparouhov, T., & Muthén, B. O. (2014). Auxiliary variables in mixture modeling: Three-step approaches using Mplus. *Structural Equation Modeling*, 21(3), 329–341.

Duncan, T. E., Duncan, S. C., & Strucler, L. A. (2006). *An introduction to latent variable growth curve modeling: Concepts, issues, and applications* (2nd ed.). Manwah, NJ: Lawrence Erlbaum.

Grimm, K., & Ram, N. (2009). A second-order growth mixture model for developmental research. *Research in Human Development*, 6(2&3), 121–143.

Kirves, K., Kinnunes, U., de Cuyper, N., & Mäkikangas, A. (2014). Trajectories of perceived employability and their associations with well-being at work: A three-wave study. *Journal of Personnel Psychology*, 13, 46–57.

Vaske, J., Ward, J. T., Boisvert, D., & Wright, J. P. (2012). The stability of risk-seeking from adolescence to emerging adulthood. *Journal of Criminal Justice*, 40(4), 313–322.

Wickrama, K. A. S., Mancini, J. A., Kwag, K., & Kwon, J. (2013). Heterogeneity in multi-dimensional health trajectories of late old years and socioeconomic stratification: A latent trajectory class analysis. *Journal of Gerontology Series B*, 68(2), 290–297.

ANSWERS TO CHAPTER EXERCISES

Chapter 2 Exercises

1. If the covariances between non-adjacent measurement occasions are higher than the covariances among adjacent time points $(t, t+1)$ negative slope variances (π_{1i}) are often found when estimating a LGCM. The covariance matrix suggests that negative variances will likely not be an issue for estimating a LGCM because the covariances are higher among adjacent time points than non-adjacent measurement occasions.

2.
   A. See Exercise Table 2.2.
   B. In general, most of the model fit indices except the RMSEA indicated that all three models (linear, quadratic, and piecewise LGCMs) have an acceptable model fit. The RMSEA of the linear LGCM was .114, suggesting that the linear LGCM does not fit the data well.
   C. *Nested model comparison (M1 vs. M2):* Compared to the χ^2 of the linear LGCM (M1), the χ^2 of the quadratic LGCM (M2) was significantly lower $(\chi^2_{\text{DIFF}} = 50.905(4), p < .001)$. This suggests that the quadratic LGCM fits the data better. *Non-nested model comparison (M2 vs. M3):* Compared to the AIC value of the piecewise LGCM (M3), the quadratic LGCM (M2) had a lower AIC value, suggesting that the quadratic LGCM (M2) fits the data better. Given the above model comparisons, the *quadratic LGCM (M2)* can be selected as the best fitting model.
   D. The following syntax should have been specified in the OUTPUT command. PLOT: SERIES=IS1-IS5(S); TYPE=PLOT3; Then, by utilizing

EXERCISE TABLE 2.2 Model Comparisons among Various Unconditional LGCMs.

Growth curve models with different time functions	χ^2	df	Model Fit Indices					
			$\Delta\chi^2$ (df), p-value	Model Comparison	CFI / TLI	SRMR	RMSEA	AIC
Linear Time Function (M1)	68.291	10			.937 / .937	.086	.114	2390.550
Quadratic Time Function (M2)	17.386	6	50.905 (4), p < .001	M1 vs. M2	.988 / .979	.041	.065	2347.644
Piecewise Time Function (M3, use T3 as a distinct time point)	26.018	6		M3 vs. M2	.978 / .964	.031	.086	2356.276

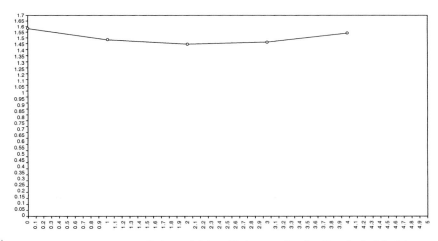

EXERCISE FIGURE 2.1 Estimated Mean Trajectory for the Quadratic Model.

the drop-down menu in the M*plus* output (Plot → View Plots → Estimated Means) the visual depiction in Exercise Figure 2.1 is found.

E. The mean growth parameters:

1.582: The average (μ_{00}) of mothers' internalizing symptoms at Time 1.

–.128: The average linear slope (μ_{10}) of mothers' internalizing symptoms. On average, a one year increase in time was associated with a .128 unit decrease in internalizing symptoms.

.030: The quadratic slope (μ_{20}) of mothers' internalizing symptoms. On average, a one year increase in time was associated with a .030 unit increase in the rate of change (linear slope) of internalizing symptoms.

F. Interpret all statistically significant coefficients related to the predictor.

.120: On average, for every one unit increase in marital unhappiness, the initial level of mothers' internalizing symptoms increased by .120 units ($p < .001$).

.059: For every one unit increase in marital unhappiness, the linear slope of mothers' internalizing symptoms increased by an average of .059 units ($p < .01$).

G. Interpret all statistically significant coefficients related to the distal outcome.

.328: One unit increases in the initial level of mothers' internalizing symptoms were associated with an average .328 unit increase in the depressive symptoms of their children ($p < .01$).

3.868: One unit increases in the linear slope of mothers' internalizing symptoms were associated with an average 3.868 units increase in the depressive symptoms of their children ($p < .05$).

Chapter 3 Exercises

1.

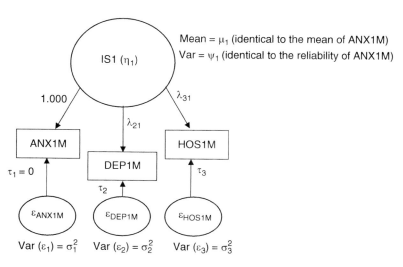

EXERCISE FIGURE 3.1 The Model Specification of a Confirmatory Factor Analysis (CFA) Model with Mothers' Three Internalizing Symptoms (i.e., Symptoms of Anxiety, Depression, and Hostility at Time 1).

Note: ANX1M was used as the marker variable. Var = Variance.

2. See Exercise Figure 3.2.
3.
- Configural invariance: Unconstrained Model
- Weak invariance: $\lambda_{21} = \lambda_{22} = \lambda_{23} = \lambda_{24}; \lambda_{31} = \lambda_{32} = \lambda_{33} = \lambda_{34}$
- Strong invariance: $\tau_{21} = \tau_{22} = \tau_{23} = \tau_{24}; \tau_{31} = \tau_{32} = \tau_{33} = \tau_{34}$
- Strict invariance: $\sigma^2_{11} = \sigma^2_{12} = \sigma^2_{13} = \sigma^2_{14}; \sigma^2_{21} = \sigma^2_{22} = \sigma^2_{23} = \sigma^2_{24}; \sigma^2_{31} = \sigma^2_{32} = \sigma^2_{33} = \sigma^2_{34}$

4. See Exercise Figure 3.3.

Chapter 4 Exercises

1. See Exercise Figure 4.1.
2.
   A. See Exercise Table 4.2.
   B. Given all available results, the *strong invariance* model (M2) is acceptable, which is the minimum level of measurement invariance acceptable for proceeding with a curve-of-factors model (CFM). Below is a discussion of each comparison.
      The general model fits across all three models: All three competing models had good model fit indices, suggesting that each model fits the data well.
      Weak invariance model (M2) vs. the Configural model (M1): As can be seen in Exercise Table 4.2, a non-significant $\Delta\chi^2$ value (11.514 (6), $p = .07$) indicated that the configural LCFA and the LCFA with weak invariance were not statistically different. The more parsimonious LCFA with weak invariance is the preferred model.
      Strong invariance model (M3) vs. Weak invariance model (M2): A statistically significant $\Delta\chi^2$ value was detected when M3 was compared with M2 (14.674 (6), $p < .05$). However, we also examined the ΔCFI value as an alternative index to compare these competing models. As can be seen in Exercise Table 4.2, the ΔCFI value between M3 and M2 was .003, indicating that the strong invariance model is acceptable (the recommended cut-point is .01).
3. The mean growth parameters of the CFM:
   1.314: The average (μ_{00}) of mothers' internalizing symptoms at Time 1.
   –.025: The linear slope (μ_{10}) of mothers' internalizing symptoms. A one year change was associated with a .025 unit decrease in internalizing symptoms.

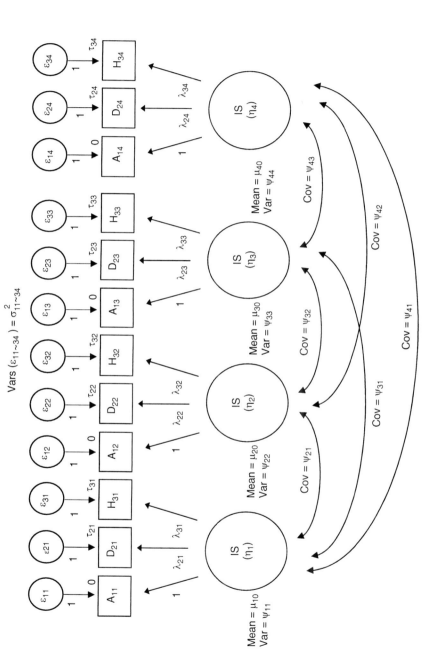

EXERCISE FIGURE 3.2 The Model Specification of a Longitudinal Confirmatory Factor Analysis (LCFA) Model.
Note: D = Depressive Symptoms. A = Anxiety Symptoms. H = Hostility Symptoms. Var = Variance. Cov = Covariance.

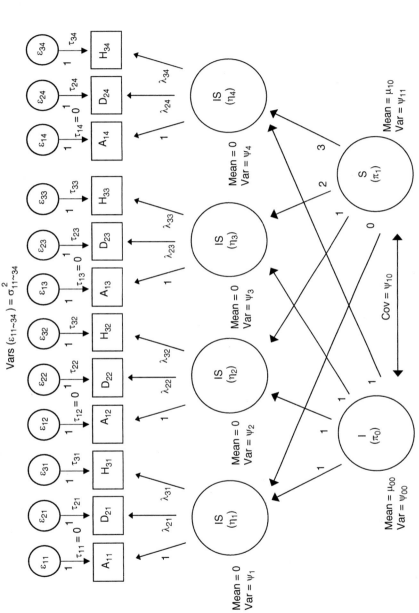

EXERCISE FIGURE 3.3 The Model Specification of a Curve-of-Factors Model (CFM) with a Linear Time Function. Note: D = Depressive Symptoms. A = Anxiety Symptoms. H = Hostility Symptoms. Var = Variance. Cov = Covariance. I = Intercept. S = Slope (rate of change).

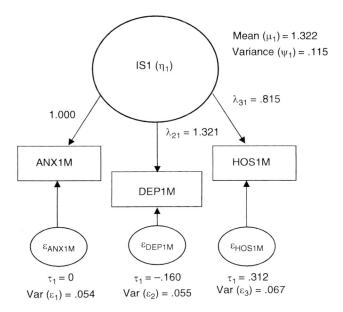

EXERCISE FIGURE 4.1   The Estimated Parameters of a CFA Using Three Manifest Indicators (i.e., DEP1M, ANX1M, and HOS1M).
Note: ANX1M was used as the marker variable.

EXERCISE TABLE 4.2 Measurement Invariance in a Longitudinal CFA Model.

	χ^2 (df)	Model Comparison	$\Delta\chi^2$(Δdf), p-value	CFI	ΔCFI	RMSEA	SRMR	BIC
Configural LCFA model (M1)	58.526 (30)			.992		.046	.022	2259.108
LCFA with weak invariance (M2)	70.040 (36)	M2 vs. M1	11.514 (6), p =.07	.991	.001	.046	.034	2233.953
LCFA with strong invariance (M3)	84.714 (42)	M3 vs. M2	14.674 (6), p < .05	.988	.003	.047	.036	2211.958

Note: M = Model. LCFA = Longitudinal Confirmatory Factor Analysis. All of these LCFA models included autocorrelated errors.

4.

    A. **.108:** For a one unit increase in financial cutback, the initial level of mothers' internalizing symptoms increased by an average of .108 units ($p < .001$).

    B. The interpretation of statistically significant unstandardized coefficients is as follows:

        The influence of financial cutback on global growth factors (π)

        .118: For a one unit increase in financial cutback, the initial level of mothers' internalizing symptoms increased by an average of .118 units after taking into account the effect of financial cutback on internalizing symptoms at T3 and T4 ($p < .001$).

        -.019: For a one unit increase in financial cutback, the linear slope of mothers' internalizing symptoms decreased by an average of .019 units after taking into account the effect of financial cutback on internalizing symptoms at T3 and T4 ($p < .05$).

        The influence of financial cutback on time-specific latent variables (η)

        .039: For a one unit increase in financial cutback, mothers' internalizing symptoms at T3 increased by an average of .039 units after accounting for the influence of global growth factors ($p < .001$).

        .057: For a one unit increase in financial cutback, mothers' internalizing symptoms at T4 increased by an average of .057 units after accounting for the influence of global growth factors ($p < .05$).

5.

    A. **1.379:** For a one unit increase in the initial level of mothers' internalizing symptoms, mothers' subsequent negative life events increased by an average of 1.379 units ($p < .001$).

    B. The interpretation of statistically significant path coefficients is as follows:

        The influence of global growth factors (π) on the outcome (mothers' negative life events):

        -5.125: For a one unit increase in the linear slope of mothers' internalizing symptoms, mothers' negative life events decreased by an average of 5.125 units after taking into account the effect of internalizing symptoms at T3 on negative life events ($p < .05$).

        The influence of time-specific latent variables (η) on the outcome (mothers' negative life events):

        1.225: For a one unit increase in mothers' internalizing symptoms at T3, mothers' negative life events increased by an average of 1.225 units after taking into account the effect of global growth factors on negative life events ($p < .01$).

Chapter 5 Exercises

1. See Exercise Figure 5.1
2. See Exercise Figure 5.2
3. See Exercise Figure 5.3

Chapter 6 Exercises

1. In general, the adjacent covariances among the same repeated indicators were higher than those between non-adjacent time points for all three subdomains, which suggests that analyzing a parallel process model (PPM), and subsequently a FCM, is appropriate for these data.

2.

    A. The PPM had a poor fit to the data (χ^2 (df) = 959.127(51), CFI/TLI = .775/.709, SRMR = .071, RMSEA = .199).

    B. The PPM with correlated errors between subdomains had an acceptable fit (χ^2 (df) = 115.650(39), CFI/TLI = .981/.968, SRMR = .052, RMSEA = .066). All between-subdomain correlated errors were statistically significant. Compared to the baseline PPM (without autocorrelated errors), the chi-square statistic was improved ($\Delta\chi^2$ (df) = 843.477(12), p < .001). However, the PPM with correlated errors within subdomains had a poor fit (χ^2 (df) = 940.562(42), CFI/TLI = .777/.650, SRMR = .069, RMSEA = .218), and most within-subdomain autocorrelated errors were not statistically significant. Although the chi-square for this model was also significantly improved compared to the baseline PPM ($\Delta\chi^2$ (df) = 18.565 (9), p < .05), we excluded this model from subsequent analyses due to its overall poor fit. For these reasons, the *PPM with correlated errors between subdomains* was selected as the optimal model.

3.

    A. *Assessing Loading Invariance (i.e., Testing a Weak Invariance Model):* As shown in Exercise Table 6.2, the chi-square test ($\Delta\chi^2$) was not statistically significant, which indicates that the model with constrained loadings (M2) was not significantly different from the model with unconstrained loadings (M1). Thus, the more parsimonious model (M2) with loading equality is preferred.

    B. **1.331:** The average (η_0) of internalizing symptoms at Time 1 (p < .001).
       -.026: The mean linear slope (η_1) of internalizing symptoms. In other words, over a period of one year, internalizing symptoms generally decreased by .026 units (p < .001).

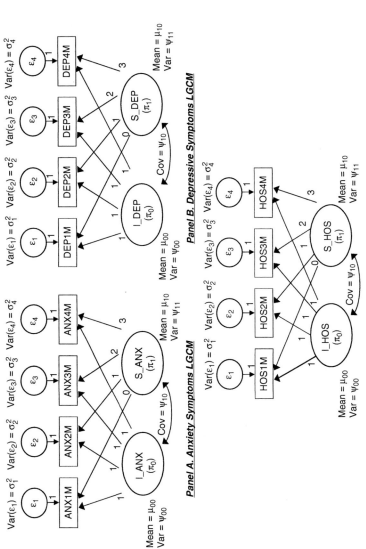

EXERCISE FIGURE 5.1 Model Specification of Latent Growth Curve Models (LGCMs) for the Three Subdomain Symptoms (i.e., Symptoms of Anxiety, Depression, and Hostility).

Note: ANX = Anxiety Symptoms. DEP = Depressive Symptoms. HOS = Hostility Symptoms. M = Mother. Var = Variance. Cov = Covariance. All intercepts ($\tau_1 \sim \tau_4$) of manifest indicators were fixed at 0 for model identification purposes.

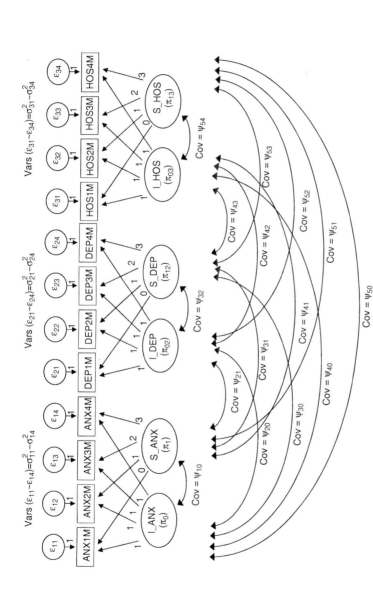

EXERCISE FIGURE 5.2   Model Specification of a Parallel Process Model (PPM) Using Three Subdomain Symptoms (i.e., Symptoms of Anxiety, Depression, and Hostility).

Note: ANX = Anxiety Symptoms. DEP = Depressive Symptoms. HOS = Hostility Symptoms. M = Mother.Var = Variance. Cov = Covariance. All intercepts ($\tau_{11} \sim \tau_{34}$) of manifest indicators were fixed at 0 for model identification purposes.

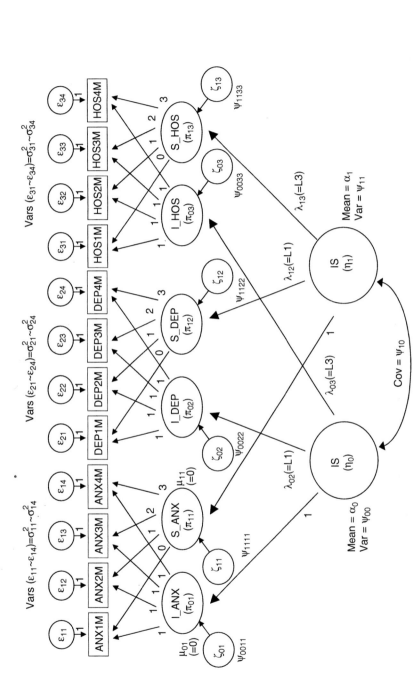

EXERCISE FIGURE 5.3 Model Specification of a Factor-of-Curves Model (FCM) Using the Three Subdomain Symptoms (i.e., Symptoms of Anxiety, Depression, and Hostility).

Note: ANX = Anxiety Symptoms. DEP = Depressive Symptoms. HOS = Hostility Symptoms. M = Mother. Var = Variance. Cov = Covariance. Assigned path names in parentheses (i.e., L1 and L3) were constrained to use the marker variable approach. All intercepts ($\tau_{11} \sim \tau_{34}$) of manifest indicators were fixed at 0 for model identification purposes.

EXERCISE TABLE 6.2 Loading Invariance (Equality) Test for a Factor-of-Curves Model (FCM).

	$\chi^2(df)$	Model comparison	$\Delta\chi^2(df)$, p-value
Uncontrained loadings (M1)	131.717 (47)		
Constrained loadings (M2)	134.872 (49)	M2 vs. M1	3.155(2), $p = .206$

4.

    A. **.060**: For a one unit increase in financial cutback (FC1_M), the initial level of internalizing symptoms increased by an average of .060 units ($p < .001$).

    B. One additional significant path between FC1_M and the initial level of anxiety (I_ANX) was detected in the conditional FCM. Below we interpret the impact of this predictor on the primary growth factor (initial level of anxiety). Then we re-interpret the statistically significant effect(s) on global growth factors after incorporating these direct regression coefficients.

      The influence of financial cutback on primary growth factor (π)

      -.016: For a one unit increase in financial cutback, the initial level of anxiety symptoms decreased by an average of .016 even after accounting for the influence of the internalizing symptoms' global factor ($p < .001$).

      The influence of financial cutback on global factor (η)

      .067: For a one unit increase in financial cutback, the initial level of internalizing symptoms increased by an average of .067 units after taking into account the effect of financial cutback on the initial level (intercept) of anxiety ($p < .001$).

5.

    A. **1.626**: For a one unit increase in the initial level of internalizing symptoms, mothers' poor health increased (indicating a decline in general health) by an average of 1.626 units ($p < .001$).

    B. There were no statistically significant paths between primary growth factors (π) and the outcome of interest after controlling for the influence of global factors (η).

Chapter 7 Exercises

1. The growth factors for the intercept and slope have statistically significant variances (unstandardized coefficients: .262, $p < .000$ and .025, $p < .01$ for

intercept and slope, respectively). These significant variances suggest the potential existence of sub-populations, or classes.

2.
   A. See Exercise Table 7.2.
   B. See Exercise Table 7.2.
   C. As can be seen in Exercise Table 7.2, all of the class solutions for the GMM-CV had negative variances. However, no estimation problems were detected for the GMM-CIs. Thus, a GMM-CV was excluded

EXERCISE TABLE 7.2 Fit statistics of the GMM-CI and GMM-CV (n = 442).

Fit statistics	2 Classes	3 Classes	4 Classes
GMM-CI			
LL (No. of Parameters)	−993.697 (12)	−910.525 (15)	−873.231 (18)
BIC	2060.489	1912.420	1856.106
SSABIC	2022.406	1864.817	1798.982
Entropy	0.975	0.980	0.954
Adj. LMR-LRT (*p*)	313.644 (*p* < .000)	157.712 (*p* < .01)	70.718 (*p* = .268)
BLRT (*p*)	330.808 (*p* < .000)	166.343 (*p* < .000)	74.588 (*p* < .000)
Group size (n, %)			
C1	45 (10.2%)	22 (5.0%)	16 (3.6%)
C2	397 (89.8%)	28 (6.3%)	23 (5.2%)
C3		392 (88.7%)	30 (6.8%)
C4			373 (84.4%)
GMM- CV			
LL (No. of Parameters)	−832.575 (14)a	−773.443 (19)a	−773.443 (24)a
BIC	1750.429	1662.621	1693.077
SSABIC	1706.000	1602.323	1616.912
Entropy	0.755	0.830	0.865
Adj. LMR-LRT (*p*)	632.290 (*p* < .001)	114.506 (*p* < .001)	0
BLRT (*p*)	653.050 (*p* < .000)b	118.265 (*p* < .000)b	0
Group size (n, %)			
C1	146 (33.0%)	21 (4.7%)	0 (0.0%)
C2	296 (67.0%)	128 (29.0%)	21 (4.7%)
C3		293 (66.3%)	128 (29.0%)
C4			293 (66.3%)

Notes. GMM-CI = Growth Mixture Model with class-invariant variances-covariances (M*plus* default model). GMM-CV = Growth Mixture Model with class-varying variances (constrained covariances to be equal across class). LL = Log-likelihood value. No. of Parameters = Number of estimated (free) parameters. BIC = Bayesian Information Criteria. SSABIC = Sample size adjusted BIC. LMR-LRT = Lo-Mendell-Rubin likelihood ratio test. Adj.LMR-LRT = Adjusted LMR. BLRT = Bootstrap likelihood ratio test. *p* = p-value. n = class sample size. C = Class. a = Inadmissible solution (i.e., negative variances of growth parameter) in classes. b = No replicated log-likelihood value (H0 log-likelihood value) of the k-1 class model.

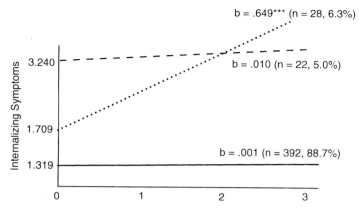

EXERCISE FIGURE 7.1 Estimated Mean Trajectories for the Potential Optimal Class Model (i.e., 3-Class GMM-CI).
Note: Unstandardized coefficients are shown. ***p < .001

from further consideration. Of the GMM-CIs, the 4-class solution was excluded due to the class sample size (the smallest class proportion was 3.6% (n = 16)). All of the model fit statistics indicated that the three-class solution was a better fit compared to the two-class solution (a lower BIC and SSABIC value, a higher entropy value, statistically significant LRTs [Adj. LMR-LRT and BLRT], and acceptable class sizes). Thus, the three-class solution was selected as the optimal class model.

D. The plot (see Exercise Figure 7.1) shows three well-separated classes, including a consistently low IS class, a moderate and increasing IS class, and a chronically high IS class.

E. The mean quality test (see Exercise Table 7.3) also provided evidence that the model has a clear classification. The intercept means were statistically different across the three classes. Also, the slope mean for the moderate and increasing class was significantly higher than the slope means for the consistently low and chronically high IS classes.

Chapter 8 Exercises

1.

A. Using the 1-step approach:
   1) Between-class variation model - see Exercise Table 8.2.

EXERCISE TABLE 7.3 Growth Parameters of the Optimal Three-Class Model (GMM-CI).

Optimal class model (GMM-CI)	Intercept		Linear Slope		Factor co-variance	Growth parameter equality (Wald)	
	Mean	Variance	Mean	Variance		Intercept mean	Slope mean
C1 (Chronic-ally high)	3.240***	.075***	.010	.002	-.001	198.136*** a, b, c	735.705*** b, c
C2 (Mod-erate and increasing)	1.709***	.075***	.649***	.002	-.001		
C3 (Con-sistently low)	1.319***	.075***	.001	.002	-.001		

Note: Unstandardized coefficients are shown. GMM-CI = Growth Mixture Model with class-invariant variances and covariances. C = Class. Wald = Wald chi-square test (df = 2).
a = Chronically high vs. Consistently low, $p < .05$.
b = Moderate and increasing vs. Consistently low, $p < .05$.
c = Moderate and increasing vs. Chronically high, $p < .05$.
†$p < .10$. *$p < .05$. **$p < .01$. ***$p < .001$.

EXERCISE TABLE 8.2 The Influence of FEP1 on Class Membership in the Optimal Model (Utilizing the One-Step Approach).

	Coefficients	OR
Chronically high vs Consistently low [a]	2.656***	14.239
Moderate and increasing vs Consistently low [a]	1.135*	3.111
Chronically high vs Moderate and increasing [b]	1.521**	4.576

Note: Unstandardized coefficients are shown. OR = Odds Ratio.
[a] Consistently low class is the reference class.
[b] Moderate and increasing class is the reference class.
*$p < .05$. **$p < .01$. ***$p < .001$.

2.656: For every one unit increase in family economic hardship, the odds of being a member of the chronically high IS class were 1323.9% (= 100 (Exp (2.656) − 1)) *higher* than the odds of being a member of the consistently low IS class.

1.135: For every one unit increase in family economic hardship, the odds of being a member of the moderate and increasing IS class were 211.12% (= 100 (Exp (1.135) − 1)) higher than the odds of being a member of the consistently low IS class.

EXERCISE TABLE 8.3 The Influence of FEP1 on Class Membership (C) and Growth Factors (πs) in the Optimal Model (Utilizing the One-Step Approach).

	Coefficients	OR
Between-class multinomial logistic coefficients		
Chronically high vs Consistently low [a]	2.500***	12.18
Moderate and increasing vs Consistently low [a]	1.059*	2.88
Chronically high vs Moderate and increasing [b]	1.442**	4.23
Added regression paths under the OVERALL option		
Intercept	.133*	
Slope	.014	

Note: Unstandardized coefficients are shown. OR = Odds ratio.
[a] Consistently low class is the reference class.
[b] Moderate and increasing class is the reference class.
*$p < .05$. **$p < .01$. ***$p < .001$.

> **1.521**: For every one unit increase in family economic hardship, the odds of being a member of the chronically high IS class were 357.7% (= 100 (Exp (1.521) − 1)) higher than the odds of being a member of the moderate and increasing IS class.
> 2) Multinomial regression of classes (involving between-class variation) and regression of global growth factors for the total sample (assessing total variation) – See Exercise Table 8.3.
> **.133**: Across all classes, for every one unit increase in family economic hardship, the intercept of fathers' internalizing symptoms increased by an average of .113 units.

B. Using the 3-step approach (R3STEP) – See Exercise Table 8.4.

C. The results are similar. The unconditional model has a high entropy value (.980), which suggests that the model has a stable classification that is not affected by covariates. Thus, both approaches estimate similar coefficients.

2.

A. See Exercise Table 8.5.

B. See Exercise Table 8.5.

C. See Exercise Table 8.5.

D. The results are similar. The reason is the same as the reason for the similarities in the predictor models (see 1.C). The unconditional model has a high entropy value, which indicates that the classification is stable when covariates are incorporated.

EXERCISE TABLE 8.4 The Influence of FEP1 on Class Membership (C) in the Optimal Model (Utilizing the 3-Step Approach).

	Coefficients	OR
Chronically high vs Consistently low [a]	2.779***	16.102
Moderate and increasing vs Consistently low [a]	1.256**	3.511
Chronically high vs Moderate and increasing [b]	1.523**	4.586

Note: Unstandardized coefficients are shown. OR = Odds ratio.
[a] Consistently low class is the reference class.
[b] Moderate and increasing class is the reference class.
*$p < .05.$ **$p < .01.$ ***$p < .001.$

EXERCISE TABLE 8.5 The Mean (or Probability) Difference for Several Approaches (1-Step, 3-Step, And Lanza's Approaches).

Outcomes	Chronically high	Moderate and increasing	Consistently low	Wald chi-square equality test (df)
One-step approach				
Global health problems (Continuous)	3.207	2.840	1.700	45.803*** (2)
Three-step approach (DE3STEP)				
Global health problems (Continuous)	3.236	2.868	1.700	53.619*** (2)
Lanza's approach (DCON)				
Global health problems (Continuous)	3.211	2.842	1.704	73.337*** (2)
Lanza's approach (DCAT)				
Global health problems (Binary)	.933	.837	.558	33.716*** (2)

Note: Unstandardized coefficients are shown. For the binary distal outcomes, probabilities for the "1" response are provided (i.e., Category 2 in the M*plus* output). df = Degrees of Freedom.
***$p < .001.$

E. See Exercise Table 8.5.

   .933: The probability that an individual in the chronically high IS class experienced a "high" level of general health problems (U = 1) was 93.3%.

   .837: The probability that an individual in the moderate and increasing IS class experienced a "high" level of general health problems (U = 1) was 83.7%.

   .558: The probability that an individual in the consistently low IS class experienced a "high" level of general health problems (U = 1) was 55.8%.

Chapter 9 Exercises

1.
  A. See Exercise Table 9.3.

  B. See Exercise Table 9.4.

  C. _SOGMM-FC_: When utilizing the GMM-CI approach, the 4-class solution was excluded from consideration as the optimal model due to estimation problems (local solutions and negative variances of growth parameters). Compared to the 2-class solution, the 3-class solution had lower BIC and SSABIC values and a higher entropy value. However, the sample sizes were not acceptable in the 3-class solution (the smallest class proportion in the 3-class solution was 2.0% (n = 2)). Given all of the model fit statistics and class sizes, the 2-class solution was retained as the optimal class model.

  SOGMM-CF: When utilizing the GMM-CI approach, the 3- and 4-class solutions were excluded from consideration as the optimal model due to estimation problems (local solutions and negative variances of growth parameters). Only the 2-class solution was extracted without

EXERCISE TABLE 9.3 Fit Statistics for a Second-Order Growth Mixture Model – Factor-of-Curves (SOGMM-FC, n = 451).

Fit statistics	2 Classes	3 Classes	4 Classes
GMM-CI			
LL (No. of Parameters)	**–867.255 (44)**	–815.431 (47)	–768.830 (50) a, b, c
BIC	**2003.414**	1918.100	1843.233
SSABIC	**1863.775**	1768.940	1684.552
Entropy	**0.950**	0.961	0.964
Adj. LMR-LRT (p)	**207.572 (p = .628)**	111.091 (p = .139)	111.938 (p = .487)
BLRT (p)	**218.894 (p < .000)**	117.150 (p < .000)	118.043 (p < .000)
Group size (%) C1	**45 (10.0%)**	9 (2.0%)	6 (1.3%)
C2	**406 (90.0%)**	45 (10.0%)	26 (5.8%)
C3		397 (88.0%)	35 (7.8%)
C4			384 (85.1%)

Notes. GMM-CI = Growth Mixture Model with class-invariant variances and covariances (M_plus_ default model). LL = Log-likelihood value. No. of Parameters = Numbers of estimated (freed) parameters. BIC = Bayesian Information Criteria. SSABIC = Sample size adjusted BIC. LMR-LRT = Lo-Mendell-Rubin likelihood ratio test. Adj.LMR-LRT = Adjusted LMR. BLRT = Bootstrap likelihood ratio test.
a = no repeated log-likelihood (i.e., local maxima). b = Inadmissible solution (i.e., negative variances of growth parameter) in classes. c = No replicated log-likelihood value (H0 log-likelihood value) for the k-1 class model.

EXERCISE TABLE 9.4 Fit Statistics for a Second-Order Growth Mixture Model – Curve-of-Factors (SOGMM-CF, n = 451).

Fit statistics	2 Classes	3 Classes	4 Classes
GMM-CI			
LL (No. of Parameters)	**-870.389 (46)**	-821.035 (49) b	-779.555 (52) a, b
BIC	**2021.905**	1941.532	1876.906
SSABIC	**1875.918**	1786.024	1711.877
Entropy	**0.948**	0.943	0.954
Adj. LMR-LRT (p)	**199.385 (p < .05)**	93.603 (p = .607)	78.670 (p =.543)
BLRT (p)	**210.260 (p < .000)**	98.708 (p < .000)	82.961 (p < .000)
Group size (%) C1	**47 (10.4%)**	29 (6.4%)	5 (1.1%)
C2	**404 (89.6%)**	37 (8.2%)	23 (5.1%)
C3		385 (85.4%)	38 (8.4%)
C4			385 (85.4%)

Notes. GMM-CI = Growth Mixture Model with class-invariant variances and covariances (M*plus* default model). LL = Log-likelihood value. No. of Parameters = Numbers of estimated (freed) parameters. BIC = Bayesian Information Criteria. SSABIC = Sample size adjusted BIC. LMR-LRT = Lo-Mendell-Rubin likelihood ratio test. Adj.LMR-LRT = Adjusted LMR. BLRT = Bootstrap likelihood ratio test.
a = no repeated log-likelihood (i.e., local maxima). b = Inadmissible solution (i.e., negative variances of growth parameter) in classes. c = No replicated log-likelihood value (H0 log-likelihood value) for the k-1 class model.

estimation problems. Although this is a limitation because it is preferable to be able to compare class solutions, the model fit indices suggested that the two-class solution was acceptable (i.e., acceptable class sample sizes, high entropy value, and statistically significant LMR-LRT and BLRT values, which indicate that the 2-class solution was an improved classification compared to a single class solution). Moreover, the same number of classes was also selected as the optimal model in the SOGMM-FC. This similarity supports our selection of the 2-class model as an appropriate SOGMM-CF classification solution. Given the consideration of all available information, the two-class model was selected as the optimal SOGMM-CF class solution.

D. See Exercise Figure 9.1.

E.

1) As can be seen in Exercise Figure 9.1, the two SOGMMs (-CF and -FC) have similar classifications. In both SOGMMs, the "consistently low IS" class was the largest class (n = 404, 89.6% and n = 406, 90.0% for the CFM and the FCM, respectively), followed by the "moderate and increasing IS" class (n = 47, 10.4% and n = 45, 10.0% for the CFM and the FCM, respectively).

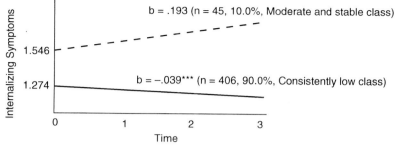

Panel A. The optimal SOGMM-FC

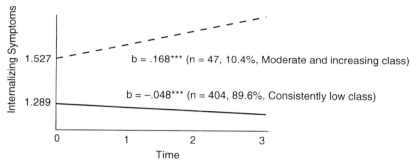

Panel B. The optimal SOGMM-CF

EXERCISE FIGURE 9.1   Estimated Mean Trajectories for the Optimal Class Model (i.e., 2-class Models).
Note: Unstandardized coefficients are shown. SOGMM-CF = Second-Order Growth Mixture Model – Curve-of-Factors. SOGMM-FC = Second-Order Growth Mixture Model – Factor-of-Curves. ***$p < .001$.

    2)   Most growth parameters also appear to be similar, but the mean of the slope factor was not statistically significant for the "moderate and stable IS" class in the SOGMM-FC, while the corresponding mean trajectory was statistically significant in the SOGMM-CF.

2.

    A.   The conditional SOGMM-FC:

        1)   See Exercise Table 9.5.

        Logistic coefficient interpretation:

        .584: For a one unit increase in financial cutback at Time 1, the odds of being in the moderate and stable IS class were 79.3% (= 100 (Exp (.584) − 1)) *higher* than the odds for being a member of the consistently low IS class after controlling for all other predictors.

        .804: For a one unit increase in family-work conflict at Time 1, the odds of being in the moderate and stable IS class were 123.4% (= 100

EXERCISE TABLE 9.5 The Results of the Multinomial Regressions for the SOGMM-FC and SOGMM-CF.

Predictors	β	OR
SOGMM-FC		
Financial cutback (FC1M)	.584*	1.793
Family-work conflict (FWC1M)	.804**	2.234
Job dissatisfaction (JD1M)	.223*	1.249
Partner support (PS1M)	-2.787***	.062
SOGMM-CF		
Financial cutback (FC1M)	.680*	1.974
Family-work conflict (FWC1M)	.674*	1.962
Job dissatisfaction (JD1M)	.184*	1.202
Partner support (PS1M)	-1.407*	.245

Note: Unstandardized coefficients are shown. OR = Odds Ratio. β = logistic regression coefficient.
[a] Consistently low class is the reference class. *$p < .05$. **$p < .01$. ***$p < .001$.

(Exp (.804) − 1)) *higher* than the odds of being a member of the consistently low IS class after controlling for all other predictors.

.223: For a one unit increase in job dissatisfaction at Time 1, the odds of being in the moderate and stable IS class were 24.9% (= 100 (Exp (.223) − 1)) *higher* than the odds of being a member of the consistently low IS class after controlling for all other predictors.

-2.787: For a one unit increase in partner support at Time 1, the odds of being in the moderate and stable class were 93.8% (= 100 (Exp (-2.787) − 1)) *lower* than the odds of being a member of the consistently low IS class after controlling for all other predictors.

2) See Exercise Table 9.6.

Coefficient interpretation:

On average, individuals in the moderate and stable IS class have higher mean scores (indicating poorer outcomes) across all three distal outcomes compared to those in the consistantly low IS class.

B. The conditional SOGMM-CF:

1) See Exercise Table 9.5 and the interpretations above for the SOGMM-FC as these coefficients are interpreted in the same manner.

2) See Exercise Table 9.6 and the interpretations above for the SOGMM-FC as the coefficients are interpreted in the same manner.

EXERCISE TABLE 9.6 The Mean Equality Test Results for the SOGMM-FC and SOGMM-CF.

Outcomes	MI/MS	CL	Wald test
SOGMM-FC			
Marital dissatisfaction (MD4M)	3.053	2.025	25.697***
Children's antisocial behavior (TAB4M)	3.261	2.130	21.424***
Global health problems (GHP4M)	3.549	2.126	216.218***
SOGMM-CF			
Marital dissatisfaction (MD4M)	3.119	2.008	36.843***
Children's antisocial behavior (TAB4M)	3.266	2.118	23.425***
Global health problems (GHP4M)	3.563	2.125	213.318***

Note: MI = Moderate and increasing class. MS = Moderate and stable class. CL = Consistent low class. ***$p < .001$.

AUTHOR INDEX

SUBJECT INDEX